Tourism, Creativity and Development

Destinations across the world are beginning to replace or supplement culture-led development strategies with creative development. This book critically analyzes the impact and effectiveness of creative strategies in tourism development and charts the emergence of 'creative tourism'. Why has 'creativity' become such an important aspect of development strategies and of tourism development in particular? Why is this happening now, apparently simultaneously, in so many destinations across the globe? What is the difference between cultural tourism and creative tourism? These are among the important questions this book seeks to answer.

It critically examines the developing relationship between tourism and creativity, the articulation of the 'creative turn' in tourism, and the impact this has on theoretical perspectives and practical approaches to tourism development. A wide range of examples from Europe, North America, Asia, Australia and Africa explore the interface between tourism and creativity including: creative spaces and places such as cultural and creative clusters and ethnic precincts; the role of the creative industries and entrepreneurs in the creation of experiences; creativity and rural areas; the 'creative class' and tourism; lifestyle, creativity and tourism, and marketing creative tourism destinations. The relationship between individual and collective forms of creativity and the widely differing forms of modern tourism are also discussed. In the concluding section of the book the contribution of creativity to tourism and to development strategies in general is assessed, and areas for future research are outlined.

The diverse multidisciplinary contributions link theory and practice, and demonstrate the strengths and weaknesses of creativity as a tourism development strategy and marketing tool. It is the first exploration of the relationship between tourism and creativity and its consequences for tourism development in different parts of the world.

Greg Richards is a partner in Tourism Research and Marketing (Barcelona) and a researcher at the Centre for Leisure, Tourism and Society (CELTS), at the University of the West of England, Bristol, UK.

Julie Wilson is an EU Marie Curie Postdoctoral Research Fellow in the Department of Geography, Autonomous University of Barcelona (Catalunya) and a researcher in the Centre for Leisure, Tourism and Society (CELTS), University of the West of England, Bristol, UK.

Contemporary geographies of leisure, tourism and mobility
Series Editor: C. Michael Hall

Professor at the Department of Management, College of Business and Economics, University of Canterbury, Private Bag 4800, Christchurch, New Zealand

The aim of this series is to explore and communicate the intersections and relationships between leisure, tourism and human mobility within the social sciences.

It will incorporate both traditional and new perspectives on leisure and tourism from contemporary geography, e.g. notions of identity, representation and culture, while also providing for perspectives from cognate areas such as anthropology, cultural studies, gastronomy and food studies, marketing, policy studies and political economy, regional and urban planning, and sociology, within the development of an integrated field of leisure and tourism studies.

Also, increasingly, tourism and leisure are regarded as steps in a continuum of human mobility. Inclusion of mobility in the series offers the prospect to examine the relationship between tourism and migration, the sojourner, educational travel, and second home and retirement travel phenomena.

The series comprises two strands:

Contemporary Geographies of Leisure, Tourism and Mobility aims to address the needs of students and academics, and the titles will be published in hardback and paperback. Titles include:

The Moralisation of Tourism
Sun, sand ... and saving the world?
Jim Butcher

The Ethics of Tourism Development
Mick Smith and Rosaleen Duffy

Tourism in the Caribbean
Trends, development, prospects
Edited by David Timothy Duval

Qualitative Research in Tourism
Ontologies, epistemologies and methodologies
Edited by Jenny Phillimore and Lisa Goodson

The Media and the Tourist Imagination
Converging cultures
Edited by David Crouch, Rhona Jackson and Felix Thompson

Tourism and Global Environmental Change
Ecological, social, economic and political interrelationships
Edited by Stefan Gössling and C. Michael Hall

Forthcoming:

Understanding and Managing Tourism Impacts
Michael Hall and Alan Lew

Routledge Studies in Contemporary Geographies of Leisure, Tourism and Mobility is a forum for innovative new research intended for research students and academics, and the titles will be available in hardback only. Titles include:

1. **Living with Tourism**
Negotiating identities in a Turkish village
Hazel Tucker

2. **Tourism, Diasporas and Space**
Tim Coles and Dallen J. Timothy

3. **Tourism and Postcolonialism**
Contested discourses, identities and representations
C. Michael Hall and Hazel Tucker

4. **Tourism, Religion and Spiritual Journeys**
Dallen J. Timothy and Daniel H. Olsen

5. **China's Outbound Tourism**
Wolfgang Georg Arlt

6. **Tourism, Power and Space**
Andrew Church and Tim Coles

7. **Tourism, Ethnic Diversity and the City**
Jan Rath

8. **Ecotourism, NGOs and Development**
A critical analysis
Jim Butcher

Tourism, Creativity and Development

Edited by
Greg Richards and Julie Wilson

Routledge
Taylor & Francis Group

LONDON AND NEW YORK

First published 2007
by Routledge
2 Park Square, Milton Park, Abingdon, Oxfordshire OX14 4RN

Simultaneously published in the USA and Canada
by Routledge
711 Third Avenue, New York, NY 10017

First issued in paperback 2014

Routledge is an imprint of the Taylor & Francis Group, an informa business

Typeset in Times New Roman by
Book Now Ltd, London

British Library Cataloguing in Publication Data
A catalogue record for this book is available from the British Library

Library of Congress Cataloging in Publication Data
Tourism, creativity and development / edited by Greg Richards and
Julie Wilson.
      p. cm.
Includes bibliographical references.
1. Tourism. 2. Sustainable development. I. Richards, Greg. II. Wilson,
Julie, 1975–

G155.AIT58966 2007
338. 4′791–dc22   2007018400

ISBN13: 978–0–415–42756–2 (hbk)
ISBN13: 978–1–138–01062–8 (pbk)

For Edna Margaret Mary Wilson

Contents

x *Contents*

Illustrations

Contributors

Vivien Andersen is a freelancer who has previously lectured at Queen Margaret University College, Edinburgh. Her research interests include cultural tourism, festivals and cultural 'exports'.

Albert Arias Sans is a PhD student at the Department of Geography, Universitat Autònoma, Barcelona, and gained his MA in Management of the European Metropolitan Regions at the Erasmus University, Rotterdam. His research interests vary from urban and cultural geography to political economy issues. He is academic coordinator and lecturer in the Masters Programme 'City management' at the Universitat Oberta de Catalunya, directed by Jordi Borja, with whom he has collaborated on many projects.

Julian Beer is the Director of Research and Innovation for the University of Plymouth. Julian is a Reader in Public Policy and prior to this appointment he was a Principal Lecturer and the Director of the Social Research and Regeneration Unit in the Faculty of Social Science and Business, as well as, more recently, a secondee to Chancellery. He joined the University in 1999 having previously worked in the private sector; he has since gained extensive experience in knowledge transfer and 'third stream' activities while working at the higher education public and private sector interface. His previous roles in the University include developing and advising on strategy, direction and policy related to Knowledge Transfer at both Faculty and wider University levels linked to the research and teaching base in the University.

Esther Binkhorst founded and manages the consultancy Co-creations S.L. (Spain and the Netherlands) and teaches at ESADE/St. Ignasi in Barcelona. She obtained her PhD at Tilburg University in 2002 for her study 'Holland, the American way'. She worked at the Centre for Research and Statistics in Rotterdam in 1994 and at the Leisure Management School, Leeuwarden in 1995.

Karin Bras studied cultural anthropology and received her PhD from the University of Tilburg in the Netherlands. Her PhD research was on the role of local tourist guides in the social construction of tourist attractions on the island of Lombok, Indonesia. She currently works as a lecturer in the Department of Tourism and Leisure Management at INHOLLAND University. Since 2005 she has participated in the Leisure Management research group at the same university in the field of cultural tourism in Amsterdam.

Paul Cloke is Professor of Human Geography at the University of Exeter, having previously held a chair at the University of Bristol and the University of Wales, Lampeter. Paul has longstanding research interests in geographies of rurality, nature–society and tourism, and more recently has also carried out research into urban homelessness and ethical consumption. He is Founder Editor of the quarterly *Journal of Rural Studies* (Elsevier) and is Editor of the new *Sage Handbook of Rural Studies* (with Terry Marsden and Patrick Mooney).

Jock Collins is Professor of Economics, School of Finance and Economics, University of Technology, Sydney (UTS). Research interests are ethnic entrepreneurship, comparative immigration, ethnic crime, ethnic precincts, ethnic diversity and tourism, cosmopolitan cities and communities, and regional and rural immigration.

Graeme Evans is director of the Cities Institute at London Metropolitan University (www.citiesinstitute.org). He led a 2 year international comparative study of 'Creative Spaces – Strategies for Creative Cities' for London and Toronto city regional authorities, and is currently evaluating city growth and creative cluster development, including cultural tourism, in London's city fringe. Key publications include: *Cultural Planning* (Routledge) and more recent chapters in *Culture, Urbanism and Planning* (Ashgate), *Tourism Regeneration and Culture* (CABI), and *Small Cities* (Routledge), with a forthcoming chapter on London 2012 for a book on Olympic Cities (Routledge).

Roos Gerritsma is an urban sociologist and lecturer/researcher in the department of Tourism and Leisure Management at INHOLLAND University, and a lecturer at the Documentary Photography Department at the Utrecht Art Academy. Roos' main research themes are: the creative city, sports, wellness and lifestyles and sociology as a tool for documentary photographers.

John Hannigan is Professor of Sociology at the University of Toronto. Hannigan completed his Bachelor's and his Master's degrees in Sociology at the University of Western Ontario. He then moved to Ohio State University where he received his Doctorate. During his time at Ohio State, he was a Research Associate at the Disaster Research Center. Hannigan joined the faculty at the University of Toronto in 1976.

Stephen Hodes is a co-founder and director of LAgroup Leisure and Arts Consulting in Amsterdam, a consultancy firm in the fields of culture, tourism, recreation and the hospitality industry specializing in project management, strategic development, marketing, trendwatching, business planning, concept development and feasibility studies. After studying architecture at the Delft University of Technology, Stephen worked for the Netherlands Board of Tourism as marketing manager and director for North America and as a consultant and partner with KPMG Management Consulting in Amsterdam. In addition to his work for LAgroup, Stephen is a board member of various cultural and civil organizations and was a lecturer in Leisure Management at INHOLLAND University until September 2006.

Patrick Kunz is a PhD Candidate, School of Finance and Economics, University of Technology, Sydney (UTS). Research interests are ethnic entrepreneurship, immigration, transnationalism, ethnic precincts, and tourism in and to urban precincts.

Robert Maitland is Director of the Centre for Tourism at the University of Westminster, London. His main research interests are in urban and cultural tourism, tourism policy, and tourism and regeneration, with a focus on capital and world cities.

Kevin Meethan is Senior Lecturer in Sociology, School of Law and Social Sciences, University of Plymouth. Kevin's research interests have focused on tourism, cultural change and globalization. His publications include *Tourism in Global Society* (2001) and more recently *Tourism, Consumption and Representation* (2006). His current research interests also include the role of culture and the creative industries in relation to tourism and regeneration.

Can-Seng Ooi is Associate Professor at Copenhagen Business School (CBS) and director of the university's masters programme in international business. His research interests include the creative industries, experience product development, place branding and tourism strategies. He has published extensively, including in the *Annals of Tourism Research*, *Tourism* and *SOJOURN: Journal of Social Issues in Southeast Asia*. He is the author of *Cultural Tourism and Tourism Cultures: The Business of Mediating Experiences in Copenhagen and Singapore* (Copenhagen Business School Press, 2002).

Richard Prentice is Reader in Tourism Management at the University of Strathclyde. His research interests include creative tourism supply (heritage tourism; cultural tourism; lifestyle tourism), experiential marketing (lifestyling), tourists' behaviour and consumption of meanings (identity; experiential consumption; imagining arts, places and periods), and public policy making (naturalistic decision making; corporate planning; policy effectiveness). Major publications include *Tourism and Heritage Attractions* (Routledge, 1993).

Crispin Raymond migrated to New Zealand in 2001 and lives in the Nelson-

Tasman region at the top of the South Island. Previously based in the United Kingdom, he worked in the arts for 25 years, first as Chief Executive of the Theatre Royal in Bath and subsequently as the founder of a specialist management consultancy that helped arts and charitable organizations with policy, management, building and funding issues. He launched Creative Tourism New Zealand in 2003.

Greg Richards is a partner in Tourism Research and Marketing (Barcelona) and a researcher at the Centre for Leisure, Tourism and Society at the University of the West of England (Bristol). He has published widely in the field of cultural and creative tourism, including books on *Cultural Tourism in Europe*, *Marketing and Developing Crafts Tourism*, *Cultural Attractions and European Tourism*, *Tourism and Gastronomy* and *City Tourism and Culture in Europe*.

Christian Rogerson is Professor of Human Geography at the University of the Witwatersrand, Johannesburg. His research foci are small enterprise development and local economic development with a special interest in tourism. During 2004 he co-edited a book titled *Tourism and Development Issues in Contemporary South Africa*, published by the Africa Institute of South Africa, Pretoria.

Antonio Paolo Russo is Assistant Professor in Tourism at the Universitat Rovira i Virgili, Tarragona (Spain), Department of Geography. He also collaborates with EURICUR (European Institute of Comparative Urban Research) at the Erasmus University, Rotterdam, where he received his PhD in Economics at the Tinbergen Institute in 2002. Previous appointments were at Universitat Autònoma Barcelona (visiting researcher) and at IULM University Milan (lecturer in Cultural Policy). As a consultant, he has collaborated with various international organizations, such as the International Centre for Art Economics, the Foundation ENI Enrico Mattei, UNESCO and the Latin American Development Bank. His research interests range from tourism and regional studies to cultural economics.

Walter Santagata is Professor and Director of EBLA CENTER, International Center for Research on the Economics of Culture, Institutions, and Creativity, as well as holding the posts of Director of the Department of Economics 'Salvatore Cognetti de Martiis'; Professor of Public Finance, Faculty of Political Science; and Professor of Cultural Economics, Faculty of Political Science (all University of Turin). His research interests include cultural economics, institutional economics, political economy and public choice.

Giovanna Segre is Assistant Professor of Public Finance and Lecturer of Economics of Culture at the Faculty of Economics, Turin University, where she received her Doctorate in European Economic Studies. She also collaborates with EBLA CENTER – International Center for Research on the Economics of Culture, Institutions, and Creativity – of Turin University. Her research focuses on welfare economics and cultural economics.

Stephen Shaw BA (Hons), PG DipTP, MRTPI, FCIT, FILT, FRGS is Director of the TRaC research centre at the Cities Institute, London Metropolitan University. His current research and publications include regeneration, built heritage and use of the public realm, especially in cities that are gateways to immigration.

Jacques Vork has his own consultancy firm, specializing in international tourism marketing and promotion. He made his career with the Netherlands Board of Tourism and Conventions (NBTC), where he was responsible for marketing, R&D, corporate strategy and public affairs. In addition to his consultancy, Jacques is an Associate Lecturer in Leisure Management at INHOLLAND University. His main research areas are the creative city and leisure and destination marketing. He is chairman of the advisory board of the Regional Tourism Office for the Veluwe (VBT).

Julie Wilson is an EU Marie Curie Postdoctoral Research Fellow in the Department of Geography, Autonomous University of Barcelona (Catalunya) and a researcher in the Centre for Leisure, Tourism and Society (CELTS), University of the West of England, Bristol. Her research interests focus on tourism, culture and urbanism; particularly the 'creative' turn in urban development and the city imagery–urban reality nexus.

Preface

This book forms part of a long-term research programme on the relationship between tourism, culture, creativity and development, supported by ongoing empirical research in many different countries. The basic research has been supported by a number of different research groups of the Association for Tourism and Leisure Education (ATLAS), including the Cultural Tourism Research Project, the Cultural Capitals Research Group, the Tourism and Gastronomy Group and the Backpackers Research Group. In particular, the numerous surveys and publications of the ATLAS Cultural Tourism Project since its inception in 1991 have provided the raw material which highlighted the role of creativity in the burgeoning cultural tourism market, and which provided a platform for later research on the role of cultural events.

Specific development projects have also helped to shape our thinking on the relationship between tourism and creativity. In particular the EUROTEX project (1997–9) funded by DGXVI of the European Commission led to the practical development of creative tourism pilot projects in Finland, Portugal and Greece. Broadening perspectives on the development of creative landscapes were facilitated through the CIUTAT project (2003–7). Work carried out for the UK Northern Way programme on cultural tourism development in 2006 helped shape our thinking about the practical application of creative development in tourism.

Over the years, a great many people and organizations have helped to support our work and to stimulate creative thought. This list is not exhaustive, but the following have played key roles in the research programme (apologies to anybody we have inadvertently overlooked):

Ajuntament de Barcelona; Salvador Antón; British Academy Small Grants Programme; Lluís Bonet; Conxita Camós; María Casado-Diaz; Ramon Cosialls; Montse Crespi Vallbona; Eduard Delgado; Carlos Fernandes; Generalitat de Catalunya; Grecotel; Antonio Guiccardo; Christa Helwig; David Sweeting and Miguel; EU Marie Curie Fellowship EIF and Experienced Researcher Programmes; Margarita Mendez; Satu Miettinen; NWO / British Council Netherlands; Leontine Onderwater; Bob Palmer; Natalia Paricio; Gerda Priestley; Celia Queiros; Crispin Raymond;

Paolo Russo; Laura Salarich; Spanish Ministry of Industry, Commerce and Tourism; Aurora Tresserras; Jordi Tresserras; Turisme de Barcelona; Universitat Autònoma de Barcelona, Department of Geography; Universitat de Barcelona; University of the West of England, Bristol; Jantien Veldman; Marisa Zanotti.

Greg Richards and Julie Wilson
Gràcia, Barcelona
March 2007

1 Tourism development trajectories

From culture to creativity?

Greg Richards and Julie Wilson

The growing synergy of tourism and culture has been one of the major themes in tourism development and marketing in recent years. Tourism destinations seeking to distinguish themselves from their increasingly numerous competitors have turned to culture as a means of distinction, and culture has increasingly been linked to tourism as a means of generating income and jobs (Richards 2001). The growth of 'cultural tourism' has been one of the major trends in global tourism in the past three decades and is still seen as one of the major growth areas for the future (European Travel Commission/World Tourism Organization 2005).

The apparent success of culture-led or cultural development strategies has encouraged more cities, regions and nations to use the combination of culture and tourism. Richards (2001) shows that the supply of cultural attractions grew faster than cultural demand during the 1990s. This has led to growing competition between destinations for cultural consumers, stimulating the creation of more distinctive and more impressive cultural developments. There is an increasing problem of 'serial reproduction' (Harvey 1989; Richards and Wilson 2006) or 'McGuggenheimization' (Honigsbaum 2001) of culture, and it can be argued that cultural development alone is no longer sufficient to create distinction between destinations.

Destinations are therefore beginning to replace or supplement culture-led development strategies with creative development. Florida's (2002) concept of the 'creative class' has been wholeheartedly embraced by a wide range of different regions seeking to capitalize on their creative resources. Creativity is also increasingly being applied to the tourism sphere. VisitBritain (2006), the national tourist office for Great Britain, has now identified 'creative tourism' as a major growth area. Questions about the development of creative tourism have also been asked in the British Houses of Parliament (Hansard 2006). The European Travel Commission/World Tourism Organization report on *City Tourism and Culture* (2005) also underlined the development potential of 'creative cities' as tourism attractions. Creative approaches to tourism are also being actively developed by commercial operators, in countries as far apart as New Zealand (see Chapter 9), South Africa (Chapter 15) and Spain (Chapter 8).

Why has 'creativity' become such an important aspect of development strategies in general, and of tourism development in particular? Why is this happening now, apparently simultaneously, in so many destinations across the globe? What is the difference between cultural tourism and creative tourism? These are important questions we will seek to answer in this chapter, before exploring different aspects of the development of creativity in tourism in the chapters that follow. The aim of this volume is to examine critically the developing relationship between tourism and creativity, the articulation of the 'creative turn' in tourism, and the impact this has on theoretical perspectives and practical approaches to tourism development.

Until relatively recently, the tourism industry had apparently made relatively little use of creative approaches to development or engaged in real innovation (Chapter 8). The majority of tourism products continued to be based on relatively static modes of consumption, and tourism remains largely based on models emphasizing traditional factors of production (such as sun, sea and sand). In recent years, however, signs of change have emerged as more attention has been paid to experience production and creativity as an element of both tourism consumption and production. In effect, there also seems to have been a shift in the power relations in the tourism production system, with the role of the consumer changing from that of mere receiver of ready-made products into the co-producer of tourism experiences. In this new scenario, the consumer is able to exercise more creativity and consumption skills, which in turn forces the erstwhile producers to be more creative in (co-)producing tourism experiences. To examine the reasons why creativity has become such a popular development strategy, it is useful to examine the reasons why culture and tourism have become so closely integrated in recent decades, and why the 'cultural turn' in tourism later developed into the 'creative turn'.

From cultural to creative production

As Evans outlines in Chapter 4, culture-led development first came to prominence in the late 1970s, as a wide range of places, most significantly large cities, were struggling to deal with the consequences of economic restructuring. The need to replace 'traditional' jobs in manufacturing (or in the case of rural areas in primary industry) led to a search for alternatives. These were invariably found in the rapidly expanding service sector, with areas such as culture, tourism, banking and finance, and research and development being among the most important. Hannigan shows in Chapter 3 how the entertainment industry provided one potential redevelopment strategy, as cities built shopping malls, theme parks and other relatively safe forms of consumption. However, as the 'fantasy city' model began to show signs of waning, culture, creativity and tourism in particular became sectors favoured by a large number of cities and regions. As well as being growth areas, they also offered the potential to develop employment and project a new image of the region to the outside world. Glasgow, Toronto, Barcelona, Singapore and Sydney are just a few of the cities examined in this volume which have employed culture in their restructuring strategies. In many cases, the development of culture and tourism went hand-in-hand, as cultural facilities became important flagships which attracted tourism, and tourists contributed the money which supported the expansion of culture. The apparently flawless symmetry of this couplet led many other cities and regions to adopt the culture-led or culture and tourism-led regeneration model. In addition, cultural and tourism bodies threw their weight behind such models (cf. UNESCO's Cultural Tourism Programme, cultural tourism developments funded by the European Union and the White House Report on Cultural Tourism in the USA). Cultural tourism arguably became a 'good' form of tourism, widely viewed as sustainable and supporting local culture (Richards 2001). Cultural tourism also became equated with 'quality tourism'; a factor increasingly important in areas which are experiencing decreasing returns from traditional forms of mass tourism (see Russo and Arias Sans, Chapter 10).

The expanded role for culture in tourism development mirrored the growth of culture as a factor of development in general. Zukin (1995) showed how the growth of culture-led development was tied to the work-ings of the symbolic economy. Culture provided the symbols, such as museums, art galleries and iconic architecture, which could be used to increase land values and stimulate business activity. This growth in turn supported employment in the cultural sector, strengthening its lobbying for more investment in culture. This produced powerful arguments to preserve the heritage of the past and expand contemporary culture in order to maxi-mize the 'real cultural capital' of places. Culture has therefore come to play an important role in distinguishing places from each other. This is increas-ingly essential in a globalizing world where place competition is fierce, and cities and regions strive to create distinctive images for themselves (Richards and Wilson 2004a).

One of the problems inherent in cultural distinction strategies is that many places adopt similar strategies (often copying or 'borrowing' ideas from one another), and therefore even 'culture' begins to lack distinction. The growth of 'serial reproduction' of culture epitomized by McGuggen-heimization (Honigsbaum 2001) and the spread of signature architecture to cities around the world makes it harder and more expensive to use material culture to distinguish places (Richards and Wilson 2006).

In the McGuggenheim marketplace, it is no longer enough to have culture – one has to have a cultural brand in order to compete (Evans 2003). The need to bundle and identify cultural resources led to the development of 'cultural quarters', 'cultural districts', 'creative clusters' or 'creative districts', where cultural and creative producers were clustered in order to generate a 'buzz'. In such a climate, the 'creative turn' became almost a logical successor to the cultural development process of previous decades.

The creative turn in culture-led development was therefore stimulated by a number of basic factors. First, the development of the symbolic economy (Lash and Urry 1994) privileged creativity over cultural products. Second, regions and cities have increasingly used culture as a form of valorization (Ray 1998). Third, the sheer proliferation of 'real cultural capital' created the need to find new cultural signs to create distinction in an increasingly crowded marketplace. Finally, places which do not have a rich built heritage or iconic architecture and therefore lack the 'real cultural capital' needed to find a new means of cultural development.

As a result, a wide range of cities and regions found the creative development option very attractive. Places which had already been down the road of culture-led redevelopment began to examine creativity as a added dimension of cultural development which could help them shine on the global stage. At the same time, places which lacked the 'hard' cultural resources to compete effectively in the cultural arena saw creativity as one of the few alternatives to cultural development.

Creative strategies therefore emerged in cities such as London and New York, which already had an overpowering store of cultural resources; in more modest 'cultural capitals' such as Barcelona, Porto and Graz; and in seemingly unlikely cultural centres such as Huddersfield, Winnepeg, Taipei and Lille (e.g. Pappalepore 2007). Whole countries and regions are now profiling themselves as 'creative'. Perhaps the first example was Australia, which in drafting its first national cultural policy in 1994, positioned itself as a 'Creative Nation', aiming to:

> increase the comfort and enjoyment of Australian life. It is to heighten our experience and add to our security and well-being, in that it pursues similar ends to any social policy. By shoring up our heritage, in new or expanded national institutions and adapting technology to its preservation and dissemination, by creating new avenues for artistic and intellectual growth and expression and by supporting our artists and writers, we enable ourselves to ride the wave of global change in a way that safeguards and promotes our national culture.
>
> (Commonwealth of Australia 1994)

The Helsinki region in Finland now positions itself as 'the most creative region in Europe', thanks to its high rating for research and development and ICT employment in a study of creativity in European regions by Florida and Tinagli (2004). In the United Kingdom, a number of regions are striving to develop a creative brand, including Northern Ireland, Yorkshire and the East of England.

The inspirations for such creative development strategies can be located in the general idea that creativity could deliver wider benefits than a cultural strategy alone. Culture, seen as relatively static and generally anchored in the past, needed creativity to inject dynamism and release the potential of people and places. The focus of many development strategies has therefore shifted from the cultural industries to the creative industries.

The term 'cultural industries' was initially coined in the 1940s by Horkheimer and Adorno (1972), who were key members of the Frankfurt School, and fiercely critical of the commodification of art. They saw the cultural industries as the producers of repetitive cultural products for capitalist mass consumption. However, policymakers later began to use the 'cultural industries' as a more inclusive concept of culture than traditional or 'highbrow' approaches which dominated discussions of public intervention in the cultural field. The first practical application of the term seems to stem from the 1980s when the Greater London Council (GLC) began using the term 'cultural industries' to refer to cultural activities which fell outside the public funding system (O'Connor 1999). This concept was taken up by other British cities and also by countries such as Germany and Australia. The emphasis of the cultural industries approach fell on the creation of employment and economic impact. As Ratzenböck *et al.* (2004: 9) argue:

> All these concepts have a pragmatic orientation in common. None of the used definitions originate from academic discourses but they are rather the result of economic policy and highly site-related concept formations. The notion of cultural industries has been very much driven by those involved in framing policy.

The transformation of the 'cultural industries' into the 'creative industries' arguably stems from the 'media boom' of the 1990s, where emerging sectors of cultural production, such as multimedia and software production, the audio-visual industries, architecture and design became increasingly hard to encompass within traditionally defined sectors of the cultural industries (Ratzenböck *et al.* 2004). The term 'creative industries' was apparently first used by the UK Creative Industries Task Force (CITF) in its *Mapping Document* (1997) which defined the creative industries as 'those activities which have their origin in individual creativity, skill and talent, and which have a potential for wealth and job creation through the generation and exploitation of intellectual property' (CITF 2001: 5).

This approach has been followed by a number of other countries and regions, but not everybody is happy with the shift from culture to creativity. For example, Finland decided to use the term 'cultural sector' rather than 'creative industries', because the latter was seen as being too narrow, and linked too closely to intellectual property issues. Koivunen (2005: 29) argued that:

> The concept of the 'cultural sector' covers creative work in different branches of the arts in the traditional field of the arts and culture, right through to distribution. The 'creative sector' is a parallel concept to the 'cultural sector', which appears particularly in the English-speaking

world and in practice often approaches the concept of the 'copyright industry' in its extent.

Similar discussions surround the relationship between creativity and innovation. Again, the Finnish approach to this question is fairly broad: 'Creativity involves producing new meaning and linking things together in new ways. Creativity is not only connected with art, culture and science, but with all forms of human activity' (29–30).

Whereas:

> 'Innovation' is the introduction of new creations or bringing the meaning of new creation into common use by society. This may happen through the *productization* [sic] of creative meaning, i.e. by the production of creative products. Creativity emerges in the right conditions and in the right operating environment. Innovations are thus consolidation processes in creative meaning combined with changes in social behaviour.
>
> (30)

Some of the contributions to the current volume make it clear that such discussions continue. For example Evans (Chapter 4) argues that some countries have deliberately avoided the creative label, and in Chapter 15 Rogerson illustrates the fact that 'Creative South Africa' is being developed as part of a wider cultural strategy.

The shift from cultural to creative industries is not merely an addition of new sectors to the former. By combining the 'arts' with industrial production, the creative industries concept also challenges previous dichotomies between high and popular culture or elite and mass culture (Cunningham and Hartley 2001). This shift might also be interpreted as an attempt to overcome the 'instrumental–intrinsic divide' in approaches to culture noted by Belfiore and Bennett (2006: 6–7). They show how the perceived 'instrumentalization' of culture has led to calls for a return to the 'art for art's sake principle' as the guiding rationale for cultural policy. Those in the cultural sector arguing for a more intrinsic approach to cultural policy say that:

> the language of government policy towards the arts does not recognize their special nature, but treats them as if they were no different from any other economic sector. It is no accident that museums, galleries and theatres are rolled up by government ministers into the one economic/industrial category – the 'creative industries'.

However, this 'instrumental' approach has proved popular with policy-makers, particularly in large cities. As Hannigan outlines in Chapter 3, creative approaches were applied to urban policy by Charles Landry and Franco Bianchini (1995) in their *Comedia* report on *The Creative City*, and

Landry (2000) later crystallized these ideas into a 'toolkit' for urban managers. The widespread influence of the creative economy approach can be judged from the recent report on *The Economy of Culture in Europe* (KEA European Affairs, 2006), which provides a review of the cultural economy policies of EU countries, most of which now include creativity in their definitions of cultural economy. The report proposes that Europe should move towards becoming more creative in order to meet growing competition from other world regions. The definitions used in this report separate the cultural and creative sectors. The 'cultural sector' includes industrial and non-industrial sectors and culture constitutes a final product of consumption, which can be consumed in situ (for example at a concert) or packaged for mass distribution (as a CD, for example). The 'creative sector' provides inputs for the production process of other economic sectors and becomes a 'creative' input in the production of non-cultural goods (such as fashion, design, etc.). In this view, culture is an 'input' to the creative sector; a raw material for creative production.

The creative class

At about the same time as the creative industries were being touted as a solution to the problems of urban redevelopment, other studies began to identify the emergence of the mobile, creative consumer. In the social sciences in general, increasing attention is now being paid to mobility (Sheller and Urry 2004), and particularly to the movements and influence of 'transnational elites' (Willis *et al.* 2002). As cities and regions compete to attract these mobile elites, there is a need to profile their members, their motivations and drivers. A number of studies have pointed out the links between mobile transnational elites and fast-growing creative sectors. In the United States, Ray and Anderson (2000) argued that the postmaterialist 'cultural creatives' accounted for about 26 per cent of the American population, and were driving the demand for creative activities and 'green' products. However, the most important boost to the application of creative strategies came with the publication of Richard Florida's *The Rise of the Creative Class* (2002), in which he argued that the basis of economic advantage has shifted away from basic factors of production, such as raw materials or cheap labour, towards human creativity. Cities and regions therefore have to develop, attract and retain creative people who can stimulate innovation and develop the technology-intensive industries which power economic growth. These creative people collectively make up the 'creative class'. A number of chapters in the current volume examine different aspects of the creative class concept and its relationship to tourism (see particularly Hodes *et al.*, Chapter 11 and Collins and Kunz, Chapter 13). At a very basic level, however, what Florida has done is to popularize the notion of mobile elites, giving them a more snappy label and linking them to specific 'sexy' economic sectors.

Crucially, Florida made the link between creative production and consumption, arguing that the 'creative class' were not just important for the production of creativity, but also its biggest consumers. This relates very closely to evidence from the ATLAS Cultural Tourism Project, which has consistently shown the important role played by those employed in the cultural and creative sectors in consuming cultural tourism (Richards 2007).

Florida also emphasizes that what is important to the creative class is the 'quality of place', which combines factors such as openness, diversity, atmosphere, street culture and environmental quality. These relatively intangible factors are now arguably more important than traditional cultural institutions such as opera houses or ballet companies in the locational decisions of creative people. One might assume, therefore, that tourists would also be attracted to such places, since many tourists are in search of 'atmosphere' and difference. As the ATLAS Cultural Tourism Research Project has shown over the past 15 years, there has been a shift away from specific cultural attractions (such as monuments and museums) towards intangible factors such as 'atmosphere', gastronomy and linguistic diversity in the attractiveness of places (Richards 2007). This also applies to the development of leisure facilities. Hannigan argues in Chapter 3 that the commercialized leisure attractions of the 'fantasy city' are no longer sufficiently engaging for the creative class, who seek instead the 'controlled edge' of urban areas in transition.

A further interesting feature of Florida's work is the link that he makes between the growth of the creative class and the search for experience. 'Experiences are replacing goods and services because they stimulate our creative faculties and enhance our creative capacities' (2002: 168). Echoing Pine and Gilmore (1999), he warns of the dangers of experiences becoming commodified and places becoming less 'authentic'. The ability of places to provide the unique experiences that attract people is seen as vital to their economic health, to the extent that 'place is becoming the central organizing unit of our economy' (Florida 2002: 224). The combination of where we live and what we do has become the main element of our identity, and this turns place into an important source of status, replacing abstract 'cultural capital' and embedding 'creative capital' more firmly into concrete places.

However, there has also been much criticism of Florida's work, mostly centring on the identification of the creatives as a 'class', and the various indicators used to measure creativity. Peck (2005) provides a detailed review of the issues raised with the creative class concept. The different types of criticism made of Florida's work can be summarized briefly as follows:

The definition of the 'creative class'

In particular, many researchers have criticized the lumping together of widely disparate activities, such as the performing arts and technological innovation into a single 'class'. In fact, it can be argued that there are many different class factions who make up Florida's creative class, many of whom have opposing interests.

There are also doubts about the inclusion of such a large proportion of the working population in the creative class. In the different contributions to the edited volume by Franke and Verhagen (2005) on *Creativity and the City*, for example, contributors quote widely differing figures for the extent of the creative class in the Netherlands. At one extreme is Florida's own estimate that the Dutch creative class accounts for 30–40 per cent of the working population, whereas Dutch scholars provide figures varying between 13 per cent (in the widest definition) to less than 2 per cent (a narrow definition of the 'core' creatives). In particular, there is much criticism of Florida's inclusion of high-technology workers in the creative class.

Issues of diversity

The minority and countercultural groups championed by Florida as creators of diversity are arguably being overtaken by commodification and gentrification. Howell (2005: 32) argues that 'bohemian' or 'countercultural' lifestyles, such as skateboarders in Philadelphia's Love Park, are 'becoming institutionalized as instruments of urban development', a phenomenon also witnessed in Barcelona's Plaça dels Angels. This is an argument taken up by Hannigan in Chapter 3 of this volume.

Binnie (2004) also points out that the producers of diversity also face problems trying to live up to their role:

> These urban transformations clearly bring benefits. But what are the costs? It seems that these are borne by those whose presence in urban space is more fragile and tentative: the disadvantaged, the poor, or those who are unable or unwilling to conform to a certain standard – not having the right kind of body and cultural capital for the new hip urban gay identity that is promoted on lifestyle programmes.

The relationship between creativity and economic growth

To some critics, Florida's suggestion that increased creativity stimulates economic development seems to be a chicken and egg argument. What comes first – employment or creative people? Wealthy places, it could be argued, are more likely to have the cultural facilities that might attract the creative class, so it might be that creativity follows economic development. As Peck (2005) notes, many of the relationships put forward by Florida are based on correlations, rather than causality.

Misreading the causes of economic growth

Some cities which score high on the creativity index have not performed as well as Florida has suggested, particularly in terms of job creation, and in some cases top 'creative cities' are actually losing population. However, Florida (2005) argues it is the quality of jobs that counts, not the number.

The indices of creativity

One of the key elements of Florida's work was the development of indicators which could be used to measure the creativity of different places and therefore produce rankings of creative places. These creativity league tables created immediate interest from cities at the top of the table, which used their high rankings to market themselves as creative cities, while cities lower down began to devise strategies to improve their rankings. Florida's analysis largely rests upon the correlations between indices of what have been described as the 'three Ts' of economic growth: talent (measured by the percentage of residents with a bachelor's degree), tolerance (percentage of the population that is foreign-born) and technology (employment in high-technology industries). It has been argued that the correlation between technology, well educated residents, foreign born population and the presence of artists is hardly surprising. The technology index tends to be highest in large cities, which also tend to attract more immigration and have a larger cultural sector. The indices also seem to underrate some places which are perceived as being relatively creative. For example, in the study of creativity in European countries (Florida and Tinagli 2004), Italy rated very low, and yet the country is famed for its design and fashion industries. It therefore seems to be questionable what Florida's indices are actually measuring.

In the realm of tourism, one would expect that cities which score high on Florida's creativity indices would also attract more visitors as they become more attractive to the creative class. As a small test of this idea, we looked at the creativity rankings produced for a DEMOS (2003) study of UK cities. In this study, Manchester ranked as the most creative of 40 British cities in terms of its 'bohemian index', above such cultural tourism icons as London and Edinburgh.

However, a rank correlation analysis of the creativity index and the change in overseas visitor numbers for 12 of these cities for which data are available indicates a fairly weak negative relationship (Table 1.1). The rank correlation between creativity and the change in tourist numbers was −0.125, indicating that cities which ranked higher in terms of creativity actually had less overseas tourism growth between 2000 and 2003. This does not necessarily mean that creativity does not attract tourists, but it may show that the indicators used by Florida to measure creativity do not say much about the attractiveness of a city for tourism. It is clear, for example, that major cities such London have been losing ground in cultural image terms

Table 1.1 Relationship between creativity and change in overseas tourism arrivals in UK cities, 2000–2003

	Bohemian Index Rank	Overseas visitor change rank 2000–2003
Manchester	1	2
London	2	12
Leicester	2	4
Nottingham	4	5
Bristol	5	10
Brighton/Hove	6	7
Birmingham	7	1
Coventry	8	7
Cardiff	9	7
Edinburgh	10	11
Liverpool	17	3
Newcastle-upon-Tyne	21	5

(Richards 2007) to smaller cities such as Liverpool and Newcastle, which both score poorly on the creativity indices.

The global economy

Florida's view of urban economies has also been criticized for lacking global scope. Friedman (2005) agrees with Florida that creative people are essential assets in the modern economy, but he argues that these people can be found anywhere. Globalization has effectively made the world 'flat' in economic terms, so that where creative people live is no longer so important as the skills they have. For Florida, cities should strive to attract creative people to move there, whereas Friedman argues that transnational firms are seeking out the creative class in the places they live now.

Wider critique of creativity

In addition to the specific critique of Florida's work, more general complaints have been made about the current vogue for creative development.

Creativity is a bandwagon

One of the most frequently heard complaints is that creativity is simply the latest in a whole line of trendy development bandwagons that cities have jumped on. This process has been stimulated by the ideas of Florida (2002), who warns that cities will get left behind if they do not become creative now.

There is nothing unique about uniqueness

As discussed above, the notion of creativity sits comfortably with the post-modern notion of identity and wanting to be different. In fact, most places actually want to be not just different but 'unique'. The problem is that there is actually nothing unique about uniqueness. If every destination profiles itself as unique, then the consumer has to go back to generic categories and themes to choose between them (or their copycat 'creative' strategies).

There's nothing new about creativity

In spite of the recent attention paid to creative development strategies, it should be recognized that the application of creativity to urban and rural development is not new. Amsterdam and Florence were the 'creative cities' of the sixteenth century, opening their doors to the 'creative classes' long before Florida. Artists' colonies have long existed and been a source of tourism development, as in the cases of St Ives in Cornwall (United Kingdom) or the Lake District (United Kingdom).

The creative turn is an academic fiction

One of the most compelling critiques of the creative turn is that it stems from an overly academic reading of society. Just because academics, as part of the intellectual superstructure of the information society and arguably members of the creative class, have decided that there is a shift towards creativity does not mean that it actually exists outside the academy. As Gibson and Klocker (2004) have pointed out, academics are part of the critical superstructure producing creative discourse, because it suits their world view. For the vast majority of the world's population, 'creativity' is actually as far out of their reach as the economic riches that it is supposed to generate.

You can't manage spontaneity

Spontaneity is an important part of creativity. The visitor and the resident alike appreciate the unexpected, which the spontaneous creative process can throw up. Cities such as Barcelona are seen as attractive because of their surprise element – the chance encounter with street art or traditional festivals (Vallbona and Richards 2007). This can be seen as the obverse of 'staging', which as Binkhorst points out in Chapter 8 relies on the McDonaldized principles of predictability and calculability. It is also ana-thema to the 'Fantasy City', which is one reason why Hannigan argues in Chapter 3 that creative strategies offer an alternative. This creates a dilemma for the creative destination – how do you maintain spontaneous creativity in a setting where the ability to deliver a predictable experience is vital in attracting visitors? Evans, in Chapter 4 of this volume, also argues that creativity cannot be branded, because brands kill spontaneity.

The idea that creativity is a freer form of cultural expression than the (re)production of culture is attractive not just to the producers of creative products, but also to many policy analysts, particularly in free market economies. As Hannigan (2002a: 3) comments: 'at the same time that regions have the power and resources to secure cultural institutions and practices through subsidization, protection, and regulation, they interfere with the ability of artists, writers, and performers to grow and adapt.'

The attractions of creativity

In spite of these criticisms, Florida's arguments continue to receive considerable attention, not least from the public sector, desperate for strategies to improve economic performance and quality of life. Creative development strategies have quickly spread to different cities, national governments and international bodies. For example, the United Nations Conference on Trade and Development recently published a report on 'Creative Industries and Development' (2004) and UNESCO launched a 'creative cities network'.

Why has Florida attracted so much attention? His arguments seem to have arrived at the right time for many cities and regions struggling with development issues in a globalizing environment. His league table ranking cities by their level of creativity appealed to policymakers who saw themselves caught up in a competitive struggle to attract employment and consumer spending. Other factors include:

- The attractiveness of creativity as a concept which fits well with the current 'fast policy' climate (Peck 2005) and neoliberal concepts of the entrepreneurial city. Creativity also seems to have a more general appeal linked to the desire for change in a new millennium and the increasing attention paid to personal development and fulfilment.
- The development of the 'experience economy' (Pine and Gilmore 1999), which seems to link strongly with the idea of the creative class. The creative class are apparently the 'imagineers' needed to develop experiences for the experience economy.
- The concept of the 'creative city' or 'creative region' also fits with wider trends in management theory, which emphasize the value of creativity as an economic resource (Porter 1998). O'Donnell (2004: 11) remarks, 'Pop management theory has, since the early 1980s, raised "creativity" to the level of an entrepreneurial imperative.' As Gibson and Klocker (2004) point out, the creativity gospel is also spread by academics (the current volume being a case in point) and celebrity 'gurus' such as Richard Florida and Charles Landry.

Richards and Wilson (2006: 15) also identify a number of reasons why creativity is now more popular than traditional cultural approaches to development:

- Culture is often associated with 'high culture', which has a traditional, staid image.
- The cultural sector is not perceived as being very flexible or dynamic.
- The creative sector is broader than the cultural sector alone, covering more sub-sectors and having a greater total value and employment impact.
- The creative sector is closely linked to innovation and change.
- The creative industries include many more aspects of visual consumption (advertising, cinema, design, fashion, video games).
- Women often play a key role in the development of the creative industries.

The dynamic image of creativity is also one of its attractions. Research of Local Authority image-making in the Netherlands (Cachet *et al.* 2003) shows that 'dynamism' is one of the most desirable images for an area to have. Creativity, which is linked closely to ideas of innovation and novelty, is therefore an attractive aspect of development strategies, rather than 'culture', which tends to be linked to tradition and a lack of change.

The emphasis on creativity can also be linked to a wider shift towards a plural vision of society in an era of increased mobility and social fragmentation. 'Culture', usually linked to existing social groups and structures, is not as flexible as 'creativity', which is process-based, fluid and apparently more egalitarian. The growing diversity of postmodern society is also seen as a resource for creativity. As Shaw (Chapter 12) and Collins and Kunz (Chapter 13) argue in this volume, ethnic diversity is a potential stimulus for creativity.

Another feature of contemporary society which dovetails well with creative development models is the blurring of boundaries between production and consumption. As Binkhorst argues in Chapter 8, involving the consumer in the production of experiences is now so important that one can talk of a process of 'co-production'. Experiences therefore depend not just on the creativity of the producer, but also the consumer. Attractions which do not allow space for the consumer to create their own experiences or help to shape the experiences produced by creative enterprises, are arguably not exploiting the full potential of 'creative tourism' (Richards 2001).

Whether one accepts Florida's arguments or not, there does seem to be a feeling among many place managers that there is a link between creativity and place performance. If the indices developed by Florida cannot identify the exact causes of this relationship, this is perhaps of little relevance for those engaged in concrete development and marketing strategies, and perhaps of greater relevance to academics. The key question for the policy-makers is whether such strategies work or not. In the end, the best advice may be to take a balanced view. A report prepared for the Department of Canadian Heritage (Donald and Morrow 2003: ii) attempts to provide such a balance, citing successful examples of creative strategies while warning against the 'dangers of a quick and careless translation of the talent model into public policies'. In particular, they argue that 'surface-level place marketing may have the potential of glossing over the essential investments required to maintain and enhance a creative city'. But if creativity is applied purposefully and carefully, it has the potential to provide many benefits for the host community.

The main problem, as Peck (2005: 744) suggests, may be figuring out just who the 'creative class' actually are, and how to deal with their needs:

> figuring out what the Creative Class wants means adopting an entirely new analytical and political mindset, and, even then, learning to accept that creatives will not be pushed around, that their behavior will be difficult to predict, and that above all they need space to 'actualize their identities'.

What is creativity?

The extensive discussions surrounding creative development and marketing strategies raise the important question – what is 'creativity'? In fact, in most of the discussions of creativity, one finds no definition of the term. This may be because creativity is seen as something 'mystical' and multidimensional and therefore difficult to pin down (Florida 2002). But there are a few key elements of creativity which tend to be widely repeated, and which also turn up in the many definitions that do exist. For example, the Oxford English Dictionary defines creativity as being 'inventive, imaginative; showing imagination as well as routine skill'. Chartrand (1990: 2) argues that '[individual] creativity occurs when an individual steps beyond traditional ways of doing, knowing and making'. These ideas suggest that creativity is above all about developing new ways of thinking and doing, which explains the close link often made between creativity and innovation.

In this sense, creativity could be applied to tourism through the development of new products or experiences; of new forms of consumption or new tourism spaces. Arguably, any form of tourism which is related to imagination, whether the imaginative capabilities of the producers or consumers of tourism, could be considered to fall within the sphere of 'creative tourism'. As Evans argues in Chapter 4, this includes a wide range of areas including design, architecture and the media, and Cloke in Chapter 2 extends the creative range to include the creativity of nature.

In fact, as the chapters in this volume show, there are a large number of ways in which creativity is interpreted and applied, both within the wider cultural development field and in tourism in particular. Among the uses of the term 'creativity' in this volume, we can identify:

- creativity as a product
- creativity as an experience
- creativity as innovation
- creativity as a marketing strategy
- creativity as an industry sector
- creativity as a social development strategy
- creativity as a landscape
- creativity as problem-solving
- creativity as a blanket term for heritage and cultural tourism
- creativity as a challenge to identity
- creativity as difference and diversity.

It is clear that there remains much discussion about the nature, scope and application of creativity, and that this debate will continue for some time to come. In the meantime, our approach in this volume is to examine the many ways in which concepts of 'creativity' have been applied in the tourism sphere, and the differing outcomes which these applications have had, regardless of the purely creative content of such ideas.

Developing creativity

In order for creative development strategies to work, there seems to be general agreement that there needs to be a critical mass of creative production, and this in turn is reliant on a sufficient volume of creative consumption. Ritzer (1999) has argued that the developing (post)modern consumer economy is driving a shift away from the means of production towards the 'new means of consumption'. The new means of consumption comprise the range of settings that facilitate people in their consumption activities, including fast food restaurants, cruise ships, casinos and museums (and increasingly the mixture of different settings). The need to 'enchant' these basic settings in order to entice the consumer entails a creative imperative, something that the Disney Corporation and other key experience producers have long understood. The difference is that now even the most experienced experience producers have to learn to engage consumers in the process of designing, distributing and performing the experiences themselves, to the extent that they become 'prosumers'.

This new landscape of prosumption offers opportunities for a range of strategies for creative tourism development, for example:

- tourism based on the consumption of creative media;
- tourism based on the creative input of the consumers themselves;
- traditional forms of tourism consumed or produced in a more creative way.

The importance of image in the new means of prosumption also gives particular prominence to the creative industries as a channel for developing and diffusing destination images. This is evident in the current trend towards 'film-induced tourism' (Beeton 2001). As Cloke outlines in Chapter 2 of this volume, film and TV productions are becoming increasingly important as tools for destination marketing: attracting tourists to locations where they are set or filmed. In some cases, film-induced tourism may relate to other areas of creativity, as in the case of films based on books, which can support the development of 'literary landscapes' (VisitBritain 2007). In other cases the pure act of filming in a particular location may generate tourism, even if the story does not relate to the 'real' (or even imagined) setting of the film, such as 'spaghetti westerns' attracting visitors to the Almería province of Spain, a frequent stand-in for the 'Wild West'.

As Vanolo (2006: 7) puts it:

> the construction of the image of the creative city lies in the building up of visual symbols – landmarks – of creativity, specific narrations, empha-sizing the creative milieu of the cities, and the location of connected functional and, above all, cultural references.

This process is, Vanolo argues, based on a range of key ideas which are increasingly utilized in creative development and marketing strategies:

- The 'buzz', i.e. scenes with people, and particularly scenes of people meeting and chatting, with a special attention towards situations of multi-ethnicity.
- The local art scene, referring both to 'official' and more 'popular' forms of art.
- Nightlife, both in the case of fancy restaurants and other places for young and trendy people.
- Public spaces, particularly natural environments and parks, together with outdoor sport situations and landmark buildings.
- Representations of high quality education, both for young people and professionals, widely considered as an essential element of urban competitiveness.

What Vanolo essentially seems to be talking about here is the creation of a 'cosmopolitan' atmosphere, which then becomes packaged in marketing strategies. The picture that emerges is one of a general shift away from tangible culture and heritage towards intangible culture and creativity. This affects tourism products of all kinds (Figure 1.1) as well as the cultural tourism field specifically (Figure 1.2). Destinations are having to learn new ways of developing and marketing tourism as the emphasis moves from traditional forms of culture to new, more diffuse resources.

The modern art of creative development therefore lies in transforming intangible elements of the culture of a place into 'experiences' that can be

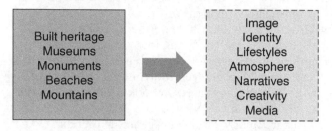

Figure 1.1 The shift from tangible to intangible tourism resources.

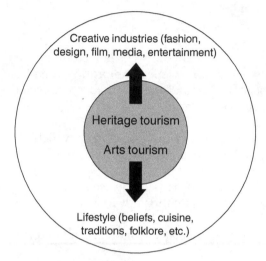

Figure 1.2 The shift from tangible to intangible cultural resources in tourism.

consumed by tourists. This is a complex process that requires the coordination of cultural and creative hardware, software and orgware:

- creative hardware – infrastructure/spaces for creative production, consumption and prosumption;
- creative software – atmosphere/ambience, fashion, quality of life, perceived diversity, 'vibrancy';
- creative orgware – sectors, industries, clusters, policies, governance.

Because many regions already have creative hardware, much attention has been paid in recent years to the development of creative orgware and software. In terms of orgware a number of cities and regions have established organizations responsible for the creative development process, such as Creative London, Vancouver's Creative City Task Force, Creative Auckland and the Cool Cities Initiative in Michigan. As Landry (2005) indicates, these organizations provide creative spaces and other support, such as financial and business advice, information and networking. These types of organizational development are spreading rapidly from one city to another. For example, a recent report on Creative New York (City Futures Inc. 2005) recommended that New York establish a centralized coordinating body for the creative industries, modelled on Creative London.

Destinations have also begun emphasizing their software, such as 'quality of life' for citizens and visitors alike. Prentice and Andersen (Chapter 6) illustrate the impact of these changes in French towns which have shifted from conservation of their historic fabric in the 1980s to the current focus on instilling pride in the residents to creating a liveable (animated) city. The city is no longer just a machine for living, but a landscape for enjoying, experiencing and tasting. This also seems to mark a shift away from the primacy of the tourist 'gaze' into other (multi)sensory realms of production and consumption. This 'sensory shift' (e.g. Dann and Jacobsen 2003; Ferrari *et al.* 2007) is also marked by the growing importance of intangible culture in cultural tourism products and marketing.

The growing role of intangible culture and creativity in tourism production and consumption raises a new challenge for destinations, because much intangible culture appears to be more footloose than tangible cultural heritage. Mobile intangible culture presents more problems in terms of protecting intellectual property than tangible heritage, as movie makers and music companies have discovered in recent years. Protecting the intellectual property bound up in intangible tourism resources requires linking creativity to place by establishing a link in the mind of the visitor between particular manifestations of creativity and specific locations. This also tends to privilege local, embedded knowledge over universal, abstract knowledge. In the past, cultural tourism has tended to be dominated by universal symbols related to high culture, such as Greek myths, the works of Shakespeare, Van Gogh or Picasso. This system was supported by the use of high culture to develop the identity of the nation state, but the rise of the new regionalism has now produced a raft of more localized, embedded cultural symbols alongside these global or national symbols.

The embedding of creativity has also allowed specific locations to accumulate and promote 'clusters' of creative activity. These clusters have become important not just in the local economy, but also as magnets for the 'creative class' and as a resource for developing a creative image (Florida 2002). Much of the current thinking about the 'creative' therefore also has its roots in the study of clusters and networks in the past 30 years (e.g. Simmie 2001). The combination of creative hardware, orgware and software can be used by cities and regions to develop a range of experiences for tourists as well as residents. Richards and Wilson (2006) summarize these combinations into three basic types of creative tourism experience:

- *Creative spectacles.* Creative and innovative activities which then form the basis of more passive tourist experiences as spectacles (i.e. production of creative experiences for passive consumption by tourists).
- *Creative spaces.* Creative enclaves populated by cultural creatives to attract visitors (often informally at first) due to the vibrant atmosphere that such areas often exude (e.g. Down Under Manhattan Bridge Overpass – DUMBO – in Brooklyn).
- *Creative tourism.* Active participation by tourists in creative activities, skill development and/or creative challenge can form the basis of tourist experiences, which can also imply a convergence of creative spectacles and creative spaces.

In tourism terms, the shift towards creativity can be seen as part of an evolution in the basis of tourist experiences (Figure 1.3). In the early stages of the development of mass tourism, the essential value of holidays for many people was the aspect of 'having' – having a holiday conferred a certain status, an intangible extension of having physical possessions such as a car, or a colour TV. As holidays became more of a normal part of everyday life, the emphasis shifted to what one saw or 'did' on holiday. This is the point at which Urry's (1990) tourist gaze became a primary discourse in the analysis of the production and consumption of tourism. However, more recently people have begun to tire of seeing an endless series of 'sights' or 'doing' a series of standardized activities.

There is evidence to suggest that in current modes of tourism consumption, the source of distinction lies increasingly in the arena of 'becoming' – moving away from having or consuming goods and services towards becoming transformed by the tourism experience itself, as Binkhorst notes in Chapter 8. In the past, cultural tourism seemed to consist of collecting

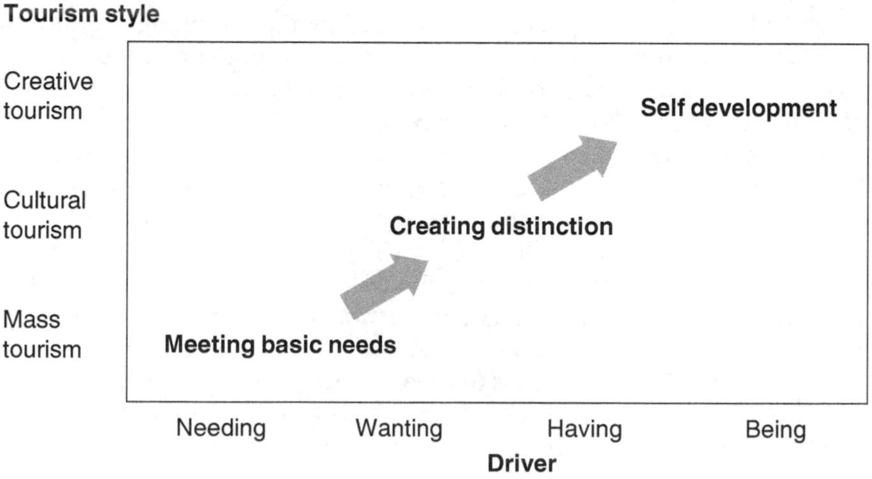

Tourism style

Figure 1.3 Changes in the drivers of tourism over time.

'must see sights', which acted as badges of cultural consumption. Now it seems that existential elements of cultural consumption, such as 'soaking up the atmosphere' are enough. Many 'cultural tourists' these days seem to want to become part of the local community and have direct contact with the everyday lives of others (Richards 2007). Wang (1999) emphasizes the growing importance of 'existential authenticity'. The idea of 'being' on holiday places more emphasis on the creativity of the tourist, rather than seeing them as passive consumers or gazers upon a series of staged experiences.

This in turn has shed light on the performative dimensions of tourism. As Edensor (1998) shows in the case of tourists visiting the Taj Mahal, the roles that tourists play become an important part not only of their own holiday experience, but also those of other tourists. In taking on different roles as tourists, people also develop narratives about themselves and their travel (Noy 2006). Travel experiences therefore become the raw materials used to develop a life biography, and travellers can also play with and shift their identities as they travel (Richards and Wilson 2004b).

In order to play their role effectively, tourists need to develop their performance skills (in other words, creativity). Highly developed tourist performance skills allow one to shift roles and become a prosumer rather than just a consumer. The destinations that are visited by these tourists can also draw upon the multiple identities of tourists to create or enhance their own identities. In the same way that tourists play with their identities, places can also take on and develop multiple identities. In the case of Amsterdam, for example, Dahles (1998) demonstrates how the city has two faces – the historic, cultural city and the liberal city linked to the Red Light District and the gay scene (see Hodes *et al.*, Chapter 11 in this volume).

The administrators of the city of Barcelona recognize that creativity discourse has a role to play in underpinning the image and identity of the city, not least by staging the 2003 Year of Design event. In terms of how this has penetrated the popular consciousness of visitors, our research indicates a close link between the tourism consumption patterns and identities of cultural tourists and the creative elements of the city. Barcelona is seen as a 'creative' city, particularly by those working in the cultural sector and cultural tourists. The city has also shifted towards intangible culture and creativity in its tourism marketing, and is now beginning to develop 'creative tourism'.

Surveys of tourists visiting cultural sites in Barcelona (Richards 2004; Richards and Wilson 2007), indicated that almost 58 per cent considered the city to be 'creative', almost the same proportion that thought it was 'historic' (Table 1.2). Perhaps not surprisingly, those who were least likely to see Barcelona as 'creative' were beach tourists, who are arguably the classic 'unskilled consumers' (Richards 1996a). But the group that was most likely to label Barcelona as a creative city were those who did not see themselves as 'tourists'. This group consisted of younger (predominantly 20–30), highly educated, professional, white collar workers with cultural occupations, who were more likely to be frequent repeat visitors, and more likely to be seeking 'atmosphere'. They also saw Barcelona as an architectural city and a multicultural or diverse city. They also tended to have made more visits to the Centre de Cultura Contemporània de Barcelona (CCCB) and the Raval than other tourists, underlining their predilection for 'creative spaces'.

Different groups of tourists clearly saw and consumed the city in different ways. Heritage tourists were the most likely to appreciate the historic and

Table 1.2 Image of Barcelona as a city by tourist type

Tourist type	What type of city is Barcelona? (% respondents)						
	Creative city	Art city	Historic city	Cultural city	Modern architectural city	Multi-cultural/ diverse city	Industrial city
Cultural tourist	64.3	62.8	66.8	78.1	51.0	51.5	18.4
City tourist	61.1	64.9	66.4	79.4	42.7	42.0	19.1
Heritage tourist	61.5	61.5	84.6	92.3	46.2	30.8	30.8
Beach/sun tourist	41.7	41.7	58.3	69.4	41.7	33.3	13.9
Business tourist	66.7	66.7	58.3	83.3	50.0	41.7	16.7
Don't like the term 'tourist'	68.3	60.0	60.0	73.3	58.3	53.3	20.0
All visitors	57.8	56.3	59.3	72.7	44.2	44.7	17.4

Source: ATLAS surveys 2004. N = 396.

Table 1.3 Correlation between images of Barcelona and its image as a 'creative city'

	Creative city correlation (Pearson's r)
Multicultural/diverse city	0.39
Modern architecture city	0.36
Art city	0.34
Historic city	0.28
Cultural city	0.27
Nightlife city	0.25
Fun city	0.25
Industrial city	0.23
Safe city	0.19
City of open spaces	0.17
Friendly city	0.17
Maritime city	0.16
Shopping city	0.15
Working city	0.12
Beach resort	0.09

cultural aspects of the city, including its industrial heritage. On the other hand, they did not notice the diversity of the city as much as 'cultural tourists' or even beach tourists. Heritage tourists were also the most likely to have visited classic heritage sites outside Barcelona, such as the religious mountain of Montserrat or the Roman ruins of Tarragona. 'Cultural tourists' visited many more sights connected with iconic figures such as Gaudí and Miró, including the Sagrada Familia, La Pedrera, Casa Batlló, Parc Güell and Fundació Miró. However, they were less likely to have visited popular cultural attractions (such as Futbol Club de Barcelona) or the town of Sitges (see Binkhorst, Chapter 8).

The image of Barcelona as a creative city was most strongly correlated with its image as a multicultural or diverse city, followed by architecture and art (Table 1.3). This seems to support the link between creativity and diversity suggested by Florida, at least in terms of city image. Creativity was also strongly correlated with a desire to live in the city (r = 0.207), which indicates that 'creatives' may also be tempted to move to places, such as Barcelona, that have a creative image.

In terms of motivation, the 'cultural tourists' were most likely to be motivated by learning new things (apparently vindicating the link between learning and cultural tourism posed by Richards 1996b) (Table 1.4). 'Atmosphere' seemed to be relatively important for most tourist types, although it scored highest with heritage tourists and those not liking the term 'tourist'. For heritage tourists the most important stimulus was 'local' culture, which seems to emphasize the fact that they are seeking 'embedded' culture. The 'non-tourists' were least likely to be motivated by sightseeing, and more frequently motivated by entertainment than by 'cultural' motives.

These figures seem to indicate close links between the creative image of Barcelona, and various aspects of the tangible and intangible culture of the city. In contrast to the analysis of the relationship between creativity and tourism at national level presented above (Table 1.1), the Barcelona example does seem to suggest that there is a direct link between creative aspects of the city and its attractiveness for certain types of tourist. The

Table 1.4   Importance of different motivations for visiting Barcelona (% very important)

Tourist type	Learning new things	Entertainment	Local culture	Atmosphere	Sightseeing
Cultural tourist	51.6	51.2	47.6	48.7	43.2
City tourist	40.6	48.0	37.5	53.9	47.3
Heritage tourist	46.2	38.5	76.9	66.7	46.2
Beach/sun tourist	38.2	55.9	41.2	57.1	44.1
Business tourist	41.7	50.0	50.0	50.0	41.7
Don't like the term 'tourist'	16.7	41.7	31.7	65.0	21.7

differential role played by creativity in city image for different groups of tourists may explain why tourist flows in general seem to be relatively unaffected by creativity ratings for a city.

Barcelona has begun to capitalize on its creative image by developing a 'creative tourism portal', which very soon after its opening in 2006 was generating considerable interest (Anon 2006: 4):

> The city council and tourism board have announced two new product lines to develop visits related to Catalan language learning, up by 60 per cent, and one to build on the success of the Year of Gastronomy, which created a new genre of visit for foodies and gourmets. Meanwhile, the recently unveiled Creative Tourism portal (www.barcelonacreativa. info) is receiving 50 enquiries per day.

This process of creative development is also being driven at grass roots level by emerging networks of creative producers and consumers, such as the 'Terminal B' website for Barcelona (www.terminalb.org).

The creative dimension of tourism is therefore no longer based on the singular production of meaning through the institutions of high culture (museums) and hierarchy (as is usually the case in cultural tourism), but also through creativity, atmosphere and narrative. The creative turn therefore implies a change in the form of production of 'culture'. Whereas culture is physically present in the museum in terms of material objects or cultural 'products', creativity as a process is intangible, often implied rather than stated, and exists in culturescapes, flows of culture and creativity across particular areas of a city or region (see Russo and Arias Sans, Chapter 10 in this volume), or the exotic and unfamiliar, crystallized into ethnic enclaves (see Shaw, Chapter 12 and Collins and Kunz, Chapter 13).

With the rise of intangible culture, there is also a more important role for 'fashion'. The atmosphere of places can change much more quickly than their physical landscapes or museum collections, and new creative elements can be added to the destination overnight. The tendency for destinations to 'borrow' from (or copy) each other also causes waves of fashionable creativity to sweep the globe, creating new nodes of creative consumption in their wake. This has been the case with many forms of festival, including the development of ethnic events such as melas, and Irish festivals, film festivals and museum nights. The creative sector is particularly useful in this regard, because creativity means new culture and new cultural forms, in contrast to 'culture' or 'heritage', which often centre on the preservation of the past and solidification of existing structures. However, developing the new, fashionable, dynamic creative product is not as simple as adding the word 'creative' to a place name. Successful creative development is a complex process which needs to recognize the delicate relationship between past, present and future, between high and popular culture, and between space and place.

Richards and Wilson (2007) identify key elements of creative development which tend to ensure their smooth incorporation into large scale urban redevelopment schemes:

- *Clustering* – Creative enterprises need a network of colleagues and suppliers, and clustering is therefore seen as providing an impulse to both individual and collective creativity.
- *Consumers* – Audiences are vital to the creative industries, and in many cases creative enterprises need to attract audiences or consumers to specific locations.
- *Co-makership* – In order to function well, creative clusters need to involve both producers and consumers in a process of co-makership.
- *Clarity* – Attracting audiences depends on a certain level of visibility within the urban fabric, and the ability of potential audiences to 'read' the creative landscape. It is also important that the creative enterprises have a certain level of permeability for the consumer.
- *Confidence* – Developers must have the confidence to invest in creativity, but trust between creative individuals is also important. Cities and regions must also have the confidence in their ability to make such developments work, and to be able to sell success.

Given this level of complexity, it is perhaps not surprising that there is an easy link between creative development and large urban areas.

The location of creativity

In much of the literature on creativity, there has been an explicit or implicit linkage of creativity and the urban (see Evans, Chapter 4). The tendency to locate creativity within the urban sphere stems from the dynamic nature of cities and their role as centres for innovation and change (Amin and Thrift 2002; Simmie 2001). The role of Florida (2002) in championing the 'creative class' has also strengthened the link between creativity and cities, because the 'creative class' according to his definition tends to concentrate in cosmopolitan, urban areas. The idea that diversity is an important resource for creativity has also tended to strengthen the links between creativity and larger cities or metropolises. Many of the chapters in the current volume also examine different aspects of creativity in urban environments, and underline the tendency for clustering and the need for large audiences which tend to favour the urban location of creativity.

Hannigan (2002a: 7) also argues that cities are 'strategic sites for research' in this respect, because they concentrate global flows of capital, information, images and people, 'the dense human relationships out of which culture flows'. The contemporary metropolis therefore constitutes a 'petri dish' in which 'the formation of new claims materializes and assumes concrete forms' (Sassen 2001: 167).

The idea of large cities as laboratories in which cultures grow from diversity means that cities are also trying to find ways to make themselves more 'cosmopolitan'. A recent report by Bloomfield and Bianchini (2003) for Birmingham City Council came up with a number of suggestions about how the city could plan for the 'cosmopolitan city'. They advocate an intercultural approach to the development of cultural facilities to achieve this. One problem they identify is the segregation of cultural facilities between ethnic groups, an effect arguably produced by the location of community facilities in the periphery. The report suggests a number of strategies to counter such problems, including relocating cultural venues to the city centre, encouraging mainstream cultural institutions in the city centre to make cultural diversity integral to their activities or adopting a very aggressive marketing strategy to promote peripherally located venues in order to achieve a city-wide appeal.

Such strategies can also be seen as part of wider attempts to develop social cohesion within cities and larger regions. This discussion has become very problematic in recent years, because of the apparent 'clash of civilizations' (Huntington 1996), and the emerging critique of 'multicultural' approaches to social cohesion which have arisen as a result. In such a climate, the concept of 'culture' becomes overburdened with meaning and increasingly difficult to apply in 'multicultural' settings. This new challenge to 'culture' has suddenly increased the appeal of 'creativity' as a relatively value-free way of allowing different cultures to express themselves. The argument is neatly shifted from cultural specifics to the general process of expression.

Expressing creativity or cultural identity is something that eventually begins to claim public space. This in turn creates new challenges for the management of that space. More and more cities have therefore been stimulated to develop creative management strategies for public space. For example, many cities in the United States have specific events departments whose job is to choreograph the creative use of the city centre. In Chicago, the Mayor's Office of Special Events is 'a city office designed to present public programs all year round' (www.cityofchicago.org/specialevents). The city is effectively 'programmed' by coordinating 'all efforts related to the event or festival including booking performances, organizing operations, facilitating sponsor partner participation and assisting in promotion'. Such creative programming efforts rely on both utilizing indigenous talent and attracting external creative performers.

The tendency of creativity to be linked to adversity is also one reason why cities, with their deprived neighbourhoods and *banlieus* (Evans, Chapter 4) are seen as nodes of creativity. As Hannigan points out in Chapter 3, the presence of a 'controlled edge' seems to be important for urban creativity. Creative people are supposed to 'live on the edge', and often perform a bridgehead function in colonizing declining areas of the city and making them attractive for others to live in, work in and visit. This is evident in the 'new tourist areas' of London analyzed by Maitland (Chapter 5) and also in the 'artistic enclaves' of other cities, such as the Jordaan in Amsterdam, the Raval in Barcelona and more directly engineered clusters such as DUMBO in Brooklyn. The 'controlled edge' concept is particularly applicable to the kind of ethnic enclaves described by Shaw (Chapter 12) and Collins and Kunz (Chapter 13), as these inner-city areas were often seen as areas to be avoided in the past, but they are now being 'pacified' for safe consumption by (largely white, middle class) residents of other areas of the city as well as tourists (Zukin 1995). The creative sector has often been at the forefront of pacification campaigns, whether in terms of 'pacification by cappuccino' (ibid.) or 'pacification by fiesta' (Richards 2007).

The development of creative strategies in cities is also closely linked to narratives of design and architecture in the contemporary city. As we have seen in the case of Barcelona, architecture, design and creativity are positively linked in the minds of visitors as part of the 'creative landscape' of the city. This linkage can be positively stimulated by the creation of events and attractions. For example, during the Gaudí Year in Barcelona (2002), a number of new attractions and itineraries related to the Catalan architect were opened in the city. This in turn tended to attract people interested in architecture and design, and the appreciation of visitors for the architectural elements of the city increased as a result (Turisme de Barcelona 2003).

However, there are also counterpoints to the prevailing urban creativity paradigm. As Cloke shows in Chapter 2 of this volume, creativity is also an aspect of rural environments, and even of nature itself. There are a number of reasons why the 'rural' has become a creative space to rival that of the 'urban':

- The rural has become a creative haven for those retreating from the city.
- The rural has long been a location for creative clusters (witness the development of rural artists' colonies and the creation of new rural museums and crafts centres).
- Nature itself is creative (see Cloke, Chapter 2).
- There is increasing de-differentiation between 'urban' and 'rural'.

In addition, the lure of a rural existence seems to have been particularly strong for people from the creative sector. As Shaw and Williams (1994) have shown in the case of South-West England, rural areas which attract tourists can also be attractive places for lifestyle entrepreneurs to relocate. Richards and Wilson (2006) have indicated that most of the existing examples of 'creative tourism' are to be found in rural areas, and Raymond (Chapter 9) illustrates some of the dynamics of these developments in rural New Zealand.

It seems that in many cases the biggest spurs to creative development in rural areas have been adversity and the relative lack of employment opportunities. However, the growing relocation of creative entrepreneurs to rural areas has also meant that these regions now face challenges of 'serial reproduction' through the increasing development of 'rural tourism'. As more people living in (or migrating to) rural areas seek new sources of income, they often turn to some form of rural tourism accommodation (such as *casas rurales*, *gîtes* or manor houses). In the European Union, serial reproduction of rural tourism products has been underwritten by subsidies from rural development programmes such as LEADER and INTERREG. Very often, however, there is a lack of business networks and other enabling resources to support entrepreneurship (Soisalon-Soininen and Lindroth 2007).

In an increasingly competitive market, rural tourism enterprises also need to differentiate themselves. Unfortunately, in the modern symbolic economy, it seems that nature is not enough (Pretes 1995). The basic natural resources of rural or wilderness areas need to be added to through narrative creation. This generates lots of rural consumption space in need of animation and consumers (Cloke 1993).

One of the solutions to the growing problems of rural tourism has been the development of creative tourism products. In the United Kingdom, rural development agencies have promoted the creation of crafts tourism, arts festivals, workshops and masterclasses (Rural Regeneration Cumbria 2006). The EUROTEX project (Richards 2005) advocated the development of creative experiences for tourists as a means of valorizing textile crafts in peripheral areas of the European Union, and similar strategies are now being applied in developing countries, such as Namibia (Miettinen 2004).

What apparently unites most of the urban and rural places that apply creativity to development strategies is the idea of adversity. Urban deprivation certainly seems to be a spur to developing creative strategies, as the cases of Huddersfield, Glasgow and Pittsburgh all testify, but also rural areas with few alternatives have turned to creativity for similar reasons. For example, the La Rioja wine region in Spain has turned to similar tactics to those employed by large cities, inviting star architects such as Frank Gehry and Santiago Calatrava to create iconic facilities for wine experiences and creative tourism. In northern Italy, the village of Viganella has turned its location at the bottom of a deep, sunless alpine valley to creative advantage. By installing a giant mirror on the hillside opposite, the village has managed to create artificial sunlight during the sunless days from November to February, and create an innovative tourist attraction in the process.

Barriers to creative development

There are also reasons to believe that it will not be easy for all destinations to develop creative tourism strategies. Although the desire for creative development may exist, there are a number of practical barriers which may intervene.

Shortage of creative skills

The development of creative attractions requires the acquisition of new skills, on the part of both planners and those providing the attraction. Are those currently employed in the tourism industry and/or the creative sector able to develop the performative, interpretive and pedagogic skills required to stage and facilitate creative experiences?

Lack of creative investment

The development of creativity also implies investment, not necessarily in physical infrastructure, but in intangible culture and orgware. In many cases, the creative sector finds it difficult to attract investment because there is a lack of visible assets or an obvious return. It is far easier to make the case for cultural 'hardware', such as art museums or theatres, since they are more widely understood as venues and more widely supported by the public sector.

Lack of creative audiences

Although the 'creative class' are now supposed to account for a large proportion of the population, it has to be recognized that many creative activities have a limited audience – effectively, most forms of creative tourism are forms of special interest tourism. There is a problem of reaching the target audience, which is often widely dispersed (as Raymond demonstrates in Chapter 9).

The cumulative disadvantage of creative locations

The fact that creativity is often spurred by adversity means that creative development may happen in areas which are at a disadvantage, for example in terms of economic structure or peripheral location (Garrod and Wilson 2003, 2004). This may hamper their tourism development, particularly where accessibility is a problem or the area has a poor image.

It is clear to us that creative development strategies in the sphere of tourism involve major challenges and pitfalls as well as potential rewards. In developing the material in the volume we have therefore attempted to provide a balanced view of the relationship between creativity and tourism. This is also clear from the selection of case study material, which illustrates successes, challenges and pitfalls. Hopefully this approach will allow readers to make up their own minds about the pros and cons of creative development strategies in relation to tourism.

Structure of this volume

This volume examines the relationship between creativity, tourism and development from a number of different disciplinary and thematic perspectives. The wide-ranging contributions are arranged into four main parts, dealing with the geographical implications of creativity in urban and rural locations, the development of creative products, the role of creative lifestyles and the creative industries.

Spaces, enclaves and clusters

This part examines the spatial implications of creative development strategies in regions and cities. In Chapter 2 Paul Cloke examines the relationship between creativity and tourism in rural environments, arguing that the concentration of research in urban environments overlooks the important role of creativity in rural spaces and in nature. He argues that tourism is one of the most important producers and commodifiers of rural space, and he identifies some of the creative activities which not only characterize rural tourism but also reconceive rurality as lived space. The creative acts of tasting, placing and performing are examined not just as means of recommodifying rural space on the part of the producers, but also as involving the performance of nature itself, as exemplified in the burgeoning field of ecotourism.

John Hannigan's analysis of the trajectory from fantasy city to creative city in Chapter 3 provides a useful overview of the underlying forces which have propelled cities and regions towards creative strategies. In particular he traces the decline of postwar concern with safety, which gave rise to the entertainment-centred fantasy city, and the concurrent rise of the creative city and the excitement offered by the 'controlled edge' of creative neighbourhoods and ethnic enclaves. He sees the search for new and more daring experiences on the part of the creative class as leading towards new products such as 'gulag guesthouses' and 'favela tourism'.

In all these developments there is a feeling that culture and creativity have become more central in the development process in general and the tourism development process in particular. In Chapter 4 Graeme Evans analyses the relationship between creative spaces, tourism and the city and illustrates the way in which cultural development strategies have been transformed into creative development strategies.

The way in which new tourism areas are created in large cities is the focus of Robert Maitland's analysis of two areas of London (Chapter 5). He examines the way in which global flows of tourists have become part of a 'cosmopolitan consuming class' that overlaps with the 'creative class' in developing and transforming city spaces for visitors and resident alike. In this sense, tourists in these new creative areas are very much in the role of 'prosumers'.

Building creative tourism supply

Creativity may abound in many different locations, but it still has to be produced and packaged for tourist consumption. This section of the book looks at how creativity is packaged and sold to locals and visitors alike. Richard Prentice and Vivien Andersen (Chapter 6) examine the way in which French towns have been creatively staged for tourism consumption. As time has passed, the goals of cultural policy have shifted away from heritage preservation towards more intangible heritage and the animation of cultural space.

In Chapter 7 Walter Santagata, Antonio Paolo Russo and Giovanna Segre examine the negative effects of tourism development on the creative sector in Venice, and argue the need for quality labels as an antidote to the problem of commodification. They also argue that there is a need for 'empathy' between the host population and tourists, in order to deliver an authentic and creative tourism experience.

In the case of Sitges – a small coastal town in Catalunya – Esther Binkhorst argues in Chapter 8 that the co-production of experiences is another possible solution to the commodification problem. However, her analysis shows that Sitges has been slow to move away from traditional beach tourism models, and still has a long way to go to develop fully fledged creative tourism.

In Chapter 9, Crispin Raymond presents a case study of the development of creative tourism in New Zealand and demonstrates some of the practical problems that arise in targeting the creative consumer. His refreshingly honest analysis of creative tourism development identifies many of the pitfalls, but also offers potential solutions to these problems.

Consuming lifestyles

Even when not actively involved in the co-production of creative experiences, tourists often consume the creative lifestyles of others. Just as the crafts content of rural areas is an attractive authentic part of the rural tourism experience, so the 'bohemian' lifestyles of students, the gay community and other 'minority groups' are important for giving places a 'creative' image and making them attractive for others who live in or visit the area (Florida 2002).

Thus, such vicarious consumption of creativity is made possible by the existence of a series of 'creative landscapes', as explained by Antonio Paolo Russo and Albert Arias Sans in the case of Venice (Chapter 10). They look in particular at the way in which students help to forge a creative landscape through their own cultural activities and the increasingly inter-cultural lifestyles that they lead.

Gay communities have also been argued by Florida (2002) to be important indicators of the tolerance and creativity of cities. This idea is tested empirically in the city of Amsterdam by Stephen Hodes, Jacques Vork, Roos Gerritsma and Karin Bras in Chapter 11. They show that a decline in the tolerant image of Amsterdam has been responsible for making the city less attractive as a destination for the gay community.

Ethnic communities have also become important resources for the development of creativity, and the presence of ethnic enclaves in particular is seen as an attraction for tourists. In Chapter 12, Steve Shaw examines the development of ethnic enclaves in London and Montreal, and demonstrates how cosmopolitanism is being developed into a form of tourism consumption in these areas. He outlines some of the policy challenges for cities in developing such enclaves, as well as some of the potential pitfalls.

Jock Collins and Patrick Kunz examine the development of four ethnic precincts in Sydney in Chapter 13, and analyze the way in which relations of production, consumption and regulation interact to create cosmopolitan sites of diversity and tolerance that are attractive for the creative classes. They emphasize that visitor reaction to ethnic enclaves is not just a function of the ethnicity of the enclaves, but also the background of the visitors. In addition they point to an emerging divide between the views of the policy-makers and ethnic entrepreneurs regarding the creation and management of such enclaves.

These chapters appear to underline the fact that groups previously invisible in the tourist economy have been pushed to the forefront in the creative economy. However, one of the big questions that has to be dealt with here is the extent to which the production and consumption of diversity (or difference) and cosmopolitanism is actually creative.

Creative industries and tourism

One of the most common forms of creative strategy is the development of the 'creative industries'. This part examines three very different examples from the United Kingdom, South Africa and Singapore. In Chapter 14, Kevin Meethan and Julian Beer examine the relationship between tourism and the creative industries in Plymouth (United Kingdom). In particular, they look at how the impact of creative industry development models can be assessed, and how public–private partnerships can be used to develop creative facilities.

Christian Rogerson shows how creative industries policy has been developed in South Africa in Chapter 15, and argues that these developments are characteristic of an awakening interest in the creative sector in the developing world. National and local policies are geared towards developing the creative industries, and using them to position South Africa and its major cities as creative hubs, for example by making Johannesburg Africa's 'fashion capital' and kick-starting the local film industry.

Chapter 16 presents an analysis of the creative industries and tourism in Singapore. Can-Seng Ooi shows that the development of creative strategies is problematic in a highly controlled society such as Singapore, not just because a high degree of control tends to stifle creative endeavour, but also because strongly imposed moral codes make it difficult to develop the tolerant, 'bohemian' atmosphere prescribed by Florida.

These case studies provide apparently different approaches to the question of creativity, but interestingly they all spring from their own problems of adversity – postapartheid reconstruction in South Africa, replacing lost manufacturing jobs with knowledge industries in Singapore and loss of defence jobs in Plymouth. In the same way that culture-led regeneration was often a response to the problems of economic restructuring in the 1980s and 1990s (as Evans shows in Chapter 4), so it seems that the response to the challenges of a postindustrial, postmodern economy lies in the creative sector.

In the concluding chapter, Richards and Wilson attempt to summarize the major themes which arise from the analyses presented in this volume, and identify areas for potential future research.

Part 1

Space, enclaves and clusters

2 Creativity and tourism in rural environments

Paul Cloke

Tourism and rural space

As several authoritative texts (e.g. Butler *et al.* 1998; Hall, D. 2005; Roberts 2004; Roberts and Hall 2001) have recently confirmed, rural areas have become increasingly significant in the (re)production of tourism over recent years. It is equally clear, however, that tourism has become increasingly significant in the (re)production of rurality, particularly in view of a clearly changing set of relationships between space and society in relation to the countryside. Mormont's (1990) classic study identified five trends in contemporary rural society and space:

- an increase in the mobility of people, resulting in an erosion of the autonomy of local communities;
- a delocalization of economic activity and the associated heterogeneity of economic zones;
- new specialized uses of rural spaces (especially related to tourism) creating new specialized networks of relations in the areas concerned, many of which are no longer localized;
- the people inhabiting rural space increasingly include a diversity of temporary visitors as well as residents;
- rural spaces now tend to perform functions for non-rural users and can exist independently of the actions of rural people.

What is evident in these trends is that each points to how conventional rural spaces have been transformed by touristic processes and practices (see Baerenholdt *et al.* 2004; Crouch 1999), such that there is no longer any simple rural space, but rather a multiplicity of social spaces which overlap the same geographical area.

In this chapter I suggest that particular practices and performances, especially those associated with tourism, are influential in bringing together how rural areas are conceived and how they are lived. With the emphasis in contemporary tourism on creativity, I want to trace some of the creative performances which not only characterize rural tourism but also reconceive rurality as a lived space.

The production of rural space

These ideas about how rural space is conceived, lived and practised are illuminated by the writings of Henri Lefebvre (1991) on the production of space. Lefebvre's contribution is to offer a framework with which to render intelligible the qualities of space which are contemporaneously perceptible and imperceptible to the senses (Merrifield 1993, 2000). The resultant 'spatial triad' sketches out three moments which in reality coalesce in fluid and dialectical fashion, but the recognition of which sets the scene for why new forms of practice and performance bring about different relations between tourism and space in rural areas. Lefebvre's triad is as follows.

First, 'representations of space' indicates that space which is conceptualized and constructed by a range of technocrats and professionals – architects, planners, developers, social scientists and so on – whose signs, discourses and objectified representations present an order which interconnects strongly with prevailing relations of production. Here then, is space as it is CONCEIVED, and as Merrifield (2000: 174) suggests 'invariably ideology, power and knowledge are embedded in this representation'.

Second, Lefebvre talks of 'representational space' – the space of everyday experience which is shaped by complex symbols and images of the dwellers and users of that space. Representational space overlays physical space, and makes symbolic use of its objects, places, landscapes. Here we are dealing with a more elusive experiential realm – space as LIVED – in which there are continual interventions from conceived space in the form of actors such as planners and tourism developers and managers actively seeking to make sense of how space is lived experientially.

Third, Lefebvre identifies 'spatial practices' as patterns of interaction which glue together society's space, by achieving cohesion, continuity and acceptable competences. These are practices through which space is PERCEIVED and given a performed identity. However, Lefebvre does not suggest any achievement of coherent spatial practices. Rather, such practices represent a fluid and dynamic mediation between conceived space and lived space, which at once holds them together, yet keeps them apart.

Lefebvre's trialectic insists that each element of conceived–lived–perceived space is related to and influenced by the others. As such, particular moments of space are unstable and need to be 'embodied with actual flesh and blood culture, with real life relationships and events' (Merrifield 2000: 175). Capitalism will tend to give dominance to the conceived realm, meaning that lived and perceived moments can be of secondary importance to objective abstractions which reduce the significance of both conscious and unconscious lived experience. However, the inevitably incoherent and fractured nature of everyday life and the notion that many performances will be more than representational (Lorimer 2005 – also see Dewsbury 2000) allow the possibility that new creative practices and performances are capable of affording a release from conceived space as well as a reinforcing of staged meanings which serve to make you know how you feel.

These ideas about the production of space have been largely eschewed by rural researchers (but see Phillips 2002) until a recent essay by Halfacree (2005) who connects formal representations of the rural, such as those expressed by capitalist interests, planners, spatial managers and politicians, with how rurality has become commodified in terms of exchange values. Practices of signification and legitimation have been vital to this process as the vernacular spaces of the rural have become variously appropriated symbolically by producers and consumers. This 'social imaginary' (Shields 1999), however, can itself become subversive as spatial practices and performances are (re)appropriated from the overarching interests of the dominant.

Changing rurality, changing commodity

In a previous study (Cloke 1993) I have sought to account for the ways in which tourism has been actively involved in the symbolic reconstruction of rural space as well as in its material reconstruction through the production of new sites, facilities and opportunities. In recent decades there has been a shift in the nature and pace of commodification in rural Britain as new markets for countryside commodities – and notably those associated with leisure and tourism – have been opened up. The production of rural space has taken new shape through the framing politics of neo-liberalism, and in particular through the specific outcomes of privatization and deregulation, which have both released rural land for new purposes and created a conception of a multifunctional countryside in which these new rural spaces can take root. Basic ideas such as a 'day out in the country' or 'a rural holiday' have begun to take on new meanings, reflecting opportunities to visit newly commodified rural 'attractions' in addition to more traditional and less 'pay-as-you-enter' rural pursuits. Consumption of and through these rural attractions has often reflected new forms of old values. Conventional concerns for pastoral idyll, history and heritage, traditions, outdoor pursuits and the like are still evident. However, such concerns are served up differently through attractions and spectacles which offer new touristic practices to participants while reproducing conventional signs, symbols and displays.

New rural commodity forms, then, suggest changing representations of rural space in that they are shaped and enabled by distinct conceptions of what are *appropriate* new land uses, sites and attractions for the countryside. The needs of changing rural production systems, notably in terms of the necessity for farm diversification, are aligned with new forms of commodity consumption which broadly uphold enduring conceptions of the rural as idyllic, pastoral, close to nature, rich in heritage, safe and problem-free, and so on. These representations of rural space are associated with new representational spaces which symbolize and present new images of what it is like to live in that space, telling us how we should feel in and about it. Equally, new spatial practices allow new rural spaces to be perceived and given a performed identity, notably through patterns of interaction involving observing and participating in event-spectacles which somehow perform rurality. Such performances often strive towards some 'authenticity' (however postmodern), but there is also evidence that staged performances of new ruralities can go well beyond the real objects and relations of the sites and buildings concerned. Here, the conspicuous consumption of the symbols on offer indicate that some new rural attractions are emphasising signs which are unrelated to the specific reality of a place, its landscape and its history. In the terms suggested by Best (1989), the society of the spectacle may be nudging up against the society of the simulacrum (Baudrillard 1983) in which rural commodities are being eclipsed by their sign-values, which can be altogether unrelated to the realities of rural space.

Creativity, tourism and the reproduction of rural space

Accepting that many rural areas have re-defined themselves as consumption spaces in which the commodifications of nature, heritage and tradition have transcended agricultural production as key signifiers of rural space, it is now possible to question what, if any, impacts the 'creative turn' in tourism has exerted on the reproduction of rural space. In terms laid out by Richards and Wilson (2006) the markers of this creative turn reflect a shift from cultural tourism to more skilled forms of consumption, resulting in reformulation of identity and subjectivity and the further acquisition of cultural capital. In large urban settings they detect the rise of creative spectacles and spaces as well as tourist participation in creative activities. It is also important to trace any such shift towards creativity in other space-settings.

Expressions of creativity in tourism in rural areas reflect some of these markers, although often in rather different clothing. Rural spectacle will often be based on particular forms of nature–society relations in which people either collude with, or pit their wits against, the non-humanity of nature (Szerszynski *et al.* 2003). Creative spaces might represent the humble multifunctionality of a village hall rather than the sweeping artistic *quartiers* of metropolitan space. Creative activities might again consist of a seemingly mundane learning to appreciate local produce rather than more glitzy skills-acquisition. Alternatively, creativity in rural tourism may suggest new ways of understanding creativity, for example in terms of hybrid nature–society performances based around eco-experience or adventure. What is clear, however, is that various practices which offer tourists the opportunity to develop their creative potential through forms of active participation in rural contexts are likely in turn to lead to further reproductions of rural space in terms of Lefebvre's triad. In particular, new creative spatial practices will begin to offer potential for gluing space together differently, and providing space with a different performed identity.

In what follows, I draw on research in the United Kingdom (Western

Scotland and Devon) and in New Zealand (Kaikoura and Queenstown) to illustrate different strands of how creativity in tourism is associated with the reproduction of rural space. Some examples are much more fully formed than others, and my account is by no means exhaustive. Nevertheless I suggest four strands of creative performance, recognizing that each is neither mutually exclusive nor necessarily mutually compatible in any particular rural place.

Tasting

Here I refer to a range of practices in which tourists (and indeed local people, for I suggest that they, too, sometimes act as tourists even when in their home locality) *taste* the creative performance of others, and in so doing develop their own creative potential and expose their identity to change or cultural acquisition. In its simplest form tasting may reflect attempts to bring cultural events and opportunities into rural space when these previously might only have been accessed in the city. In Devon, for example, the *Nine Days of Art* project is an artist-led initiative which provides a trail around venues throughout rural areas of the county to view jewellery, textiles, ceramics, painting, sculpture, photography and printmaking both inspired by the rural/coastal environments and placed within them, in the creative spaces of homes, studios, galleries, village halls, hotels, country houses and even outdoor locations. The project not only provides residents and tourists with the opportunity to appreciate and purchase rural 'art', but has sparked off several subsidiary packages whereby people can learn to 'do art' for themselves. The *Villages in Action* project, subsidized by local authorities and the Arts Council, brings a programme of performance arts (theatre, comedy, music and dance) to 63 village halls in South Devon. The taste of performance is thereby brought into rural space, where it encourages existing and new performative potential and credentials.

Tasting, more literally, can also involve opportunities to sample and get to know local food and drink, including the production and presentation. Tasting occurs at very different scales – New Zealand's Marlborough region, for example, is becoming sufficiently geared to tourism to suggest a highly organized wine tourism industry. Elsewhere, however, tasting local food and drink forms the basis of more limited spectacle and site-specific attraction. The Mull and Iona Community Trust, for example, organizes an annual *Taste of Mull and Iona Food Festival*, bringing together its food and whisky producers, restauranteurs and shopkeepers in a frenzy of events and opportunities for tourists to get into Mull's food culture. The invitation to 'come and share our food with us' is combined with unmistakeable messages about food, place and identity:

> We are more aware than ever before that 'we are what we eat'. We need to know *what* we are eating and *where* it comes from. Food from Mull

and Iona is food from farmer to consumer – Real Food – food you can put a face to!

The localized identity of rural food offers tasters the knowledge credentials to be more discerning, sustainable and identity conscious. Other facilities emphasize this point. Sharpham Vineyard in Devon invites the tourist to 'learn the fine art of making wine and cheese' and, perhaps subconsciously, to learn from their visit about the credentials to be gained from more discerning food and drink choices in the future.

It might be argued that tasting represents a passive form of cultural tourism. Rather, I suggest that it involves practices and performances which develop creative knowledge, intuition, capacity and skill. Interaction with art exposes the rural tourist to signs which both reinforce the rural idyll and contest that idyll with more dystopic interpretations of countryside. Music in the village hall permits the learning and honing of taste, as does food and drink tourism. In so doing, rural areas become replete with creative spaces and practices which are capable of performing the rural differently.

Placing

Another form of creative performance comes with the interaction of rural tourists with imaginative creative performances. For some time now the representation of rural space has commandeered literary figures and narratives – Heriot Country, Lawrence Country, Robin Hood Country, Lorna Doone's Exmoor and the like – to convey key significations about particular places, their history and their heritage. By visiting these places, tourists may not only learn more about particular authors and their literary imaginations via interpretative attractions and commodities, but they are also enabled to re-read the original texts with these 'real' rural settings in mind. Placing imaginative texts is therefore a performative practice which in turn contributes creativity to the place(s) concerned.

The use of rural locations in mass media programming and filming has added new dimensions to the creative performativity of placing. While clearly contributing to the conception of some rural spaces as film sets, and thereby helping to reconstitute such places as tourist magnets – the village of Goathland in the North York Moors, for example, is the location for the popular television programme *Heartbeat*, and is now the most visited site in the National Park (see Phillips *et al.* 2001) – visitors will inevitably have to engage in imaginative performances to perceive film-space experientially. Travellers on '*Harry Potter's*' railway in West Scotland have to work hard to connect tourist scene with film scene, as do visitors to New Zealand as *Lord of the Rings* country, especially given the digital enhancement deployed in such movies which performs radical transformation to rural locations. Such places do, however, offer tourists the opportunity to develop their creative potential through place participation.

A good example of place participation is the recent popularity of Tobermory on Mull in terms of its status as the filming location for the childrens' television programme, *Balamory*. Regular streams of young children (and their parents and grandparents) take the ferry from Oban and the bus to Tobermory so that they can explore the various brightly coloured houses which are used in the programme. The Island's tourism managers have been, in this case, wary of representing place identity in terms of the imaginative texts associated with Balamory. Although providing visitors with a Balamory map, tourist authorities have avoided elaborate representations of space for a number of reasons, including: copyright restrictions prevent further commodification of the brand; local people can react negatively to tourist congregations outside private dwellings; child-oriented day tripping is less valuable economically than other target tourist groups, and may indeed conflict with the longer-term target clientele who are attracted by representations of wildness and nature (see below). Nevertheless, the placing of Balamory in Tobermory has created a place-spectacle which is performed via a dutiful trudge around a mapped network of sites.

While creating and touristically re-creating Balamory might be thought of as a relatively undemanding performance of cultural tourism, a series of new projects to place performative arts projects in places of natural beauty are intended to make more serious demands on tourists. Perhaps the best known of these is a Glasgow-based arts organization's (NVA) project combining drama, music, history and night-hiking on the Trotternish Peninsula of the Isle of Skye. Two hundred participants each night are taken on a torch-lit walk across rugged terrain. Aspects of the landscape are illuminated by lightscapes and soundscapes: 'It ravishes the senses with a fantastic and beautiful right-time journey ... above all, it makes us think and feel the relationship between humankind and nature' (*Observer*, 10 July 2005: E8).

Participating in this 're-placing' of part of Skye offers tourists cultural distinction based on restricted opportunity in an elite landscape. However, this, too, can be seen as a touristic performance of placing – a creative interaction with imaginative texts in a rural setting.

Performing creatively

The 'creative turn' in tourism brings immediately to mind the upsurge of opportunities for tourists to learn new skills and undertake recognizably creative activities. Rural tourism is now replete with such opportunities. Any self-respecting country house hotel now offers add-on creativities: learn to cook country-style, to fish, to shoot, to ride, to engage in various forms of environmental art and so on. To some extent these offerings represent the capitalist imperative of conjuring up new commodity forms with which to refresh traditional businesses, and while some of these creative activities present obvious connections with traditional conceptions of rural space, other attractions – murder mystery weekends, learn to play tennis, writing or music master classes and so on – reflect more of a society of the simulacrum than any particular fusion with rural signification.

I want to suggest, however, that offering tourists the opportunity to develop their creative potential will not be restricted to a series of obviously 'country' cultural activities. The creative performance of 'being rural' has moved on in many parts of the developed world. An example of alternative creative performance can be found in the adventure tourism industry of New Zealand (Cater and Smith 2003; Cloke and Perkins 1998, 2002). Here an amalgam of particular circumstances – not least the 'outdoorsness' of New Zealanders, the technological innovation of jet boats and bungy technology, and the willingness of government to license adventure tourism businesses in elite environmental sites – have resulted in first, key brand-leading firms (Shotover Jet and A.J. Hackett), then an explosion of follow-on operations, providing opportunities for adrenaline-fuelled adventurous pursuits in rural areas of New Zealand. Engaging in bungy-jumping or white-water rafting may appear simply to be a creative spectacle, with the active involvement of the few providing exciting events for others to watch. Our research suggests, however, that the performance of adventure contributes a number of highly creative aspects to the spatial practices and representations of rural New Zealand, which has become signified and experienced in terms of cultures of adventurousness. In particular, participants in adventurous activities report a sense of having to overcome their own fear as well as having to overcome the forces of nature. Not only do they win the cultural credentials of having successfully completed the activity concerned, but in so doing they ask questions of their own identity and often claim to be different after the event. At the same time, the actual performance of, say, bungy-jumping, defies representation. Post-jump interviews encounter jumpers struggling to put into words what they have just experienced. It is a kind of performativity in which although the actual process is staged, nevertheless the unfolding event is entirely immanent, and resistant to representational signification. In these ways, the creative performance of adventure appears to shift the conceived–lived–perceived register of rural space into different directions from those traditionally defined.

One further disjuncture from the performance of rural idyllism is also worthy of mention. The post-productivist countryside is beginning to play host to a range of activities and practices which sit somewhat uneasily alongside traditional conceptions of the production of rural space. For example, alongside the farm parks, potteries and farmers' markets of contemporary rurality there are a growing number of paintball sites offering opportunities for 'adrenaline pumping' 'full-on combat'. Although military uses of the countryside are familiar, they are often closed off and always non-participatory. Here, a different form of creativity is emerging – one where skilled consumption is significant, identity formation is challenged and the cultural capital among particular group cultures is assured – yet one which is dystopic to the production of spaces of rural idyll.

Performing interactively

Here, I suggest that the 'creative turn' in rural tourism needs also to encompass how tourists interact with the creativity of nature. Rural studies has in recent years embraced the idea of 'hybrid geographies' (Whatmore 2002) of nature–society relations in which relational ensembles of humans, non-humans (whether living or not), discourses, technology and so on are taken seriously as the formative networks of agency (see Cloke and Jones 2001, 2004). This philosophical movement requires nature in its multifarious guises to be recognized as a co-constituent of rural places, and while this is no easy task given the anthropocentric assumptions of most social science, particular nature–society relations can be used to signify the wider point. Thus instead of regarding nature as a backcloth to rural tourism we can begin to ask questions about how nature performs interactively with humanity in tourism, and vice versa.

At the simplest level, creativity can engage with particular facets of nature in the development of new touristic opportunities. An example of this is the *Mull and Iona Tree Festival*, a 2-month celebration of trees combining exhibitions, workshops and trails through which visitors can learn of the ancient tree alphabet of the Celts and its associated iconographies. The festival represents an innovative fusion of creative appointments, and fits into the calendar of special events to attract tourism to Mull.

Another tactic of tourist operators is to upgrade and re-commodify classic activities of rural tourism so as to render them more guided, informative and skilled, thus providing them with enhanced credentials. In New Zealand, the representation of particularly scenic hikes (or 'tramps'), such as the Milford Track and the Routeburn Track as 'Great Walks', has increased the apparent skilled consumption involved, and heightened their popularity. Elsewhere in the country, walks that have always been available have been similarly reproduced. Bell's (1996: 42) account of walking up to Fox Glacier in the West Coast of New Zealand provides an excellent example:

> The experience, which I'd remembered from previous times as an easy walk, had now been mediated for these tourists by converting the placid (non-paying) traveller to a (paying) alpine adventurer, with all the correct garments and equipment required for such an adventure. By augmenting (artificially) the scale and danger, in effect they become participants in a recapitulation of nineteenth-century experience of the vast and sublime. Literally, the fifteen-minute jog in track shoes has been transformed into a half-day guided 'expedition'.

These mediative alterations to the tourist experience can be seen simply in terms of re-commodification for greater exchange value, but alongside such motivations there is a tacit acceptance that nature itself can perform in articulate ways to challenge, to induce embodied response, to make difficult and so on.

In some cases, the specific performances of particular creatures is central to the creativity of rural tourism. The burgeoning reputation of Mull as 'the best place in Britain to see wildlife' is underwritten by the iconic performative manoeuvres of eagles soaring, whales surfacing, puffins pausing to stand by their burrows and otters nervously fishing. While tourists can connect with these performances independently, increasing numbers are willing to pay for the services of an experienced guide to lead them on their 'wildlife safari' and to pass on essential skills and knowledges.

Perhaps one of the ultimate connections with nature's creative performances can be seen in the astonishing growth of New Zealand's ecotouristic whale-watching and swimming-with-dolphins activities. Research centred on the South Island tourist town of Kaikoura (Cloke and Perkins 2005) suggests that it is the majestic 'blowing' and 'fluting' of whales, and the playful acrobatics of dolphins which provide tourists with magic moments of connection with these cetaceans. Moreover, such encounters are often presented as moments of unfolding, affording release from the staged and conceived spaces of the tourist industry. In a very real sense, then, the embodied performances of cetaceans lie at the heart of Kaikoura's growth as a tourist town. Some tourists are fortunate enough to achieve creative interactions with cetaceans in the wild. For others, just visiting the town and soaking up the representational space signifying the performance of these creatures is enough. In assessing the creativity of cultural tourism in rural areas we would do well to recognize the co-constitutive hybridities of human and animal performances.

Conclusion

In one sense, the significance of creativity in rural tourism can be represented in terms of heralding a new turn in the touristic commodification of rural areas. However, when set against Lefebvre's ideas about the production of space, the break-out of creativity needs to be assessed in terms of any resultant practices which are capable of changing the ways in which rural space is performatively brought into being. The research reported on here suggests that care needs to be exercised over the potential conflation of rurality and creativity; creativity is multifaceted and reflects far more about rural space than just the 'obvious' connections with 'rural' cultural activities and skills.

In general, rural space is increasingly *conceived* as a commodity form, in which rurality is reproduced both as an object of desire and as a stage on which to perform. Much of the apparent creativity of rural tourism deepens the relationship with rurality, and therefore deepens its desire as a place of performance. Some performances enhance the traditional appeal of the rural; others appear to re-conceive rurality as a new space for adventure, entertainment, spectacle and the like. Equally, rural space is increasingly *lived* in terms of new portfolios of symbols and images by which dwellers and users experience and make sense of the rural. Many of these messages signify the apparent need for a greater cultural engagement with rurality, involving new skills of observation, understanding and embodied participation. Thus rurality is becoming signed as a place in which to perform (although performances range across different degrees of passivity and activity) and the creativity *of* the rural emerges through creativity *in* the rural. The *perceived* space of the rural thereby begins to demonstrate new cohesions played out in terms of practices which provide opportunities for visitors to dwell, to reside (albeit temporarily) and to be involved. Involvement ranges from a post-tourist sense of dealing easily with the inauthentic, to a seeking out of new elements of the 'authentic', but throughout this range new performed identities are exhibited, some of which are more than representational in their affordance of release from orthodox conceived space.

Thus, it is clear that creativity in rural tourism is resulting in a range of different practices and performative spaces in which the identity and subjectivity of the tourist can be reformed and enhanced. In some cases, for example in interactions with particular aspects of nature, these practices have the capacity to reinforce perceptions of how traditionally conceived and lived space is played out in rural settings. Alternatively, new creative practices may be regarded as dystopic, and can certainly present conflicting social demands on a particular geographical space – as the example of Mull illustrates. In these latter cases, creative touristic performance does pose serious questions about the production of rural space. New ways of bringing together how rurality is conceived and how it is lived out can involve a release of new kinds of perception, as well as a retrenchment of existing ways of being made how to feel. It is in the thrilling and exciting credentials of adventurous and environmentally interactive rurals as well as in the skilled consumption of tradition and heritage, that the creativity of rural space is being discovered in contemporary tourism.

3 From fantasy city to creative city

John Hannigan

In his seminal article, 'From Pilgrim to Tourist – or a Short History of Identity', the eminent social philosopher Zygmunt Bauman (1996) detects a fundamental ambivalence in the tourist condition. The tourist, Bauman says, is the prototypical postmodern citizen, the successor to the vagabond and the flaneûr of earlier periods of modernity. Tourists, he observes, periodically become restless – as the 'joys of the familiar wear off and cease to allure' (1996: 29). To cope, they become conscious and systematic seekers of new and different experiences. At the same time, the tourist is rarely willing to sever the umbilical cord to everyday life. Consequently, the touristic experience is characterized by 'a profusion of safety cushions and well-marked escape routes'. In a phrase, 'shocks come in a package deal with safety' (1996: 29–30).

Fantasy city

In the years following World War Two, the United States was beset by a discernible tension between comfortable conformity and a desire for 'safe adventure'. Haunted by the battlefield ghosts of traumatic experience, veterans and their families craved familiarity and social convention. Yet, Americans also displayed a growing hunger for excitement and experience beyond the commercial utopias represented by backyard swimming pools and automatic washing machines. This was almost certainly traceable, in part, to lingering memories of the 'exotic' locales encountered during overseas service in Japan, North Africa and the South Pacific.

The leisure merchants of the 1950s and 1960s exploited these conflicting tendencies by constructing fantasy landscapes which reconciled the desire to escape from the iron cage of conformity with the wider corporate project of creating an aggressively marketed, family-oriented national consumer culture in America. Central to this was the manufacture of a sense of 'risk-less risk' – where one may take chances that aren't really chances (Hannigan 1998; Nye 1981).

One important example of this was Disneyland in Southern California where visitors could experience foreign cultures, historical events and classic fairly tales, all wrapped in a totally controlled and protective bubble. Today critics almost uniformly disparage the Disney theme parks as bland, plastic and inauthentic, but in the 1950s this was much less evident. As Doss (1997: 180) points out, the art and architecture of 'Fantasyland' in the original Disneyland fused postwar enthusiasms for the 'bizarre, the eccentric, the grotesque, the unconventional and the unrestrained' with 'deep-felt desires for safety, security, restraint and direction'. Furthermore, the adventure rides and attractions at Disneyland provided what Rojek (1993: 205) has called the 'recurrence of reassurance' – after a few well orchestrated thrills, the pirates, bandits and other villainous characters are inevitably defeated, thereby restoring a sense of social order.

Another leading mid-century example of riskless risk in action is the creation and marketing of the Las Vegas 'Strip'. In an era when the mainstream United States was allegedly under attack from mobsters, Communists and other subversive elements, Vegas was promoted as an exotic but absolutely safe getaway where middle class vacationers could enjoy some perfectly harmless casino gambling, *Folies Bergère*-style floor shows and Hollywood lounge acts. Drawing on a wealth of archival sources – promotional literature, oral histories, newspaper and popular magazines – Schwartz (2003) demonstrates how a holiday on the Strip was pitched as being a harmless interlude of 'sin without guilt' well removed both geographically and ethically from the racketeering, corruption and addictions that were threatening the integrity of hometown America (85–86). Going on a jaunt to Las Vegas constituted a kind of 'moral time out', but one that was strictly monitored and regulated by the state.

A third example of riskless risk from the 1950s and 1960s is what has come to be known as 'Tiki culture'. Also referred to as 'Polynesian pop', it encompassed an impressive menu of artifacts, design elements and leisure activities: themed bars and restaurants, backyard luau parties festooned with torches, exotica music, South Seas-inspired vernacular architecture (motels, apartment buildings, bowling alleys, roadside stands), invented tropical rum drinks, amusement parks and an impressive outpouring of Tiki artifacts – mugs, souvenir menus, matchbooks, swizzle sticks, postcards and carved idols. While it flourished most robustly in the balmier climates of Southern California, Florida and Hawaii, Tiki fever also reached the Mid-West and spread into Canada. Trader Vic's, a faux Polynesian-themed supper club that was popular in the 1950s and 1960s, had franchises in Boston, Chicago, New York, Detroit, Kansas City, St. Louis and Vancouver, as well as foreign outposts in London, Munich and Havana. Martin Denny, a pioneer of 'exotica music', broke into the Top 5 on the Billboard charts in 1959 with his stereo re-recording of 'Quiet Village', and appeared on network television on popular series such as 'The Dinah Shore Show' and 'American Bandstand'.

Tiki culture conjured up a version of riskless risk that was, at one and the same time, more respectable than Las Vegas but also more sensually charged than Disney (although Walt himself opened the 'Enchanted Tiki Room' at Disneyland in 1963). Sven Kirsten (2000: 39–40), the foremost contemporary 'connoisseur' of Polynesian pop, suggests that Tiki culture represented a type of liminal zone in the 1950s where the 'man in the gray flannel suit' could escape from the social and moral restrictions of a nation obsessed with plastic and chrome and embrace a makebelieve world which promised guilt free sex, exotic culinary experiences and a new 'suburban savage identity'. Tiki culture offered the suburban middle class a way of reconciling modernism with a romantic quest for authenticity. Even as Tiki incorporated such modernist elements as A-frame architecture and cool jazz, it also promulgated an invented and largely fictional religious mythology featuring primitive deities and rituals associated with Tiki idols. As is generally characteristic of romantic consumerism, this was clearly self-illusory, just as visitors to the Disney theme parks today suspend belief self-reflexively in the authenticity of the simulated world which they encounter there.

By the mid-1960s, the triumph of the counterculture movement created a new, riskier culture of leisure that tested the limits of escape experience in a more concerted and formidable manner (see Rojek 2000: 147–53). In one popular musical anthem from 1967, erstwhile travellers to San Francisco were advised to 'be sure to put some flowers in your hair' because 'there's going to be a love-in there'. This, of course, meant an experiential cocktail of consciousness-altering drugs, free sex and psychedelic rock music that made the exotica of Tiki, Disneyland and Vegas seem rather tame.

Surprisingly perhaps, a quarter of a century on, the seemingly defunct liminal escape hatches of the 1950s had evolved into an even more extensive fusion of consumerism, entertainment and popular culture (Goldberger 1996) that once again pivoted on the axis of 'riskless risk'. This new 'fantasy city' (Hannigan 1998) is constituted by a mix of casinos; themed restaurants, pubs and hotels; virtual reality arcades; megaplex cinemas; convention centres; sports stadiums and arenas; and branded retail 'shoppertainment' outlets, most of which are modelled on some combination of Disney and Las Vegas. Fantasy city development is aggressively themed and branded, deals in the marketing of iconic images and is 'solipsistic' in that it isolates sports and entertainment complexes from their surrounding neighbour-hoods (Hannigan 2006: 201).

A major economic impetus for building these urban entertainment desti-nations was the fiscal crisis triggered by the migration of manufacturing plants and jobs offshore, and the ensuing deindustrialization of cities. This was most dramatic in the 'rustbelt' region of the United States, but it also severely impacted on other locations, for example in the North of England. While some communities eventually rebounded, most notably by rein-venting themselves as high technology centres, most remained desperate to find some other way of reviving the local economy. Even as this was trans-piring, Las Vegas kept topping annual compilations as the fastest growing city in America. In particular, the proliferation of themed casino hotels in the city – Treasure Island (pirate adventure), Luxor (ancient Egyptian), New York, New York (the 'Big Apple') – seemed to offer up a failsafe template for a successful urban future through 'spectacular consumption'. This was reinforced by a barrage of heavily attended seminars on urban entertainment offered by the Urban Land Institute, the International Council of Shopping Centers and other industry groups where fantasy city development was heavily hyped as the way ahead for cities.

A second economic force contributing to the growth of urban entertainment destinations was the shifting nature of retailing itself. Dominant two decades earlier, by the mid-1980s conventional North American shopping centres had begun to seriously lose their appeal. In part, this was due to the competitive threat posed by 'big box' stores (Wal-Mart, Costco, Staples), online retailing and 'off-price' malls where overruns and seconds of high fashion brands such as Polo/Ralph Lauren, Saks Fifth Avenue and Jones New York were offered at a discount. To bring back the crowds, some retailers retrofitted and expanded, adding megaplex cinemas, themed restaurants and sporting goods retailers with in-store attractions such as rock climbing and fishing. Others opted to build completely new 'super regional' malls that combined 'value retailing' (selling brand-name goods at prices below those offered by department and speciality stores) with leisure and entertainment activities. Even traditional retailers such as department stores responded by making the activity of shopping more entertaining, for example by embracing 'experiential retailing' wherein shopping is transformed into a themed retail experience complete with interactive exhibits (Hannigan 2002b: 27–28).

After initially high expectations, fantasy city development has faltered. In large part, this is a matter of too much competition and too little product differentiation. With nearly every city of any size opting to build its own festival marketplace, sports and entertainment complex, and waterfront revitalization project, the tourist market has splintered. And, rather than resulting in a net increase in per capita entertainment expenditures by local residents, this has more often than not led only to a redistribution of consumer dollars. For example, megaplex cinemas with their glitzy exteriors, stadium seating and a lobby full of faux Hollywood memorabilia frequently only prosper by cannibalizing moviegoers from older, smaller screens in the same theatre chain.

Finally, the novelty and sense of excitement that fantasy city destinations initially promised have dwindled. As Hayward (2004: 190) explains, the 'gloss of these "cathedrals of consumption" and other so-called fantasyscapes is wearing thin'. Urban visitors are increasingly becoming bored with the 'Vegas aesthetic' and with activities such as dining in a themed shopping mall food court. One might even go so far as to conclude, Hayward (2004: 82) asserts, that 'the landscape of spectacle, semiotics, pleasure cultural diversity', contrived as consumer playgrounds for the middle class 'had become little more than a regeneration industry'. In similar fashion, Russo and Arias Sans (Chapter 10 of this volume) note that an 'industrial model' of redevelopment by spectacularization which has been embraced by private developers in the 'fourth generation metropolis' has watered down the urban experience, inspiring more culture-aware and experienced visitors to turn away in search of more appealing areas of the city that offer more original experiences. 'The search is now on', Hayward (2004: 191) states, 'for places of excitement and "calculating hedonism", somewhere where a "controlled suspension of constraints" or "controlled sense of decontrol" can be purchased, experienced and played with'.

On the wane in North America, fantasy city development has flourished in other parts of the world. The most dramatic example of this is in the Persian Gulf state of Dubai. Until recently, Dubai was a little known emirate on the Persian Gulf with only modest petroleum reserves. In a strategy designed to diversify Dubai's economy, its rulers financed an international airline and established a major presence in ocean port management. Next, they shifted their attention to tourism and real estate development. Among the iconic structures already built or under construction are the Mall of the Emirates, known globally for its indoor ski slope; an artificially constructed island in the shape of a palm tree that is visible from space; and the Burj Dubai tower, which will instantly become the world's tallest building when it is completed in 2009.

Creative city

While the entertainment economy sputtered as a motor of development in North American cities, a new *eldorado* emerged among city planners and politicians attempting to secure advantage for their communities in the global marketplace. Rather than casinos and convention centres, 'culture' and 'creativity' were now to be the new saviours of cities. As Miles and Paddison (2005: 833) observe, within a space of just over two decades, 'the initiation of culture-driven urban (re)generation has come to occupy a pivotal position in the new entrepreneurialism' and 'the idea that culture can be employed as a driver for urban economic growth has become part of the new orthodoxy by which cities enhance their competitive positions'. In particular, cultural development has become intertwined with neoliberal 'place marketing' strategies whereby older industrial cities seek to 're-present' themselves to the world as centres of artistic excellence, cutting-edge style and globalized consumption (Hannigan 2007).

The term 'creative city' initially rose to prominence in the 1990s in public policy circles in the United Kingdom, most notably through the wide diffusion of an influential report (and later a book) authored by the cultural consultancy *Comedia. The Creative City* (Landry 2000) is a call for imaginative action in the development and running of urban life and offers up a clear and detailed toolkit of methods by which 'our cities can be revived and revitalized' (sleevenote). Although the arts are valued as an important constituent, the concept was meant to convey a more general strategy for urban regeneration and growth:

> In the interurban competition game, being a base for knowledge intensive forms and institutions such as universities, research centres or the cultural industries has acquired a new strategic importance. Future competition between nations, cities and enterprises looks set to be based less on natural resources, location or past reputation and more on the ability to develop attractive images and symbols and project these effectively.
>
> (Landry and Bianchini 1995: 7)

The creative city approach differs from other forms of culture-led regeneration in several ways (see Hannigan 2007: 68–69). First of all, unlike other strategies for urban revitalisation, the target population is not chiefly composed of either investors or tourists. Rather, the goal is to attract a specific fragment of the middle class that is highly valued as representing the 'messiah' of urban economies. This 'creative class', as the influential American author and urban consultant Richard Florida (2002) has called them, combine technological skills with the ability to think in an original fashion. The hope is that the spectacular success of Silicon Valley in the 1990s can be replicated in cities that have been struggling in the wake of de-industrialisation. Thus, computer animators, website designers, video producers and software engineers are thought to be the type of migrants that cities need to attract in order to plug into and profit from the new knowledge economy.

Smart cities, in the Florida model, are those that actively encourage the growth of a lively, demographically diverse arts landscape. Unlike previous formulations, Florida's version does not emphasize the number of jobs created in the cultural sector itself; instead he focuses on the positive effect this has on luring 'creatives'. The link between artists and economic growth, Florida explains, is not causal per se but rather assumes the form of a 'creative ecosystem' that radiates a culture of tolerance and open-mindedness, thereby stimulating creativity among knowledge workers. By contrast, the type of rapid growth found in tourist dependent cities such as Las Vegas is deemed to be inferior because it is fuelled by the creation of less desirable service and construction jobs (Hannigan 2002b).

Second, the urban environment and lifestyle characteristic of the 'fantasy city' are said to hold minimal appeal for this new cohort of 'creatives'. Rather than professional sports, large museums and art galleries, gambling casinos, megaplex cinemas and hallmark events, they prefer neighbourhood art galleries, experimental performance spaces and theatres, small jazz and dance clubs, and independently operated coffee shops and cafés. Therefore, it makes no sense to invest huge sums of money in temples of spectacular consumption and *grands projets culturels* (see Evans, Chapter 4 of this volume, Evans 2003), such as the Guggenheim Bilbao Museum, whose appeal lies elsewhere.

Third, the 'riskless risk' that suffused the landscapes and activities of Disneyland and Las Vegas holds little potency for this new generation of creative workers. Rather, they seek out experiences that are infused with *controlled edge*. Controlled edge is the hip cousin of riskless risk. Like the latter it implies a 'safe adventure', but one that is grounded in a countercultural or bohemian milieu. Elsewhere, I have defined controlled edge as:

> the process whereby bohemian culture is captured and made safe for gentrifiers, tourists, art collectors, suburban day trippers and other middle class consumers by a cadre of real estate entrepreneurs, leisure merchants, fashion designers, restauranteurs, record producers, television and movie directors, casino czars, tourist operators and advertising agencies.
>
> (Hannigan 2007: 73)

While controlled edge thrives in many corners of contemporary culture, it is especially evident in what have come to be known as 'neo-bohemias'. A *neo-bohemia* is a formerly marginal urban area characterized by derelict industrial spaces, inexpensive rents and a visibly 'gritty' street panorama. In the first instance, it tends to be 'invaded' by a wave of penniless young artists, filmmakers and musicians in the early stages of their careers, some of whom have been squeezed out of adjoining neighbourhoods by gentrification. When not working, these neo-bohemians spend a considerable amount of time hanging out together at local 'dives' – bars, greasy spoon restaurants and clubs. It is not long before word spreads that this is where to find the 'cool' pulse of the metropolis. Soon, art gallery owners descend, followed by proprietors of secondhand clothing boutiques and small cafés. Inevitably, the neighbourhood is 'discovered' by lifestyle magazines, Lonely Planet-type tourist guidebooks and MTV, who create a 'buzz'. Before long, there are regular weekend lineups of architects, fashion editors and other urban professionals at local clubs and dining spots. Along with them come Starbucks and other retail chains, as well as property developers who set about converting lofts and other cheap real estate into luxury housing. Typically, the director of sales and marketing for a new condominium project in Toronto called 'The Bohemian Embassy' was quoted in a newspaper advertising feature on 'New Dreamhomes and Condominiums' as claiming: 'There is a sense of being on the leading edge and setting down roots in a neighbourhood beginning to bubble with a heady mix of artists, musicians, designers and with-it young professionals.'

In the early years, the neo-bohemia constitutes what Richard Lloyd (2006: 161–66) has described as a *creative milieu*. By this he means that local spaces 'attract relevant types of creative workers, such as artists, performers, and musicians, and they provide social conditions needed for the nurturing of talent and the ongoing creation of cultural producers' (162). Not only do these spaces make a social life possible for individuals who typically work in an isolated setting, but also they are invaluable in providing an environment for creative collaboration and for linking artists to audiences.

Lloyd is cautious in endorsing Richard Florida's ideas about the natural affinity between 'edgy' artistic producers in neo-bohemian areas and the 'creative class' that allegedly seek them out for both economic and aesthetic purposes. In his ethnographic study of Wicker Park, a neighbourhood on Chicago's Northwest Side that was transformed into a neo-bohemia in the late 1980s and early 1990s, Lloyd found some high technology firms did indeed conscript local artists into their ranks of employees and subcontractors, but the latter felt rather alienated, experiencing 'long hours, mediocre wages, and extraordinary vulnerability' (68). He implies that the 'diversity dividend' (Rath 2002) attached to Wicker Park was not primarily a matter of creative class entrepreneurs being altruistically attracted to neighbourhoods that were culturally open and socially tolerant, as Florida claims. In addition to their usefulness as low cost labourers, the Wicker Park locals were 'avatars of cool' (Lloyd 2006: 243), that is, catalysts for the creation of a hip local scene.

Once that scene was established, however, it was the merchants of 'controlled edge' – property entrepreneurs, venue operators and large corporate players such as Nike and MTV, that marketed it to the creative class as a source of subcultural capital. At the end of the day, Wicker Park became 'not only a target of gentrification but also a bohemian-themed entertainment district where patrons are not starving artists but rather affluent professionals' (Lloyd 2006: 69). Clearly, then, the creative city was not as different as one might have thought from the fantasy city template for urban regeneration.

Tourism and the creative city

As Richards and Wilson note in the first chapter of this volume, the 'creative turn' that has been so pervasive in recent strategies for urban regeneration is also increasingly visible in new approaches to tourism. Earlier in this chapter, I noted the deepening sense of *ennui* related to fantasy city developments and the concomitant demand for what Hayward (2004: 191) describes as 'places of excitement' and 'calculating hedonism'. In similar fashion, contemporary travellers are said to be 'more interested in enriching their lives with experiences than being passive consumers of entertainment' (Singh 2004: 5). This search for experience has assumed a variety of forms.

At the more exotic and unconventional pole, new experiences, trends and practices include 'extreme adventure tourism' (Buckley 2004), 'thana-tourism' (tourism and travel associated with death) (Seaton and Lennon 2004) and travel to atrocity heritage sites (Ashworth and Hartmann 2005).

No longer confined to African safaris or Californian wine tasting tours, today's boutique travel repertoire reproduces the everyday worlds of political prisoners, the homeless and third world slum-dwellers, but without fully exposing consumers to the dangers attached to the real thing. Some recent examples include:

- *street retreats*: inner city safaris in which altruistic, middle class consumers hand over hundreds of dollars each to spend 36 hours begging, eating in soup kitchens and sleeping underneath highway overpasses;
- *gulag guesthouses*: where tourists spend a night in a dark cell once inhabited by prisoners of the *Stasi*, the secret police in the former East German state;
- *favela tourism*: where visitors to Brazil are selectively exposed to conditions of poverty in a Rio de Janeiro shantytown.

At the less extreme pole is *creative tourism*. Richards and Wilson (2006) take this to mean – inter alia – empowering individual tourists to produce their own narratives and experiences, with tourism providers supplying the creative raw materials. Most examples of this (perfume making, porcelain painting experiences) can currently be found in rural or peripheral areas, they say, but nothing precludes doing this in a metropolitan venue.

Creative tourism both parallels and differs from the 'creative city', as profiled by Richard Florida and his acolytes. On the one hand, both reject the *administered consumption* of the fantasy city model of urban development in favour of a more *collaborative consumption*. Both put creativity and innovation front and centre as key components of cultural consumption and urban economic growth. Both recognize the superiority of experience over passive forms of leisure. At the same time, Florida's model has an undercoat of managerialism that the creative tourism strategy does not.

If creative urban tourism is to succeed, it needs to meet two tests. First of all, it must allow the tourist a true measure of creative input. That is, it must be more than an enriched mini-course in cooking or arts and crafts. One of the most extensive, and ultimately unsuccessful, attempts to mix tourism and learning is the 'Disney Institute'. The Institute attempted to market itself as a hands-on learning vacation, but has since become transformed into an experiential learning programme for corporate clients. Second, it must figure out a way to integrate the tourist into the 'creative milieu' that Lloyd identified as being central to the bohemian scene in Wicker Park. This is very difficult to do, insofar as cultural producers constitute a subculture that actively resists invasion by outsiders, whether they be tourists or gentrifiers.

4 Creative spaces, tourism and the city

Graeme Evans

The tour bus picked us up outside of the designer hotel in Manhattan. Commuting office and shop workers, tourists, police and road diggers mingled in the chaos of downtown traffic. Across the Williamsburg bridge we stopped to pick up our tour guide for the day, Angel Rodriguez, from an unsalubrious building covered in layers of posters, graffiti and grit. He was a Latino musician, a salsa drummer from the Bronx who proceeded to give the tour group the background to the area – 'Bronx is burning' (arson attacks on tenement blocks by landlords); old jazz and dance club haunts; Fort Apache ('the movie'), the now rebuilt district police station self-styled to defend itself against the 'natives' (i.e. Black/Hispanic); graffiti art of local rap stars; the massive American Mint building covering four blocks, where 65 per cent of all US dollars were once printed, and now housing two community schools, artist's studios and employment schemes; the local penitentiary with 12 year olds kept in shackles – before arriving at our destination, *The Point*. Here the graffiti boys' operation base – once the crew that covered the New York subway trains and led to the Mayor's zero tolerance regime – has now gone 'legit', working for large advertising firms and department stores in Manhattan on large-scale shop displays and billboard art.

This was the 'Creative Spaces' tour, which had followed itineraries in London, Barcelona, Berlin and Toronto. Creative spaces, creative tourism, the creative class abroad? If the 1990s were about expansive cultural tourism, then the twenty-first century is about the so-called creative city, the confluence of production and consumption, signalling the burst of the passive tourist bubble by 'cosmopolitan engagement' (Evans and Foord 2005). Creative tourists then – technically and of course economically, tourists, but also insiders, creative business tourists and educational tourists, marking a growing incidence of cultural tourism motivated by 'knowledge economy transfer'.

Figure 4.1 Graffiti – rap artist 'memorial', the Bronx, New York.

Cultural tourists and the creative class

Some time ago, MacCannell claimed that tourism was the 'cultural component of globalization' (1976). Today creativity and the creative class are
advocated as the panacea for economic growth and the arbiter of urban
taste. Where they go, we want to follow; where they create and mediate, we
want to be part of the scene. We are all creators now – even, especially, 'on
tour'. Ten years ago too, tourism was fêted by advocates and academics, set
to become the largest global industry by the new millennium, fuelled, literally, by commercial aerospace and transport growth, and also by conflating
urban 'visitor' activity and flows which had previously been hidden from the
tourist gaze (and statistics). Perhaps because tourism has become so ubiqui-
tous and fragmented, no longer confined to the western generator countries,
its distinctions have become ever more blurred with other human movement and motivations.

Moreover, the urban conurbations that fed mass tourism have increasingly been on the receiving end, as cities become destinations and the sites
of intense cultural exchange and experience – physical and virtual. Major
city tourism arrivals now exceed that of entire countries. The cultural and
creative industries, in their new and old guises, have taken over as the future
economic and symbolic global industry, of which tourism is now only one

element, along with the Internet and the consumption and transmission of a widening range of cultural goods and services (Scott 2001).

In the cross-national survey of cultural tourism carried out by ATLAS in selected European heritage sites (Evans 1998; Richards 1994, 1996b), a significant proportion of visitors, not surprisingly, were well educated and employed, but over 10 per cent were also cultural aficionados – working or studying in the arts and museum worlds. Today's creative tourist includes creative producers, artists and designers and the associated brokers – including trade/event producers, creative product organizations and networks, 'critical curators' (Harding 1997), and an increasingly discerning tourist. The latter includes students and consumers who are also the path-finders for this new form of creative tourism trail, supported by mainstream and specialist press and media where place, production and consumption are seamlessly and hungrily presented and packaged.

The creative tourists therefore join a global elite of culture vultures: international architects, artists, curators, designers, government officials, sponsors, global foundations such as Getty, Guggenheim, Thyssen and touring blockbuster exhibitions. These include efforts at cultural diplomacy and rapprochement – for example Mexico, Iran, Turkey and numerous Chinese and Islamic art exhibitions – cultural tourism as a tool of foreign policy. The city itself is a growing subject of the art museum exhibition, notably Century City, the inaugural Tate Modern show and earlier 'reflective' city exhibitions at MACBA, Barcelona; Pompidou, Paris; and URBIS, Manchester (Evans 2003).

Richard Florida's reworking of the 'Creative Class' (2004, 2005) can also be viewed as a postindustrial version of the cultural milieu which has been evident in the heyday of culture cities for thousands of years (Hall 1998). This is also explained through the role of avante garde and bohemian movements which attracted artists from outside the city, including from overseas (Wilson 2003). Spatially this has also been represented in terms of cluster theory popularized by Porter (1995), but harking back to Alfred Marshall's economic theory of industrial agglomeration (1920) and the tacit exchange of innovation and knowledge – 'something in the air'. The 'Creative Class' is now measured through a range of quasi-scientific rankings using indices for determinants of 'creativity' (Florida 2004; Florida and Tinagli 2004). These include openness, tolerance, population diversity (e.g. Gay Index), counting workers in subjectively chosen 'creative occupations', R&D spending, and innovation measured by proxies such as the number of patents issued.

Problems with this approach to branding human cultural capital include the use of the word 'class', which is viewed as regressive and divisive (Nathan 2005; Peck 2005). The growth model of a creative cluster – Silicon Valley, California – survives on a Hispanic underclass servicing a footloose and detached industry enclave. Disconnected culturally and symbolically from their city's creative hubs, industrial park-style concentrations of creativity ignore the conditions and appeal that the more sustainable creative clusters have long established. Few tourists will see the attraction of such sites, and the creative spirits are likely to resist relocation and higher rentals. The creative scene will stay in Berlin's Kreuzberg and Mitte districts (see below) while its cosmopolitan edge and low cost opportunities – which have attracted club, music and visual artists (and MTV) – persist.

Cities with high creativity rankings therefore also score highly on inequality – the widening divide between those on high and low incomes, evident in cities such as San Francisco, Austin, New York, Washington, Boston (Florida 2004), and in London (Evans *et al.* 2005). Many cities therefore increasingly reject the creative group as a class apart, but seek quality of life and places 'for all' and to retain a cultural business and heritage, as opposed to a creative class and industries approach to economic development. This includes mid-West US cities as well as developing ones which see the importance of the protection of cultural identity (language, heritage) against world trade in creative content and products. Here the term 'creative industries' is resisted in favour of the 'cultural industries' (e.g. Hong Kong – Evans *et al.* 2005).

From city of culture to creative city

As cultural tourism and its variants become diverse but at the same time standardized, and the cultural dimension to mainstream tourism expands, the imperatives of maintaining distinction and promoting tourism in postindustrial cities has led to the renewed process of city cultural branding (Hankinson 2001; Kavaratzis 2004). Cities that are most successful offer both consumption and production, heritage and contemporary culture, as well as a cosmopolitanism that cannot easily be replicated or imported. Bilbao will not achieve this diversity with a monocultural edifice (Baniotopoulou 2000; Evans 2003), nor Singapore, 'Global City of the Arts' (Chang 2000a; see also Ooi, Chapter 16 of this volume), nor Shanghai, where western-style regeneration is 'sapping the city's own creativity' (Gilmore 2004: 442; Wu 2000). New facilities alone do not create a creative city, only human cultural capital and interaction evolving over time (Evans 2005a).

This aspirational move from the capital of culture emerging in the late 1980s/1990s (Zukin 2001), towards a creative city (Landry 2000; Nichols Clark 2004) can be characterized as a shift from singular cultural branding (Evans 2003) to city spaces which depend on creative diversity and tension more than predictability (Table 4.1).

A national or place-based association with cultural creation has of course been the source of cultural tourism campaigns by individual destinations and tourist boards. These have promoted the largely passive tours to performing and visual arts festivals, theatres, museums and galleries, and heritage venues, and have been visualized by stereotypic images, itineraries and productions (London's West End/Broadway musicals, 'opera-lite'):

Table 4.1 From cultural branding to creative spaces

Hard branding the culture city >>	>> Creative spaces
• Museums and heritage tourism	• Cosmopolitan culture
• Cultural districts	• Creative production and consumption
• Ethnic quarters	• Creative clusters
• Entertainment cities – Times Square, Potsdamer Platz	• Creative class – new Bohemia
	• Cultural trade and art markets
• Competitive advantage	• Comparative advantage
• Pilgrimmage and literary trails	• Showcasing the designer city
• *City of culture*	• *Creative city*

arguably a democratization of culture, or just 'postmodern grand tourers' (Evans 1998). Politically and regionally, this is also represented by designated Cities/Capitals of Culture more often producing marginal and unsustained cultural impacts that are notably devoid of creativity – artists are not engaged or opt out, for example Glasgow 1990, Cork 2005, Liverpool 2008 – a tourist event, as opposed to a cultural development focus (Evans 2005a; Palmer Rae Associates 2004; Richards and Wilson 2004a).

Exploiting creative arts and culture for tourism has therefore been an established destination marketing tool which has included the familiar and the emergent. From literary tourism and trails; film and television locations; architecture tours and branding (Lasansky and Greenwood 2004) – for example *Gaudí* (Barcelona), *Macintosh* (Glasgow), *Mies van der Rohe* and *Frank Lloyd Wright* (Chicago), *Grands Projets* (Paris); to Ethnic Festivals, for example Notting Hill (London) and Caribana (Toronto) Carnival Mas; and 'Quarters', for example *Little Italy* (Manchester), *Little India* (Singapore), *Currytown* (Bradford), *Banglatown* (East London – see also Shaw, Chapter 12 of this volume) and *Little Africa* (Dublin).

However these cultural tours have little connection with creativity or contemporary spaces. Fictional representations, for example Sherlock Holmes' 'house' in Baker Street; Braveheart in Scotland (shot in Ireland), and the low cost alternatives to Victorian London which cities such as Prague now offer film producers, are some of the obvious candidates. The iconic architecture presented by Guggenheim Bilbao (and its franchised collection) has little resonance with Basque art or artists. Here, declining visitors, now predominantly tourists, have seen low cost flights cut back and high maintenance and upgrades to the building itself. UNESCO World Heritage Sites, on the other hand, provide a staple attraction which is expanding into cities (e.g. Maritime Greenwich and Liverpool), but this is also a diluting and weakening 'brand', and national/local status is often preferred (Evans 2002; Smith *et al.* 2005) while the global imprint is minimized as a result.

Whose culture, whose heritage?

Problems of past association also arise, for example Auschwitz (a German Nazi camp in Poland); historic francophone Québec (British garrison and battlefield); Spanish Town, Kingston ('colonial capital' of Jamaica); and most recently in Belfast, Northern Ireland. Here, the Maze prison, now emptied of its IRA and Unionist terrorist prisoners and symbol of the sectarian divide, is destined to be razed and replaced with a 42,000 seater sports stadium. Tagged the 'terror dome' – this facility, over 10 miles from the city centre, with questionable viability and 'need' and little community consultation (officially both parties support the prison's demolition), is also the historic site of the troubles and heroes/villains such as IRA 'terrorist/ freedom fighter' Bobby Sands, whose hunger strike ended 25 years ago with his death in the H-block prison hospital. The Northern Ireland Tourist Board has however come out against the development. What all of these have in common is a lack of cultural value (or at least, awareness and community ownership) and connection to the creative city and its current identity(ies).

Ethnic festivals and quarters can suffer from staged inauthenticity – as is often the case with ethnic enclaves and the post-migration branding of such areas after the original ethnic community has long since moved to the suburbs (e.g. Greek and Portuguese Town, Toronto; Little Germany, Bradford). Major ethnic festivals have also been the victim of their success but also of a form of institutional racism, being re-sited from their neighbourhood roots and sanitized into parks and islands away from the downtown residential areas (Notting Hill, Caribana).

Whose cultural identities should be celebrated is also becoming fraught as so many migrant groups lay claim to space and event programmes – New York has had to limit its ethnic festival weekends and street closures, so Bosnians will have to wait until the Italian community diminish enough to lose interest in their festival space! (Evans 2001) Cultural space can therefore be contested, in short supply and subject to commodification. Creative space can more successfully negotiate difference, exhibit temporal as well as spatial (and symbolic) dimensions and be less place-bound. It can also resist, as well as drive and mediate, consumer culture.

Rationales for creative spaces policy

Reasons why city and other agencies are developing creative spaces strategies and interventions which focus on creative industries (production) rather than cultural tourism (consumption) can be gauged from an international survey carried out as part of a joint study undertaken for London and Toronto (Evans *et al.* 2005). The prime policy rationales for over 200 such interventions at city level were justified in terms of economic development, and by physical regeneration and infrastructure (buildings, transport, IT).

However a significant number were based on tourism and events as well as city branding through culture. Heritage was the least cited policy rationale.

In major cities, the clustering effects of creative industry production and consumption are of course exaggerated by the location of national administrative and arts and cultural organizations and venues (e.g. museums, theatres, media), national and international headquarters, higher education and research establishments (including art, design and language schools) and the hospitality and recreational facilities which mutually support them. This generates a seasonally spread market of business, educational and visiting friends and relatives (VFR) tourism, a dynamic cosmopolitan urban culture and diasporas, and consumption – eclectic shopping, food, arts, entertainment and subcultures – as well as mainstream 'city break', arts and heritage tours.

Smaller towns and cities can also maintain a niche creative base and profile, often on the back of annual or periodic events such as festivals and historic associations. Tours and routes can also be developed in order to link otherwise remote and uneconomic visitor attractions and destinations (e.g. Mississippi and Montana – Hannah 2006). However although these can serve as successful cultural touristic attractions, they seldom diversify or grow into all-year-round creative spaces, even when they are cutting edge, for example art biennales and influential gatherings of artists and curators, such as Documenta in Kassel, Germany. Their creative groups are less established and often come from the 'outside' (e.g. curators, artistic directors – Evans 2003).

This is important for towns and cities which see the potential for importing and tempting a creative class with loft studios and café culture. Key features of the more successful creative city and interventions include public space, design and showcasing and the quality of the whole visitor experience.

Public realm and urban design

Creative cities – particularly those recognized and emulated – pay as much attention to the public realm and city environs as to cultural buildings and visitor attractions. The city as art gallery or design centre is therefore an ambitious but shrewd strategy which benefits residents, workers and tourists, improving quality of life, access and mobility – the first to suffer under mass cultural tourism. This includes sculpture and art and architectural installations in squares, parks and gardens – from traditional statuary to commissioned and temporary artworks, for example London's Trafalgar Square, Christo's decorated gates in Central Park, New York, and war monuments such as memorials to Vietnam, Washington and the Holocaust, Berlin (Figure 4.2).

Barcelona was famously awarded the RIBA Gold Medal for architecture in 1999, not for an individual architect or building, but for the quality of

Figure 4.2 Holocaust memorial, Berlin.

urban design for the whole city. The city has served as an exemplar for its waterfront and street art, from Montreal to Madeira. Its own cultural quarter and regeneration of the Raval district and MACBA art museum was in turn modelled on the Beauborg redevelopment in Paris (Balibrea 2001), which hosts the Pompidou grand projet culturel, one of the first explicit locations in an area subject to regeneration and gentrification based on a major arts and visitor-led facility (Evans 2003). Most visitors of course do not enter the (charged for) gallery exhibitions or library, but congregate in the surrounding square, cafés and foyer area. Travelling up and down the (now enclosed) external escalators is no longer free since the Pompidou's makeover, which has further separated culture from the tourist (Evans 2003; Silver 1994). In contrast Barcelona has invested substantially in public art and 'outdoor museums' with over 1,000 sculptures created in city spaces, including installations by Miró, Lichtenstein and Calatrava.

Public events

Barcelona has also led the way with festival/event programming linked to city re-imaging and expansion, harking back to earlier exhibitions and fairs (1888 and 1929 EXPOs). The 1992 watershed Olympics have been built upon by themed festival 'years', exploiting cultural brands and community

development, for example a Literature Festival celebrating the four
hundredth anniversary of Don Quixote with a literacy campaign and new
library projects in the city (Evans 2006) and the Year of Food and Cuisine
event (Figure 4.3).

Transport by design

Despite its reach and diminishing cost, transport has gone from a benign to
an increasingly negative aspect of tourism, rather than an enjoyable part of

Figure 4.3 Year of Food and Cuisine, Barcelona.

the travelling experience. Creative use of transport spaces and interchanges has animated and improved safety in city transport (in contrast to international airports). Art on the Metro in Stockholm presents the largest underground art gallery (Figure 4.4); on the London tube network Platform Art and Poetry, and design-led stations and facilities using artists and architects (versus engineers) have created landmarks and attractions which lift the travelling experience out of the mundane. Examples elsewhere include (Norman) 'Fosteritos' in Bilbao (Figure 4.5) and Anish Kapoor's sculpture entrance in a Naples metro station.

Well-designed and legible wayfaring/signage, and creative lighting and routes, also combine to animate and improve safety and surveillance in exposed pedestrian walkways, and encourage walking at night.

Showcasing creativity

A city's cultural life and art and design constituencies have in the past largely bypassed the touristic commodification processes. The art market and the design industries – which together form the new economies of postindustrial cities – have served a commercial and elite world and imperatives which have not made them 'public' or subject to display. Trade exhibitions and conventions; private galleries and shows (including student shows) have been the established marketing and exchange fora. The growth of art

Figure 4.4 Graffiti art on the Metro, Stockholm.

Figure 4.5 Belles Artes Museum, Bilbao.

museums, 'collectables' and the celebration of industrial and contemporary design has however brought this world into that of the tourist, as well as generating its own system of creative tourism. Creative interpretation and invention is not limited to art and artefacts, since heritage too, as Kirshenblatt-Gimblett maintains: 'is a mode of cultural production [which] depends on display to give dying economies and dead sites a second life as exhibitions of themselves' (1998: 7).

However it is beyond the walled museum and heritage site that the actively creative city is to be found, where art market, design house and student show co-exist with club culture, fashion and 'events' – not featured in, or at least one step ahead of, any tour brochure or guide. Major design fairs have existed for many years, attracting higher spending visitors and a production chain far wider than most cultural tours. Milan's mega furniture, textiles and fashion fairs now spread out over two massive fiera sites linked by metro and with many fringe events more akin to the Edinburgh Festival. Competitive fashion weeks (and student fashion shows) in Paris, Milan, London, New York and Berlin grow more like rock concerts and film premieres combined, and generate a buzz of media coverage and place-making that few, if any tourism campaigns can muster, or importantly, maintain on an annual basis. London's designer shows attracted 150,000 in 2004/5; its film week 114,000 (with 3.5 million website hits), while London

Fashion Week was trailed a week before the event itself in New York with the Anglomania exhibition at the Met celebrating 30 years of British Fashion – underlining the global phenomenon of cities exhibiting cities.

The international coverage guaranteed to these glamour events is therefore valued as highly as if not more than the visitors themselves. As the national tourist office remarked: 'In terms of raising our profile these events have been huge. To have a magazine as highly acclaimed as *Vogue* promoting Britain is fantastic exposure. It is all about popular culture seeing Britain as very cool right now' (VisitBritain cited in Whitworth 2006: 5). With no sense of irony (or history) less than 10 years earlier *Vanity Fair* produced its Cool Britannia issue, presaging a short burst and then backlash against political and institutional associations with 'cool'.

Culturepreneurs

Berlin has become the first German city to be appointed a City of Design by UNESCO. In 2006 the Berlin Senate designated the city's design sector Create Berlin, modelled on Creative London and emulated by Creative Toronto, New York and Sheffield. One of the key urban and cultural developments in post-reunification Berlin has been the emergence of a new hybrid of artists and entrepreneurs, so-called culturepreneurs. Germany's new capital has been suffering under continuous socio-economic crises requiring individualized marketing strategies and balancing unemployment and self-employment in cultural production. Declining public funding forced many artists and designers to open up their professional practice towards corporate firms, and new forms of project-based cooperation. These groups have provided the foundation for Berlin's dynamic creative and club 'scenes' and a critical mass of creative talent (Lange 2005).

The growth of trade exhibitions as cultural events is a phenomenon which links existing creative production and cultural tourism around major and fringe city sites. In Berlin, for example, annual fairs and conferences include ECHO and POPKOMM,[1] WOMEX Music Fair and fashion fairs such as Bread & Butter, Premium and Spirit of Fashion – its Fashion Week attracted over 60,000 visitors in 2006. The 10 day Berlinale is currently one of the most prestigous film festivals, with 13 cinemas and over 13,000 seats attracting 400,000 ticket buyers in 2005. The first art biennale held in 1998 attracted over 80,000 visitors: now several times that number attend. The €30 million culture programme in the build up to and during the 2006 World Cup has featured football in Fashion Week and in galleries – the Martin Gropius-Bau gallery filled with football shorts, videos, fan memorabilia and a mini-pitch laid out in the main gallery. The 2006 World Cup promotion is being used to lever a larger image campaign, including an exhibition of 70 artists from 20 countries, a football opera, business campaigns and worldwide football road shows, and the first ever Olympic-style opening ceremony to the World Cup itself.

The symbolic Love Parade that relaunched in 2006 was started by a local DJ in 1989 on the fall of the Berlin wall. In 1990 2,000 people came to dance in the streets, two years later 50,000, and by 1995 there were 300,000. By then the route was changed to accommodate what by 1999 were 1.4 million people and over 50 floats. The city organizes a host of sponsored cultural events around the parade – art shows, operas, clubs, films – seeing this event as a draw for youth culture with the hope that the visitor (average age, 21 years) will like it so much that they will come back. The Parade is now a self-sufficient cultural event – largely 'free', but 800,000 people spend on average £69 (€102).

Open studios – Hidden Art

On a smaller scale, the Berlin Design Mai Festival is organized by a local society which is coordinated by seven voluntary members. The initiative started out as a magazine – now 130 open studios participate over a two-week period in May each year. Some locations provide a venue for several design presentations, while a showroom offers a retail opportunity to purchase direct from designers/creators. The central festival venue is the Forum in Berlin-Mitte, with an auditorium for workshops, lectures and presentations. Design Mai is an international as well as a Berlin event. In 2005 over 12,000 tickets were sold, the Design Mai website received 6 million hits.

'Open studios', where artists and designer-makers open their otherwise closed workspaces to the public, have taken off in London. Hidden Art began in 1994 as a small event to promote quality work by over 40 East End artists/designers. A decade on, it has developed into a unique network which has promoted and supported over 1,800 designer-makers and built links between the creative and manufacturing industries. Open studio events have now spread to several east London boroughs and to creative hubs across London. Design exhibitions, once 'trade only', have now transformed into cultural events – Designer's Block, 100% Design, Art shows (the annual Frieze Art fair receives 50,000 visitors), and fringe and 'off-piste' events with art and design students, music and club gigs and publishers extending the scope and duration of these former product shows. These are now international events that spawn other creative production showcasing.

Architecture too, which has resisted museumification, has followed the open studio format with biennales combining open architecture studios with installations, lectures and children's events. In its first year (2004), London's Architecture Biennale attracted over 25,000 people over the opening weekend to what would have normally been a deserted area of the city (Figure 4.6). In 2006 the biennale attracted 75,000 visitors and included a route across London celebrating 'change', with Norman Foster driving sheep across his Millennium Bridge (Figure 4.7). 'Open House', similarly taking place over a weekend each year, gives access to working heritage and other buildings of architectural interest. In these ways, the everyday culture of the city is celebrated and accessed (Lefebvre 1974), in contrast to the increasingly deleterious experience at tourist and heritage sites. This more democratized built environment offers users 'the freedom to decide for themselves how they want to use each part, each space'. As Hertzberger (1991: 170) suggested: 'the measure of success is the way that spaces are used, the diversity of activities which they attract, and the opportunities they provide for creative reinterpretation'.

Conclusion

Creativity and tourism are uneasy bedfellows and in many respects their interests conflict – the latter tending to display an institutional culture that dislikes diversity and distinction, risk and resistance – key elements of contemporary urban culture. The 'creative city' is also paradoxical, potentially self-destructive, but unlike 'cultural capitals' (e.g. Paris, Venice) also self-renewing. Over-planning in terms of branding and itineraries also tends to kill the experience and be anti-creative. As independent travel overtakes package tours, visitors organise themselves, leaving space for risk and spontaneous behaviour – the first step to creative tourism: 'if you plan it, if you destroy all the spaces in between, you also destroy some of the energy that gives the nerve to the city . . . you have to create spaces where things are

Figure 4.6 London Architecture Biennale 2004: grassing over the city.

Figure 4.7 London Architecture Biennale 2006: sheep drive through the city.

more anarchistic and unorganized, otherwise it all turns into a mall' (Walberg 2006: 4).

Cosmopolitan cities have blurred edges and centres, fluid images and identities (Hannigan 1998), and are as much a part of wider inter-city, regional and intra-regional clusters and networks. Spatially they offer a more sustainable distribution and diversity of cultural and visitor activity. A creative city arguably would have little or no need for tourist boards, and would resist being branded and designated 'creative' or 'cool'. This would avoid the crass branding and perversely isolated process of city tourism promotions. In 2005 the city of Toronto, attempting to reposition and reimage itself after 9/11 and SARS events seriously damaged its tourism trade, launched a Toronto Unlimited campaign and strap-line. The same month, London had presented a similar PR commissioned campaign, London Unlimited. Fortunately the London Mayor rejected this campaign (being 'unlimited' also means exposing oneself to being sued for debts, without limit). Toronto's more savvy neighbour, Montreal, on the other hand, rejects the cultural or creative city tag. One of the first year-round festival cities, culture here is more integrated. A city design commissioner mediates the city's image and urban design quality, with annual awards to firms and landlords who compete with office and retail design projects and public art. In turn these add to the city's 'creative tour' and quality of life. Very creative.

Creative spaces also depend on 'habitus' (Beck 2002; Lee 1997), and in economic terms, comparative rather than competitive advantage. More attention (apart from good governance) might be paid to investigating the hidden cultural assets of incumbent communities, a place's current and past heritage and distinctiveness, rather than imagining a creative class gathering over the horizon ready to descend into a cultural wasteland with cool urbanity. As Hall maintains, successful culture cities retain their creativity only by 'constantly renewing themselves. Or rather, cities don't do that; their people do in a particular creative (or innovative) milieu' (Hall, P. 2005: 5). Celebrating and accessing creative spaces can therefore offer an alternative and arguably a more sustainable approach to city culture for host and visitor.

Note

1 ECHO rivals the Brit Awards as the music industry's second most important accolade after the Grammy; POPKOMM is the international business and communication platform for the music and entertainment industries.

5 Tourists, the creative class and distinctive areas in major cities

The roles of visitors and residents in developing new tourism areas

Robert Maitland

The tourism of daily life

For many years, I have lived in Greenwich, one of London's most popular tourist destinations outside the centre of the city. About a year ago, an Argentinean café opened close to where I live, and well away from the tourist hotspots. This was unexpected, and I was eager to try it. When I did, I found not just very good Argentinean food, but many Argentinean customers, though I had not realized there was such a community in my neighbourhood. Globalization has meant not only more flows of more tourists around the world, but that the rest of the world increasingly flows to us as products, images and fellow residents – particularly in major cities like London (Franklin 2003). Yet much of the discussion and writing about tourism still sees it as a separate activity, undertaken for leisure purposes in 'exotic' resorts or well defined tourism districts in cities, and detached in time and space from everyday life and work. In this view, tourists are seen as comparatively passive consumers of a product that the city creates for them, exploiting whatever resources it can marshal.

In this chapter, I will draw on research in the Islington and Bankside areas of London to argue that tourism cannot be so neatly separated off – at least in major cities. Rather than being a disconnected activity, tourism and tourists are integral to the way that some areas of the city are changing. In these areas, tourism has been 'built into the architectural and cultural fabric' (Deben *et al.* 2000; Terhorst *et al.* 2003). Many tourists are connected to the city, as they are there on business or are visiting friends or relations (Maitland 2006) or are simply frequent visitors, and use it in sophisticated ways. It is not clear how even in principle they should be differentiated from other 'city users' (Martinotti 1999), including mobile professionals who change jobs and location frequently. I will suggest that we can see a 'cosmopolitan consuming class' (Fainstein *et al.* 2003: 243) that overlaps with the 'creative class' (Florida 2002), so that 'tourists' – along with 'residents' – drive the growth of areas that add to city amenity and extend its range of creative spaces.

Producing tourist places: if we build it, will they still come?

The dominant account of recent development in tourism in major cities takes a supply side perspective. It sees former industrial cities, and former industrial districts in polycentric cities exploiting their heritage and cultural resources, revalorizing places and creating new attractions in order to bring in visitors. There are convincing accounts of how cities are 'converted' into centres for consumption in which tourism plays an essential part (Judd and Fainstein 1999). Tourism is seen as having both material and symbolic effects (Harvey 1989); it drives economic and physical change directly but also contributes to the reaestheticization and revalorization of places (see for example Clark 2003). Tourism development has often tended to take the form of a zone that is planned for tourism (Judd and Fainstein 1999), in which a range of new attractions is provided to bring in visitors. Typically these include flagship museums, galleries or other arts venues, entertainment offers such as a casino or an aquarium, leisure shopping, branded bars, cafés and restaurants, and frequently a convention centre or other business tourism facilities, and take the form of a tourist bubble (Judd 1999) or urban entertainment district (Hannigan 1998): isolated physically and symbolically from other parts of the city. For early pioneers such as Baltimore, this approach was at least in part a response to the poor image of cities in the 1970s and 1980s. Former industrial cities, and indeed many areas in world cities such as London and New York, were seen as unattractive, unfashionable and probably unsafe, so there was a need to focus on particular areas that could be reconfigured and reimaged.

The success of some of these early endeavours prompted emulators in other cities, who sought to apply the same blueprint, leading to 'serial reproduction' of planned tourist spaces (Fainstein and Gladstone 1999; Jones 1998), offering a similar range of attractions in 'placeless' environments (Entriken 1991; Relph 1976; Smith 2007). Proliferation reduced competitive advantage and cities sought new forms of differentiation. 'Cultural quarters' aim for more locality-based development, drawing on local business and culture, and are less reliant on international brands (Montgomery 2003, 2004). Alternatively, 'iconic' structures intended to symbolize changing character and to provide a memorable image can be inserted in areas designed for new tourism development. The Guggenheim museum in Bilbao is the most celebrated example, and is widely credited with changing the image and boosting the economy of the city. However, both approaches have themselves been criticized for in practice fostering standardization and destabilizing specific locality characteristics (for example Chrisafis 2004; Honigsbaum 2001; Richards and Wilson 2006; Sudjic 2005).

In short, the process of converting cities to tourism has a number of common features. It tends to be based on planned development, to attract visitors to 'tourism zones'. It sees visitors as largely passive consumers of products marketed to them by the city. The city creates these through reconfiguring or revalorizing its own (limited) heritage, culture and other resources or through investing in new attractions to produce a tourist bubble, a cultural quarter or a city icon. And it relies heavily on models that policymakers think have proved successful elsewhere, so that ironically, even attempts to draw on specific local characteristics tend to result in the creation of standardized tourist spaces (Judd 2003b). This is not surprising. As Richards and Wilson (2006) point out, cities find copying good ideas a safe strategy. The problem is that borrowing 'good ideas' means the repetition of similar models, and this creates more competition among commodified cities.

The impetus for these tourism initiatives is the need to reconfigure cities in the face of rapid change. Yet comparatively little attention is paid to how they link in with wider processes of development in the city. Equally, there is a substantial literature on neighbourhood renewal which tends to ignore the role of visitors even though they are often seen as important economic contributors to regenerated places. Hannigan (2004: 13) argues that creating successful new quarters or districts is 'increasingly a privileged approach to urban development' but one on which there is a lack of good research on 'who patronizes them and with what effect'. He argues that we need to bring together tourism with other development drivers, and consider the demands of visitors along with other city users.

What do visitors and other city users want?

An essential starting point is to acknowledge that in many cities tourism is not a separate activity that occurs only in particular defined locations at particular times (Franklin and Crang 2001); it is much more pervasive, and part of the very fabric of the city (Terhorst *et al.* 2003). While this point of view is becoming more widely accepted in the literature, its implications are too rarely explored. The regeneration of many cities produces new built environments that may be attractive to visitors or may be intended to attract tourists to exploit the economic rewards of a visitor economy. Yet visitors are not seen as having an active role in the process of regeneration. Indeed, there is little research to explore what visitors want. Although the tourist experience – in other words the tourists themselves, their perceptions and what they enjoy – is at the heart of the activity of tourism, it has received comparatively little attention. There is little research on urban tourism generally (Page 1995; Shaw and Williams 2004), and particularly on visitors' experience of cities (Selby 2004). Most research on tourism in cities has been through quantitative visitor surveys, but as Page (2002: 113) complains, these: 'may have provided rich pickings for market research companies [but have] . . . all too often been superficial, naïve and devoid of any real understanding of urban tourism'.

He goes on to argue: 'The tourist experience of urban tourism . . . is a complex phenomenon, a frame of mind, a way of being, and, above all, more complex to researchers than a simple series of constructs which can be measured, quantified and analyzed quantitatively'.

Little is known of the characteristics of the visitors it is assumed will be attracted to planned or commodified tourism areas, and still less about others who might seek different urban experiences. And the focus on the supply side discussed above extends even to studies of visitors' demands. As Selby (2004) says, many studies look at the production of places rather than their consumption, and thus frame questions and enquiry from a producer's point of view, rather than that of the consumer.

While some attempts have been made to move away from conventional typologies to focus on emerging forms of tourism (for example Poon 1993; Urry 1990), Wickens (2002: 834) argues that 'attempts [that] have been made to subdivide the tourist by various typologies . . . neglect the tourists' voice'. Despite recent work to investigate the characteristics and experience of visitors to particular areas in cities (Hayllar and Griffin 2005; Maitland 2006; Maitland and Newman 2004) there is still 'limited research material . . . focussing on the experience of tourist within [tourism precincts and] . . . an understanding from the tourist's perspective has been a neglected dimension' (Hayllar and Griffin 2005: 518).

There is a need for more sophisticated analysis that draws on the visitor perspective. As McCabe (2005) notes, tourists are not as dumb as they are often portrayed, and there is no simple traveller/tourist dichotomy. Franklin (2003) argues that people learn new skills of interpretation as visitors, which makes them better tourists, since they can better interpret what they see. We might equally argue that in their everyday lives, some visitors have learned the skills of interpretation and the signs and markers (MacCannell 1976; Urry 1990) that allow them to recognize and navigate areas not designed as tourism zones.

This brings us to the ways in which tourism demands can overlap those of other city users, and the links between tourism and other city development. Discussion of the broader processes by which cities and their economies are being reconfigured has drawn attention to the role of amenities in attracting mobile elite workers – part of what Fainstein *et al.* (2003: 243) term 'the cosmopolitan consuming class'. Such people are highly mobile, drawn from middle and upper strata in the developing as well the developed world, and work in industries such as marketing, the professions, finance and producer services – all of which require high levels of education. The importance of amenities came to the fore in 2002 when Richard Florida's book *The Rise of the Creative Class* created a significant impact among politicians and professionals in the USA. Florida pointed out that conventional theories of urban development had stressed the role of capital investment, including in physical infrastructure such as highways. They have been generally superseded by more recent analyses that emphasize the importance of human capital in driving city growth. However, that begs the question of what attracts high quality human capital – that is, well educated, creative, talented workers –

to particular cities. Florida considers that such people – his 'creative class' – are attracted to places that offer a combination of amenities and tolerance of difference. Amenities are widely defined and include a range of non-market attributes – low crime and good public transport as well as physical and cultural amenities such as good architecture and design, art galleries, and bars and restaurants; all of which might, of course, be expected to be attractive to tourists also.

Florida challenges conventional opinion on how people view the different elements of their lives and argues that they do not see the city where they live as a worksite, separated from places they would visit for leisure or holidays. Instead, they seek locations that provide the amenities they require, along with work opportunities. They will make location and career decisions to reflect their pursuit of amenity and desire to experience quality places. He sees attractive cities as offering high quality of place, which has three dimensions – 'what's there (the built and natural environment); who's there (the diverse and stimulating people that make a community interesting); and what's going on (street life, café culture, arts etc)' (Silver *et al.* 2006: 6). Different mixes of amenity and place qualities will imbue different localities with particular character – or 'distinctiveness' (Maitland and Newman 2004). Florida's views coincide with Lloyd's (2002) notion that some residents behave like tourists in their own city – but we should perhaps also stress that they mean it is often the touristic qualities of amenity that make many residents choose a particular place to *be* their own city.

Broader processes of city regeneration aim to make places attractive – including to the 'lifestyles and consumption practices of managerial, professional and service class' (Bell and Jayne 2004b: 3). Indeed there has been a substantial criticism of this style of urban renewal that harnesses the 'aspirations of middle class professionals' (Slater *et al.* 2004: 1142), to produce gentrified areas. Renewed parts of the city may have a base in gentrification or in the 'new economy' (Hutton 2003), or cultural clusters (Mommaas 2004). Such changes may result from large scale remodelling and rebranding or more organic processes (Hannigan 2004). In the case of new industry quarters, new media industries, for example, are characterized by small or single person firms who meet in local bars and restaurants to gain intelligence on opportunities in open, flexible and precarious labour markets (see Pratt 2004). The resulting cityscape of young talented workers of the creative class, together with fashionable bars and restaurants, may prove irresistible to other city users, such as tourists.

Changing cities meet tourist demands in ways that go far beyond the development of defined tourist zones. Improving amenity and gentrifying areas to meet the demands of the middle class professionals of the cosmopolitan consuming class/creative class might be expected to attract tourists among a range of different city users. Research in London was designed to explore this.

'New tourist areas': Islington and Bankside

The research examined the experience and perceptions of overseas visitors to Islington and Bankside. These are areas that have not formed part of traditional tourist itineraries, and neither has been planned as a tourist precinct, although Bankside is contiguous with established tourist areas in central London. Both have been subject to a process of regeneration and gentrification that involves upmarket housing, offices, restaurants, bars and shopping.

Islington is well connected to but separate from the main tourist locations of central London. It has been subject to a long process of gentrification for a period of some 30 years, and behind the main roads residential areas divide into either high value private Georgian and Victorian streets or mass housing blocks of the 1960s and 1970s. The area has very good public transport connections, to central London in particular.

Although a number of initiatives had sought to develop tourism, there was no comprehensive plan, the build up of tourism has been gradual and no major new tourism attractions were developed. There has been considerable public investment in renewing cultural assets (for example the Sadler's Wells and Almeida theatres) and substantial private investment in a bustling Islington nightlife that has seen the opening of numerous new bars and upscale restaurants. There is a rich mix of land uses, and loft style apartments and studio workspaces have been developed in former industrial buildings.

Tourism-related policy in Islington has a broken history. There were visitor-related initiatives from the early to mid 1980s, one result of which was the creation of Discover Islington, an independent not-for-profit organization whose mission was to develop tourism to create economic activity and foster civic pride; it was set up in 1991 and closed in 2001. However, tourist development has not resulted from comprehensive planning by either public or private sectors, in contrast to the process of creating a tourist bubble. It has been estimated that over four million visitors spent an estimated £105 million (€153 million) in the London Borough of Islington in 1998 (Carpenter 1999).

The Bankside area of Southwark lies between established tourism nodes (Westminster/the South Bank Centre and the London Bridge/Tower Bridge area) and includes two new flagship (or iconic) attractions – Shakespeare's Globe Theatre and the Tate Modern. There are other galleries and theatres in the area, and rapidly developing speciality food shops, bars and restaurants particularly focused around Borough Market. Along the river, warehouses have been converted to loft apartments, in many cases let to companies to accommodate overseas employees on temporary assignment. There is a significant population of office workers – some in modern towers, others in smaller spaces in converted warehouses. Further south, and away from the river, development appears to have been driven by apartment and office conversions, often for the 'creative industries'. The regeneration partnerships responsible for the area sought to spread the benefits and opportunities from congested locations north of the river to Lambeth and Southwark in the south, and tourism was part of that policy. A key element was improving pedestrian links across the river and a new pedestrian bridge creating a route from St Paul's to Tate Modern was opened in 2002. So while tourism has been a significant element in development, it has been just one part of a wider process of regeneration that included major office and residential building, and designing a place commodified for cultural consumption (Teedon 2001) which some see as an image-based and opportunist response to a process of property speculation (Newman and Smith 1999).

Visitor experience of Islington and Bankside

The research was intended to explore the characteristics of visitors to Islington and Bankside, their perceptions of the area and what they liked or disliked. It began by surveying overseas visitors to investigate their characteristics and the appeal of the area for them. It has been more fully reported elsewhere (Maitland 2007; Maitland and Newman 2004). Visitors in the surveys differed from overseas visitors to London as a whole. They were older, had more experience of visiting London, and made use of friendship networks in deciding on the areas they wanted to visit. These characteristics were more pronounced in Islington than in Bankside. This is unsurprising: Bankside now has good connections to established tourism beats and is more accessible – physically and symbolically. Visitors were asked what they liked and disliked about the areas. While there was a range of responses, it turned out that in both areas, the appeal was less to do with individual attractions, and much more about a distinctive built environment, sense of place or 'leisure setting' (Jansen Verbeke 1986). In Islington, visitors liked physical and cultural elements (e.g. architecture and the cosmopolitan atmosphere) but also the landscape of consumption – the range of shops, bars, cafés, restaurants and clubs. In Bankside, visitors referred to the architecture, the river, views, sense of history and atmosphere, and these aspects of sense of place were more important than the major attractions of Tate Modern and Shakespeare's Globe. It seemed that for these visitors at least, the appeal of the area lay in those qualities that make it a distinctive place, and that apparently mundane elements of vernacular architecture, and shops and cafés, could constitute attractions. Perhaps the appeal of the familiar and everyday (Maitland 2000) was part of the draw.

Exploring visitors' experience required further qualitative research. Semi-structured interviews with visitors to the areas allow investigation of the complexities of visitors' perceptions and motivations and can yield data that are rich and holistic, locally grounded, focused on ordinary events in a natural setting and with an emphasis on lived experience (Miles and

Huberman 1994). The research is continuing, but an initial round of interviews was completed in summer and autumn 2005 and forms the basis for the discussion below. It draws on 20 interviews with overseas visitors whose ages ranged from early 20s to mid 60s. They originated from a wide range of countries from around the world including western, central and eastern Europe, North and South America, Australasia, India and Africa, but interviews were conducted in English. Most were in professional occupations – architect, IT consultant, doctor, academic, teacher, language teacher, developer, management consultant, student – so could be seen as current or future members of the cosmopolitan consuming/creative class. They were often from or had spent considerable time living in cities.

They were approached on the street but interviews took place nearby in hotel lobbies or a Tourism Information Centre. These lasted up to 30 minutes. A topic guide derived from the first round of research was used to steer discussion. The interviews were recorded and subsequently transcribed. The transcripts were reviewed in the light of the themes thrown up by the earlier research to examine interviewees' perception and experience of the area. A number of themes emerged and are discussed below.

Narratives

Visitors interpreted the areas differently. For some, Bankside was a continuation of London's well established heritage landscape. They talked about easy pedestrian links from well known attractions like Tower Bridge or St Paul's, views across the river to heritage buildings, and the new flagship attractions of the Tate Modern and Shakespeare's Globe. 'I'm interested very much in historical places . . . I started from Westminster Abbey, then crossing the Westminster Bridge . . . it's very fascinating to walk alongside the Thames river . . . there's St Paul's cathedral'.

But for others, the appeal was an opportunity to get away from London's familiar touristscape – 'we did not feel like going to Buckingham Palace and get 10,000 people. We like to go to out of the way places where there aren't so many people catering to the tourists.' For these visitors, Bankside and the south side of the river seemed much more *normal*.

Some interviewees emphasized the history and heritage, the specific 'Englishness' of what they saw – 'one thing I really admire in English people is that you keep your history . . . London Bridge, Tower Bridge, we studied that in Brazil'. However for others, the story was one of modern buildings, new construction and development that could be found in other cities – 'it's modern. But in a way it's too clean . . . maybe the old story is gone'.

There was not a single dominant narrative. Rather, visitors created their own narratives – their own 'city' – from the variety of raw materials available in the areas. The raw materials included conventional elements such as heritage and historic buildings, but for many visitors, it was the ordinary, everyday qualities of the areas that made a difference. This is discussed further below.

Complex motivations

Visitors and their perceptions could not be accommodated within simple tourist typologies. On a very basic level, the purpose of their visit to London was far more complex than the 'business', 'holiday', 'visiting friends and relatives (VFR)' categories of tourism statistics. Combined purposes such as VFR and holiday, business and leisure, holiday and education were common, and combined in varying proportions.

Discussion of what tourists do is still often enveloped in assumptions about broad typologies – 'mass tourist' or 'backpacker'; 'traveller' rather than 'dumb tourist' (McCabe 2005). These imply uniform behaviour on the part of particular broad categories of visitors. Once a dumb tourist, always a dumb tourist, while 'travellers' aim to explore and avoid sightseeing in the well known 'tourism zones'. However, the interviewees showed much greater subtleties and variations in behaviour. Those who relished Bankside or Islington as 'quieter, less crowded, more off the beaten path' nonetheless had often planned or had already enjoyed visits to other parts of London that are firmly on the tourist trail – tourism attractions and familiar landmarks including Buckingham Palace, The Tower of London, Covent Garden, Oxford Street, the British Museum, the London Eye, Piccadilly Circus and St Paul's – 'things you can't really come to London and not go to see'.

Others behaved differently on different trips, visiting 'big places' such as the British Museum on their first visit, but subsequently enjoying an area that 'not many people know'. Visitors switched between experiences – 'sometimes it's nice to be in a touristy area but I like to change venues every day'. The fondness and affection that many interviewees had for London came across strongly in the animated way they talked about places they liked. As was the case in the earlier visitor surveys, many were frequent visitors to London and felt they knew the city quite well, but wanted to know it better. This could be linked to their connections. Many had friends in London, or were over on a business trip, or had family who were living here temporarily or had lived and worked in the city at an earlier stage in their life. In some sense, they 'belong to the city they were visiting as well as to the locations of their primary residence' (Hoffman *et al.* 2003: 243).

Perceptions of the areas

Architecture, buildings and street patterns were seen as important in both areas. For some visitors, they seemed to function mainly as a general backdrop for the visit. For example, Bankside could be seen simply as 'a nice place ... we like the buildings ... history ... the way the buildings are built . . . the style in which the buildings have been made'. Islington was 'a lot cleaner, nice and tidier than Clapham or sort of down that way'. These interviewees apparently liked what they saw, but rarely had more to say about details or specifics. We would expect many visitors to well managed tourist bubbles or cultural quarters to have similar reactions to areas that have been planned and reconfigured for their benefit.

But other visitors were looking at and assessing the areas more carefully. Some made much more detailed appraisals of the nature of the architecture and how it compared with other parts of the city: 'This part of the city, well, they're doing construction to make it more modern but it looks like there's older buildings from the late nineteenth century, early twentieth century . . . Some of the buildings seem like they're about a hundred years old. They haven't renamed them at all or tried to keep them in a current state. They're just trying to upkeep the brick and the whatnot from a lot of years ago which is nice.'

Perceptions of the physical aspects of place were seen as embedding history, character and distinct qualities. 'I think just the scenery is really unique, and just that it's interesting, all sort of cobbled streets, and tall, well, not too tall building, but just buildings right on the side of the street, like not much of a sidewalk or sort of open area, and just how it's all, try and sort of achieve it's all really wind [sic] and curvy, and winding. [The buildings] . . . they've got tons of history to them, and I'd love to know more about them. Like they're old and unique, and just very different.'

Some visitors looked with care, examining particular design details. One commented on spending time walking around Islington looking at the different designs of door lintels in the houses. For another it was the fascias of shops that were attractive – 'The windows of the shops, how they are designed. . . . I really like these things.'

Distinctiveness and everyday life

In describing the areas, interviewees often saw them as 'not touristy' and contrasted them with familiar tourist spots (which they might nonetheless visit on their trip). Bankside was seen as less designed for tourists with only 'two little theme parks . . . the Museum of War and the London Dungeon'. However, these interviewees had visited Borough Market, Southwark Cathedral, the Roman ruins and the Golden Hind and were planning to take a look at the Globe. These are all promoted in tourist literature about Bankside, but from these visitors' point of view, they were not 'touristy'.

Bankside was described as 'really more like kind of off the beaten track. Not the big touristy attractions'. The interviewee felt that it told her about the city, in a way that visits to conventional attractions would not. 'Museums are museums and they're all interesting but museums are anywhere and I like to see more what the city is actually about. . . . The London Eye looks cool but it's just a big tourist attraction as well . . . you get to know a city more by seeing the little, not so touristy things.'

Another interviewee felt that 'if you want to know London, you have to walk the streets where the tourists don't go'. The opportunity to observe or partake of everyday life seemed central for some interviewees. One liked the area around the Angel because 'people are more relaxed, they sit down, drink coffee, and they talk a lot because it is more relaxed they have a little conversation with you'. 'Meeting the locals' is a familiar element in what some visitors seek from their experience. But here the interest goes beyond that and encompasses the mundane details of everyday life: 'I remember I went to Tesco to buy things, well it was an incredible experience because how people buy their food, the people wear different type of clothes'. Another interviewee pointed to the importance of the everyday and her enjoyment of it – 'people going about their daily tasks . . . I just see how people go about their day, as they would'.

Part of the appeal of Bankside was that 'people were just out and about, normal Londoners just doing their thing, as opposed to where all the tourists go'. The routines of daily life took on significance as part of a tourist experience. 'You can actually see the real London and see how people work. You can see their offices, them working in there. So you can't miss this . . . you can walk by . . . and see them typing on their computers . . . It's kind of cool.'

This 'real London' was not a romantic heritage construct. Visitors knew that the people they were seeing were workers, often affluent – 'yuppies, maybe' – and cosmopolitan. And the appeal of everyday life of course also included consumption. Cafés, bars, places to eat and interesting shops were mentioned frequently, and many of the interviewees were staying in branded hotels.

Synergy: some tourists, some residents

The interviewees had little in common with stereotypes of dumb tourists v. travellers, or conventional typologies of business or leisure visitors. Their accounts of their experience and perceptions of the areas are revealing and show that many are sophisticated users of places, conscious of switching between more or less 'touristy' activities, as one would expect of 'post-tourists'. In effect, they experience not a uniform 'London' but a series of different 'cities'. This allows us to explore how tourist experiences overlap or converge with those of residents and workers.

First, while some interviewees saw the areas mainly as another part of London's tourist landscape, others seemed to be discriminating users of the city, with tastes and preferences that converge with those of other users. They were capable of creating their own narratives, appreciative of different facets of places and they found pleasure in what they saw to be the everyday life of the city. Many had a sophisticated view of the areas' architecture and buildings and the people they thought used them. They were often very familiar with other cities through their own places of residence and through travel, and were comfortable in exploring both Islington and Bankside. In both cases, these were seen as places that were safe and easy to navigate, certainly during the day and evening. In part, this reflects the changing role and perception of cities more broadly. In some cities at least, the negative images of the 1970s and 1980s meant that the safe tourist zones – tourist bubbles – which had to be established to attract visitors have been replaced by cities as the focus of modernity and consumption.

Second, while conventional attractions in the two study areas or elsewhere in London often formed part of visitors' itineraries, they valued much less recognized experiences too, which they often saw as indicating the 'real London'. In contrast to some major tourist attractions that were what you would expect to find in any city – 'museums are anywhere' – the opportunity to be in an area with city residents and workers going about their daily routines was a valued experience. The humdrum occurrence of seeing an office worker at his or her keyboard became a point of interest, and made the area and the experience distinctive. It was the combination of the physical qualities of the place along with the presence of aspects of everyday life that was important. This could lead us into debates about the nature of the 'real London' and the extent to which visitors' experience was 'authentic', especially in the carefully designed Bankside. But as Wang (1999) has pointed out, authentic experiences can arise from the visitor's feeling that they are authentic – existential authenticity. Here, it seems that for many visitors the presence of locals confirmed that they were having an experience they had created for themselves, rather than one scripted by the tourist industry.

Third, though distinctive, the version of 'everyday life' the areas offer is a very particular one – in parts of the city that have been through a substantial process of gentrification. Gentrification has been part of a process by which cities are recast to meet the needs of dominant groups in the face of rapid economic and social change. As pointed out earlier, regeneration efforts frequently mean transforming places to meet the requirements of middle class professionals for improved amenities. As Martin (2005) shows in a discussion of neighbourhood change in London's Notting Hill, the requirements of the dominant groups may differ from those of other residents. In Notting Hill, more working class residents saw local shops and markets in functional terms, as convenient places to buy what they needed. For more middle class residents, they were about the 'character of the place' – 'what this area of London is all about' and 'epitomising its cosmopolitanism' (77). Some of them acknowledged and in some ways regretted the loss of small local shops, but saw that there was an upside in terms of amenity – 'you do get better restaurants' (81).

Balancing potential loss of character with welcome improvement in amenity is difficult, and applies equally to Islington and Bankside. If, as

Molotch *et al.* (2000) argue, the distinctive 'character' of a place is the product of connections between 'unlike' elements, and the way economic, social, physical and intangible elements are 'lashed-up', then it is vulnerable to being degraded as some of those elements change. But for the moment, we can see a clear overlap between what (some) visitors and (some) residents and workers want from the areas. A combination of aesthetic values, inherited bricks and mortar, arts and cultural activities, and opportunities for a varied range of consumption in bars, restaurants and shops provide a level of amenity that appeals to city users of the cosmopolitan consuming class whether they are tourists, workers or city residents. For some tourists, as for Florida's creative class, 'what's there, who's there, and what's going on' combine to form 'high quality of place' and amenity.

This suggests that we need to move beyond conventional analysis that focuses on tourism in cities from a supply side perspective, and when it does consider visitors, tends to take a limited and stereotypical view of who they are and what they want. This research suggests that we cannot simply see visitors as passively consuming a tourist product that has been constructed for them from the city's heritage, culture and other resources. It is clear that there are visitors who want to construct their own experiences from the raw materials that the city provides – a mix of amenity, consumption opportunities, and everyday life of a particular sort. We can see them as members of the cosmopolitan consuming class/creative class – we might amalgamate the term to 'cosmopolitan creative class'. The amenities and attributes of the areas these visitors value also appeal to city residents and workers of the same dominant class. For visitors, the presence of cosmopolitan/creative locals marks out the kind of area they want to spend time in – at least for part of their visit. For cosmopolitan/creative locals, we can see that visitors – from the same dominant class – add to character and cosmopolitanism. Thus gentrification transforms areas and constructs aesthetics and amenities that are valued by city users with similar tastes. Some of them are currently residents, some currently visiting, but their shared demand and tastes combine in the transformation of the areas. As Judd (2003a) says, in some ways mobile workers and consumers demand and therefore help create similar qualities of place wherever they are. Rather than attempt to distinguish host from guest, it may be more helpful to see gentrification as a process that satisfies the demands of a dominant cosmopolitan creative class composed of a variety of city users – temporary migrants, more long term residents, business visitors on short term assignments, leisure visitors, those visiting friends and relations, and so on. Conflicts arise not so much between 'host' and 'visitor' but between that dominant cosmopolitan creative class and a poorer local population who see the areas transformed. This research has examined two areas of London, but we would expect to find similar processes at work in other gentrified parts of the city and in other cities too – for example Amsterdam's Jordaan, Berlin's Prenzlauer Berg or Barcelona's Gràcia.

From this perspective, we can see (some) visitors seeking areas that strike a chord because of their combination of amenities and opportunities for consumption, and distinctive locality characteristics (Maitland 2007). In doing so, they contribute to a process of gentrification and environmental upgrading that develops the kind of creative spaces that attract the skilled and talented workers that cities require. Tourism is thus an integral aspect of urban change, not simply one development option. Looked at like this, some tourists are contributing to the development of amenities that enhance the city, not simply consuming them. The problems of serial reproduction do not disappear, as Judd implies. But if, for these visitors, the attractions of an area are a blend of amenity, consumption opportunities and everyday life, then there is the possibility of avoiding some of the standardizing effects of tourist bubbles, planned quarters and iconic cultural facilities – though this will depend on a subtle, sensitive and limited role for tourism planning.

Part 2

Building creative
tourism supply

6 Creative tourism supply
Creating culturally empathetic destinations

Richard Prentice and Vivien Andersen

Introduction

This chapter has twin objectives. First, it seeks to conceptualize creative tourism supply, and to offer a vision of a market yet to be fully made by suppliers. Products are classified and a structure to facilitate creative tourism supply is offered. Second, the chapter looks at a case example of culturally based creative tourism supply, and seeks to draw out issues from this. The case example is taken from France, and considers the *Villes et Pays d'Art et d'Histoire* (VPAH), a label which can be translated as *Towns and Localities of Art and History*. This case study is chosen as it shows the operation of a locally led initiative to foster creativity through visiting, both by tourists and others. Particular attention is given to leadership roles within the municipal organizations that take charge of the VPAHs and to the use of local essentialisms as tourism products.

Much of this chapter results from fieldwork undertaken in France in 2002 and 2004. Provincial France is a diverse country (Braudel 1986), and the fieldwork design sought to reflect this. Interviews were held in four French *régions*: Bretagne and Pays de la Loire in the north, and Provence-Alpes-Côte-d'Azur and Languedoc-Rousillon in the south. All four *régions* are major tourist-receiving regions (Peyroutet 1998), and differ not only in their location, but also economically, socially, demographically and in terms of social malaise (Champsaur 1997). Within the four *régions*, in-depth semi-structured interviews were undertaken with *animateurs* in nine towns and with officials in three *Directions Régionales des Affaires Culturelles* (DRAC) offices. To maintain the confidentiality of the respondents, the towns are labelled N1 to N3, and S1 to S6, and the DRAC offices, D1 to D3. Three of the towns were in the north (N1 to N3), and six in the south (S1 to S6); two DRAC offices were in the north (D1 and D2), and one in the south (D3). Interviews were undertaken in French and tape recorded.

From a supply perspective, creative tourism is about *facilitating becoming*: that is changing tourists, either temporarily through hedonistic sensation or more permanently through enhanced cultural capital. Quite literally, from such a perspective the appropriate criterion of success is that a tourist becomes a different person through the experiences facilitated. This is rarely a sudden conversion of personality; usually it is an incremental change, adding to a person's cultural capital of experiences. Just as people's identities change over time, as tourists their identities may also change through being creative. As tourism resources are commonly used by residents as well as tourists, a similar process of change can also be used to engage residents and to enhance their qualities of life. In an era in which social and cultural inclusion is becoming prominent as a policy objective, creative tourism may be used to facilitate inclusion at a range of different levels. At the most general is that of a common humanity; at the most specific, that of engaging with a local community or heritage. Equally, creative tourism may be used as a means of cultural exclusion: extolling the ideology and culture of a dominant group to the exclusion of others. As such, from a supply perspective, creative tourism can not be value neutral. Rather it depends explicitly or implicitly on the policy objectives underpinning provision.

Creative tourism supply challenges conventional conceptualizations of production and consumption. Traditionally, it has been assumed that suppliers produce products which consumers use, primarily for utilitarian benefits. For example, package tours are produced by tour companies, and are sold to holidaymakers to escape to the sun and hedonistic activity. The utilitarian benefits purchased are accommodation in the sun, the company of like escapees, a bronzed body to tell everyone that a holiday in the sun has been had, and fun. A tourist purchasing such a product is expected to contribute little to it, other than possibly choosing when to opt out of gregarious activities. Mindlessness as relaxation is the proffered style of enjoyment.

In contrast, creative tourism supply is different to traditional approaches to supply and demand, both in conceptualization and delivery. Creative tourism supply starts from the stance that *creativity* rather than mindlessness is the objective of being a tourist, at least for part of a tourism trip, and possibly for all of a person's time away from home. Indeed, this may apply to some package tourists, who may be re-conceptualized as wanting choice and the opportunity to develop interests. For these tourists the package is merely a vehicle to enable engagement in specialist interests that a destination may cater for, but for which the package was not necessarily intended. More generally, sports and cultural facilities become objects of supply, encouraging tourists into driving, walking or visiting traditional towns, for example. Journeys for experience and discovery are proffered. Commonly, this reorientation involves spatial relocation of activities, away from the littoral and inland. Catalunya is a case in point, with the inland mountains and towns becoming alternative offerings for tourism consumption (Prentice 2005a). It can also mean a re-orientation of supply to regions that traditionally have attracted few external tourists, and which may have negative imagery as destinations (Prentice 2002).

Creative tourism facilitated

Creative tourism is an emergent term. Many in the tourism industry still cling to an activity-based breakdown of their offerings: for example, festivals, sports tourism, clubbing, shopping, museum visiting, theatre going. The problem with a classification like this is that it implies that activities are relatively discrete, and that, for example, museum visitors are distinct from theatre goers and from shoppers. Such a classification also focuses on the supply of formal attractions, rather than informal facilities such as everyday life, townscape, seascape and landscape. An activity-based conceptualization of a destination offering also tends to give prominence to the tourism industry as a stakeholder, rather than tourists as customers. In other words, a supply-driven approach tends to evolve, rather than a market-driven strategy. In an industry characterized by many small scale suppliers who undertake little, or no, market research, conservatism can dominate thinking. Creative tourism supply, in contrast, often requires more radical thinking.

Key to creative tourism supply is the recognition that tourists (and others) contribute to their own product, by their cultural capital, expertise, experience, emotions and the like (Bourgeon-Renault 2005; Cunnell and Prentice 2000). Products are effectively bespoke, the antithesis of the standardized tourism of package holidays. All that suppliers can do is to *facilitate* experiences for tourists and suggest meanings: they can *supply* neither. This process is shown in Figure 6.1, and can be termed *creative co-production* (Prentice and Andersen 2000).

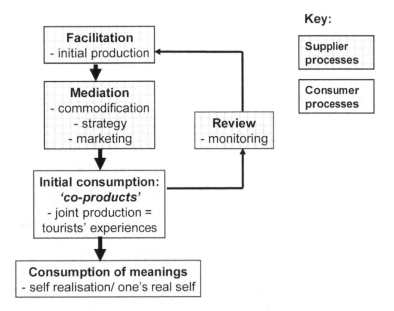

Figure 6.1 Co-production in supply.

From such a perspective, suppliers need to think of the utilities, experiences and symbols they can use to proffer their tourism offerings as suggestions to drive consumption. These are the *three levels of consumption* that define tourism benefits to tourists. Suppliers thereby offer guidance to consumers as to what is worthy of their attention and engagement: a cafeteria-like array of information from which tourists and others may select, and create a bespoke itinerary. In consequence, tourists face an implosion of information, which both destination theming and personal experience can help manage. But only theming is within the potential control of suppliers. Through providing context, implosion can be managed (Prentice 1996). However, with attraction interpretation and promotion commonly distinct from destination branding, destination theming is commonly conspicuous by its absence. Indeed, destinations rarely systematize their unique selling points (USPs) into those needed to match the three levels of tourism consumption they need to be addressing. *Visit Scotland* is a notable exception in this regard, and has promoted Scotland for several years using a 'brand wheel' that in effect emphasizes answers to four questions:

- 'What does the product do for me?'
- 'How does the brand make me feel?'
- 'How does the brand make me look?'
- 'What emotional benefits do I get?'

However, the brand wheel is essentially a destination positioning model, rather than an 'integrated destination positioning–attraction interpretative' model. Prentice (2006) has suggested that a generic systematization matching the three levels of consumption would involve:

- unique utility selling points (*UUSPs*)
- unique experiential selling points (*UESPs*)
- unique symbolic selling points (*USSPs*).

An advantage of this systematization is that it can underpin integrated destination positioning–attraction interpretative models. In terms of culturally based creative tourism, in Prentice's system UUSPs and UESPs lead to what may be termed *culturally* or *heritage augmented destinations* (*CHADs*), and USSPs lead to *culturally* or *heritage empathetic destinations* (*CHEDs*). It is a stance of the present chapter that such USPs need to be developed in order to effectively promote creative tourism; that from a supply perspective, destinations need to become CHADs and CHEDs if they are to be successful as creative tourism destinations. This in effect means two important changes are needed in destination promotion for creative tourism. First, that USPs not only have to be specified, but also need to be thought of across the three levels of tourism consumption. Second, that structures need to be developed to integrate attraction, place interpretation and promotion.

Figure 6.2 outlines the type of structure needed to deliver CHEDs and it is also useful in structuring the delivery of CHADs; the difference being that integration is not essential for the latter. All too often, however, services are commonly independent of destination positioning, rather than a consequence of it. Likewise, reductionism is common: marketing is frequently reduced to advertising, and the latter in turn to brochure and poster design and distribution. Effectiveness is in consequence also reduced as context is lost. Compared to what commonly happens, Figure 6.2 implies the extent of back-office activity needed to develop themed creative tourism destinations. It is an aspiration. Equally, the French VPAHs offer a halfway house to the achievement of this aspiration, based upon the promotion of conserved authenticity.

Academic thinking about creative tourism has changed in emphasis over the past two decades. Indeed, academics have commonly used other terminology to describe creative tourism: notably sports, heritage, cultural and lifestyle tourism. More recently festivals and industrial tourism have been added to the list. Sports tourism has been a consistent theme over the past two decades, but some the other terms have changed in prominence during this period (Prentice 2005b). During the mid-1980s heritage tourism became prominent, a decade later, cultural tourism, and today, lifestyle tourism. Each incorporated the latter, expanding the conceptualization of what resources were needed to facilitate creativity, and how consumption was driven (Figure 6.3). Heritage tourism has been concerned with the supply of museums, heritage centres, monuments and the like; cultural tourism with theatres, public art and daily lives; lifestyle tourism with festivals, boutiques and cafés. Not only has a destination's offering been expanded, but the

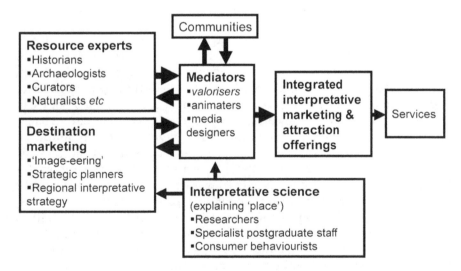

Figure 6.2 Idealized structure needed to develop CHEDs.

Figure 6.3 Changed academic conceptualizations of creative tourism.

focus of promotion has changed from encouraging formal learning, to encouraging informal learning, to encouraging play and self-classification. Cultural and heritage attraction interpretation has likewise changed from emphasizing what might from its affinity to the objectives of school visits be termed 'fieldwork', to more informal 'journeys for experiences', and now to celebration. Needless to say, once everything has been celebrated and celebration ceases to offer a USP, another focus is likely to be found. It is not that earlier foci are lost; rather they are displaced in prominence as the concept is enlarged.

This sequence of enlargement tended first to emphasize past creativity, and more recently creativity to be found in the present. Monuments, museums and heritage centres are essentially past-focused attractions, although serving contemporary purposes. Particularly through performances and attention to the informal, cultural tourism added the contemporary to the historical; lifestyle tourism has tended to add the transient. What is absent in this conceptualization of supply is a *future* orientation. Industrial tourism has begun to counter an emphasis on the past and the now, by offering insight into how technology is likely to change, as well as how it is now and has changed. But this is a fledgling sector, more developed in some European countries than others. Commonly, retailing, fantasy and industrial heritage are being promoted as industrial tourism. France, with a cultural emphasis on modernism and economic nationalism, has developed a fledgling but true industrial tourism sector over the past decade. Equally, how far such 'industrial' visitors are tourists rather than residents is questionable, and the social changes needed to make the 2010s the era of *future orientated* creative tourism are great. Not least would be the continued erosion of the distinction between recreation and work, as industrial tourism may be seen by many as an extension of work.

A more likely future in the sequence of enlargement is for digital technology to expand the period of *engagement* in tourism. Traditionally, engagement in tourism has been equated with actually visiting a destination. A pre-visit stage has been considered as preparation for engagement, and a post-visit stage as memories of engagement and recommendation to others. But remote access enables engagement to begin prior to a visit, and continue after it. Indeed, engagement may prompt visiting, or repeating, through the creation of empathy with a place, people, product, period or art form. Equally, remote accessibility could act as a substitute for visiting. Much depends on the extent to which future 'tourists' will demand authenticity and sincerity as a means to creativity. Authenticity and sincerity can only be experienced through actual visits. If demand for experiences of this kind continues, destinations seeking to facilitate creative tourism will need to ensure that their offerings are both authentic and sincere, and not unnecessarily staged or superficial. In this scenario, the 2010s will be an era of *engaged creative tourism*. Engaged creative tourism offers destinations both an objective and a means. The objective is one of an appropriate product offering while the means is the scope and convergence of digital technologies. Not only do these technologies offer accessibility and information management to consumers, they enable suppliers to integrate offerings using common platforms with content tailored to themes. In short, with the appropriate will, digital technologies enable the building of CHEDs, thereby operationalizing Figure 6.2. The market is there to be made by suppliers.

Building creative tourism supply: the French *Villes et Pays d'Art et d'Histoire*

Through its emphasis on cultural policy, France provides examples of the development of culturally based creative tourism. With the country's traditional reliance on cultural tourism as its inbound tourism USP, this might be thought unexceptional. However, the VPAH initiative originated as much for domestic as for inbound tourism, and with France currently trying to diversify its domestic tourism market, initiatives such as the VPAH take on added pertinence. This pertinence is increased by concerns about cultural cohesion and inclusion. Public concerns over immigration from North Africa and culturally excluded and alienated young people have driven many cultural and urban policy initiatives of the past 15 years. Progressively, the recognition of cultural plurality has replaced objectives of assimilation with those of insertion and integration. Multiculturalism in turn sets challenges for *what* essentialisms are to be presented as French in initiatives such as the VPAHs. Implicitly, these issues also raise the question of *whose* essentialisms should be presented.

Valorization as cultural policy

Post-war, successive French Presidents have emphasized the importance of cultural exports to France's place in the world, and internally, as a means of development. Internally, policy has emphasized French identity, a *vitalisme culturel* (cultural essentialism): cultural and economic policies in France having been described as two sides of the same coin (Forbes 1995). Cultural essentialism has frequently been equated with cultural elitism. This has been reflected in the personal lead of successive French Presidents in their *Grands Projets* and *Grands Travaux* (big projects and works), monumentalist projects in Paris or other large cities, usually concerning buildings for 'high' cultural uses (Cahm 1998). The VPAH initiative has been quite different, and embraces a more popularist concept of French essentialism. It has been assisted by the *volontarisme* (voluntarism) that now dominates French policy. Effectively, a broader vision of culture has been defined (Davies 2004) including a proliferation of *mémoires plurielles* (multiple histories) as heritage, or *patrimoine*. Popularist concepts of essentialisms have commonly impelled localist concepts of essentialisms, implicitly challenging the hegemony of French centralism.

Volontarisme has been expressed in terms both of *déconcentration* (deconcentration) and *décentralisation* (decentralization). The former is the spreading of central services regionally; the latter, their devolution and power of decision to local authorities (Bodineau and Verpeaux 1993; Pontier 1996). Decentralization involved localities in *la mission de diffusion culturelle* (the task of diffusing culture) (Lugbull 1999). Outcomes were no longer elitist. Initiatives included a new museology of community inclusion, *projets culturels de quartier* (cultural projects in deprived parts of towns) and the VPAHs. Deconcentration was effected in the Giscard d'Estaing era in cultural affairs as an experiment by creating regional directorates. Known as *Directions Régionales des Affaires Culturelles* (*DRACs*) (Regional Directorates of Cultural Affairs), their role was enhanced in 1992 in a drive towards *territorialisation* in cultural infrastructure improvement (Poirrier 1998). Their role is to act as the *relais de terrain* (territorial staging post) between central and local levels. However, the DRAC system has been criticized as being overly reactive and procedural, and for failing to set a lead in regional cultural development (Routenberg 2003). Localist interpretations of essentialisms are thereby encouraged, with the DRACs as central outposts in the regions.

Valorisation is the process of assigning value to things in order to convert them into resources. Policy in France has traditionally separated the *conservation du patrimoine* (heritage conservation), from *formation* (professional training), *création* (production of new works), and *démocratisation* (widening of access) (Loosely, 1998). Local *valorisation* potentially reconstitutes these four activities as integrated product development. This equates to the integration shown as the intermediate output in Figure 6.2. *Valorisation* is broader than Cohen's (2004) *commoditization*, and is a social value that is central to the VPAH scheme. It became significant in France in the 1980s in fostering urban social inclusion (Poirrier 1998). This has involved, first, the *sensibilization* of the local population and tourists to something of value; that is, developing awareness and feeling about something, initially the urban form, as valuable. This has involved its identification, signalling, researching and contextualization in time and place (Lamy 2003; Ory 2003). Second, *valorisation* has involved *mobilization*, encouraging and assisting the local population into identifying their heritage. This is the community input of Figure 6.2. Implicit here is an attempted democratization of culture. However, *valorisation* has also extended into conservation, interpretation, representation, retailing and symbolic consumption; including the enhancement of living culture by valuing local know-how, through the provision of a market for artisan products (le Reun 1999).

The process of *valorisation* has important creative tourism implications for France. First, through domestic tourism it can enhance the appreciation by French residents of regional French and potentially immigrant identities in their many forms. As such, it is ultimately a process of identity consolidation, and implicitly it is about the valuing of multiple identities. It is central to a plural concept of the State. Second, *valorisation* in creative tourism is ultimately about cultural exports, as it can communicate French culture to international tourists visiting France. If applied elsewhere, *valorisation* offers like opportunities to develop CHADs and CHEDs.

Villes et Pays d'Art et d'Histoire

In 2004 there were 130 towns designated as VPAH or the earlier *Villes d'Art* (VA). Because of its reliance on local initiatives the scheme is very varied, both in emphasis and objective. It was initiated by the *Caisse Nationale des Monuments Historiques et des Sites* (CNMHS) (National Commission for Historical Monuments) in 1985, as a successor to the VAs. The latter had been initiated in the 1960s to develop a network of heritage towns for tourism promotion. Currently the scheme is overseen centrally by the Ministry of Culture and Communication.

The origins of the scheme in the CNMHS are important in terms of inertia, as this was the agency responsible for the conservation and interpretation of historic buildings and other monuments. The reference point of CNMHS has been heritage as material and built form. The VA scheme was conceived likewise in built heritage terms, and originally defined art as local pre-twentieth-century urban architecture. This focus had relevance for many provincial French towns which had been bypassed by post-war redevelopment and modernization, as it valued their previously neglected built form, and gave impetus for its conservation and renovation. The successor VPAH designation is broader in its definition of architecture and market, but in concept has remained essentially one of *buildings*. While the two designations have co-existed for 20 years, the older is being phased out and all its members were required to convert or apply for conversion to the new scheme by the end of 2005, and become either a *Ville d'Art et d'Histoire* (VAH) or, if the designation includes surrounding villages, a *Pays d'Art et d'Histoire* (PAH).

The intended focus of the VA/VPAH scheme has shifted centrally from a solely tourism objective to include social work and education; and from interpretation primarily for visitors, to awareness and identity formation among local groups. In other words, the scope of facilitation of creativity and awareness has been broadened in terms of audience. Of local groups, children and young people, and those in deprived urban areas feature prominently. Recognizing the informality of contemporary cultural tourism, workers on the fringes of the tourism industry are also targeted to enhance professionalization. However, the promotion of sincerity has not as yet become a central objective beyond the welcoming approach central to customer care.

Originally, the primary focus was on a professionalization of the reception and mediation services for tourists in historic towns. This focus has markedly expanded. Emphasizing greater local integration than before, today, members of the VPAH network commit themselves to seven objectives:

- To both valorize and present their heritage in all its components; to promote the architectural features of this heritage; and create a permanent interpretative centre (*Centre d'interpretation de l'architecture et du patrimoine*).
- To sensitize the local population to their urban environment and to the local architectural and landscape features, through diverse activities such as themed tours, signage, workshops and conferences.
- To introduce young people to the subject of architecture, urbanism and heritage, and establish a permanent workshop to that end.
- To offer scheduled quality guided tours to tourists, both general and themed.
- To facilitate the communication and promotion of architecture and heritage to diverse audiences, employing the VPAH logo; and to produce publications and posters and Internet pages which conform to a standard format.
- To appoint qualified staff, primarily the *animateur du patrimoine* (heritage animator). The towns also agree to appoint trained guides.
- To develop training programmes for public sector employees, as well as for interpreters in the tourism industry, social services and associations.

Animation has been seen as central to the VPAHs, as it is to Figure 6.2. Literally, *animation* is to bring something alive, and was traditionally associated with *animateurs* as entertainers enlivening children's holiday camps. In cultural applications the consistent feature is the use of a person, or persons, as *animateur*. The *animateur du patrimoine* is required to work with all the local service departments such as culture and leisure, social services, community projects, education, buildings and works, and conservation areas. As the *animateur* needs the resources of others, decision networks become important. *Animation* has also been seen as central in community work among excluded groups, and in terms of socio-cultural *animation*, making communities aware of their rights and helping their members to realize them (Poyraz 2003). In the VPAH the *animateur* is the mediator central to Figure 6.2. *Animateurs* are key to making the system of integrated supply work in VPAHs.

Animateurs *and cultural instrumentalism*

The nine *animateurs* interviewed were all highly educated, holding post-graduate degrees. Some held multiple postgraduate qualifications. In contrast to the French civil service tradition of generic administration, the *animateurs* all had a professional background. Although their initial back-grounds varied, history or history of art had been common areas of study for first degrees. However, despite their higher educational background, the *animateurs* seemed most at ease in describing what they did, rather than *why* they were doing it. They were functionaries rather than policymakers.

With one exception, the *animateurs* defined cultural instrumentalization through the VPAH scheme as townscape conservation. In view of the origins of VPAHs this might be thought unsurprising, and was operational-ized either as the conservation of old buildings in town centres, or the fostering of local pride to conserve this heritage. However, the broader implications of the VPAH title, *art* and *history*, had not extended thinking; nor had French debates about collective memories (Davies 2004; Gombault and Eberhard-Harribey 2005). One exception was at S3, where gardens also featured prominently. However, gardens are 'built' in the sense that they are constructed, planted and managed. Instrumentalization tended to reflect the policy emphases at the time of the towns' designation, with contractual agreements used as the local reference point. The *animateur* of S5 explained this: 'We are not affected by the new demand to give first priority to the local population, because our agreement pre-dates this requirement' [in translation as for all interview text cited].

The conservation focus was particularly prominent in towns designated before the 1990s. Any shift locally towards fostering local pride was justified as much in terms of encouraging property owners to appreciate the contri-bution of their buildings as historic townscape as it was to sensitize a wider population to heritage issues. Town S3 had been re-designated in 1991 as a VAH, having previously been a VA. A conservation ethic continued:

> Restoration work in this district has been awful in the past. People have not known what materials to use, what colours, and the walls have flaked after a couple of years. Now the guides can explain to them . . . what materials to use.

(*animateur* of S3)

Similarly, S4 had been designated in 1986 as a VAH. The *animateur* explained changes in instrumentalization as increased selectivity and broadening of scope in conservation, rather than greater social inclusion: 'There was so much reconstruction to be done after the war . . . Now we have to be selective, some things are not worth conserving. And we also have to preserve and explain the more ordinary buildings.'

In contrast, S2 was re-designated much more recently, in 2003, as a VAH. The *animateur* described an extended mission, brought about in response to social problems:

> The town centre suffers from a lack of love from the local residents. They even are afraid of coming into the old town in the evening, especially certain areas . . . The old town centre has been decaying and neglected for a long time, with dirty buildings . . . I shall be attempting to bring life back to the town. This will also bring back the shopkeepers.

As such, it would be a mistake to think that central changes in French policy have necessarily affected how the *animateurs* saw their core task locally. This was still primarily to give value to historic townscape. The fostering of creativity flowed from their conservational stance, rather than the prompting of conservation.

The principal benefit of designation was also seen as the recognition of the quality of the local townscape, or potential quality once conserved. The designation was seen as a label asserting a town's *commitment* to townscape conservation. The label projected an image locally, and nationally. The latter was seen as important both for tourism promotion and for gaining local and central resources. Indeed, some towns had achieved plural labels and thus plural recognition, notably S6. Resources were in turn important in assisting conservation, and for professionalizing tourism services, through training guides. In one case, this professionalization had extended into producing multi-lingual interpretative panels. These panels were justified as a demonstration of commitment to enhancing local pride in townscape, and encouraging locals to feel that their everyday and often dilapidated environment was important.

Valorizing local essentialisms as interpretative product

Valorisation was defined by the *animateurs* as a process of creating and enhancing public awareness and engagement with both the local built environment and associated heritage. The built form was the principal reference point for *valorisation*. Asked explicitly about *valorisation*, the *animateurs* gave very similar replies. For example, 'Above all, [it is] the physical maintenance of the buildings. If I want to do any "animation" with exhibitions the building has to be presentable' (*animateur* of S2). And, 'Valorization is to bring out the heritage quality of a place' (*animateur* of S3).

Both the creative tourism and wider educational missions flowed from this reference point. 'You can have a beautiful building, clean, but if you do not make the building speak, you stand in front of it without seeing it' (*animateur* of S2). Such sentiments were almost universal: 'Valorization is . . . everything directed at the public: explanation and education about a monument' (*animateur* of S5), and the 'Communication to residents of ongoing works; to teach them to respect their own property, their own city' (*animateur* of N3).

Museums as depositories of memories were often marginal to the activities of most *animateurs*. Museum directorates were frequently described as elitist and lacking interest in popular heritage. They could even be seen as rivals in *valorisation*, with the *animateurs* in contrast categorizing their own approach as new and relevant. As such, the integration of product implicit in Figure 6.2 was only being partly achieved. The selection of built form through *valorisation* flowed variously from the *animateurs'* training, what seemed self-evident, or from community choice. Once again, Figure 6.2 was being approximated to, but only in part.

Reflecting the diversity of France, localism rather than a national French essentialism was commonly a reference point of the *animateurs*. Likewise, localism was the basis of the UESPs and USSPs they were developing and promoting. Contexts were at most regional in extent:

> In order to develop the cultural aspect of the town it [the VAH] will have to focus on attaching itself to its Provençal roots. The Cote d'Azur is more touristy, about leisure. I believe that by first of all sensitizing and teaching the residents to look at and understand the architecture from a local point of view, they will then be able to apply their knowledge to a wider sphere; regional, national and wider still.
>
> (*animateur* of S2)

A local focus was commonly effected though through the *mobilisation* of local people. 'During the first years it was important to inform the local population that it was a service for them. In the museums you never saw anyone. Locals don't go to the museum, only tourists' (*animateur* of S3). Cultural engagement had been an outcome of this work. For example, community involvement in the production of interpretative materials and documentary sources was a common strategy of the *animateurs*:

It was the same when I produced all the interpretation boards around the region. I didn't do it by myself. I asked all the local people in the villages to get together and discuss for a year what they wanted to include on the boards. They brought in their own files and records.

(*animateur* of S5)

Such an approach emphasizes community engagement, but potentially at the expense of being supply-driven and unreflective of any wider interpretative strategy of a VPAH.

Although their main programmes were often delivered to an undifferentiated public, *animateurs* were also aware of multiple heritages. The *animateur* of S2 noted, 'We have to identify all the various communities . . . We already offer a tour of all the religious institutions in the town.' In S5, the *animateur* distinguished between her town tours and those which were either themed or in the villages. Town tours were largely for tourists, the others were largely for locals. In N3 efforts were directed towards engaging the residents of the peripheral housing estates. In both S2 and S5 immigrant communities were being targeted:

We also work with the Social Centre, people with problems, new immigrants who speak little French. We work with them and show them the town: how a French town functions, the history of architecture, and the social evolution. There are Moroccan, Algerian, and Tunisian agricultural workers throughout the region.

(*animateur* of S5)

Implicit here was the notion of a traditional French local essentialism with which immigrants were to be encouraged to engage. None of the VPAHs had programmes interpreting recent immigrants' cultures to others as part of the scheme. The focus on architectural heritage is one explanation, as the interpreted buildings pre-date recent immigration. Equally, unintentional social exclusion might be another, through effectively defining Frenchness as deriving solely from the so-called *Français de souche* (French people of old stock).

The *animateurs'* sub-categorization of *valorisation* is summarized in Figure 6.4. Environmental quality as built form and mediation were universal components, but the inclusion of collections was more ambiguous. Town S3 was a notable exception, with collections being developed as a resource to aid conservation. Local people had been encouraged to 'bring in photographs and documents. We now have nearly 10,000 photographs . . . We act as a resource centre . . . and people come to find out about the origins of buildings' (*animateur* of S3). Mediation was sub-divided into explanation and education, the former essentially for adults, the latter usually for children and young persons. *Animation* was defined as involving human beings as interpretative media. This could take various forms: guides leading walks,

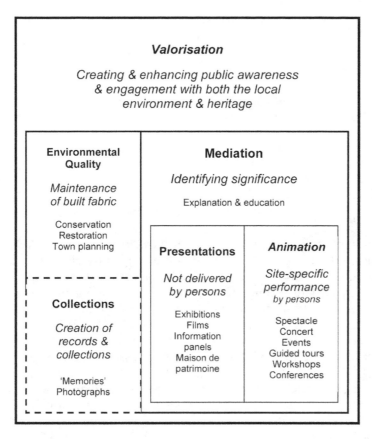

Figure 6.4 *Valorisation* and *animation* in practice in the *Villes et Pays d'Art et d'Histoire.*

workshops, spectacles or concerts. As the *animateur* of S4 said, 'Any kind of performance is *animation*'. Digital interpretation was limited to websites describing activities and buildings. Not all VPAHs had these, and they seemed to be approaching the era of engaged creative tourism with very traditional tools.

Instead of interpretative strategy, functional leadership characterized the *animateurs'* description of their key task of *animation*. The latter was defined effectively in terms of the skills they considered essential. Besides their educational background, practical multi-skilling in leadership was emphasized. At S3 this was described as 'dynamism and tenacity', at N2 as 'co-ordination and negotiation are the essence', and at S1 as 'The ability to communicate; and diplomacy. And a passion to stimulate the team.' The *animateur* at the recently re-designated S2 elaborated on the skills she thought necessary:

[Multi] competences are indispensable, particularly polyvalence. You have all the cultural baggage: art, history of art, communication skills (to various audiences, internal and external), guiding, plus all office routines including accounting, secretarial. . . . learning by doing. It is more important to be a subject specialist first.

As such, the *animateurs* animated their bureaucracies as much as their publics. They were professionals who effected wider managerial and leadership roles, and to whom decision networks were essential. How interpretative strategy originated and evolved was less clear. Instead of strategy, mayoral preference, incrementalism and opportunism characterize VPAH processes locally subsequent to designation.

Decentralization

Despite cultural affairs in general being deconcentrated in France, VPAHs are decentralized. Past and present mayors were frequently cited by the *animateurs* as initiators or sustainers of heritage policy. Mayoral support was both psychological and economic. The reasons cited for mayoral interest were usually those of economic development, although in N3 cultural education was given as the initiating objective. DRAC officials were not assigned a generic initiating role. Instead, they were seen in the mainstream tradition of the French civil service as generalist and essentially responsive administrators. One partial exception was noted in S1, where a *former* DRAC employee was credited with having convinced the mayor of the time to focus on heritage as a means of economic regeneration through tourism.

Despite being representatives of the centre, the promotion of a French rather than local essentialism seemed at most secondary to the DRAC officers. They were hesitant in targeting towns to bring forward VPAH plans, and preferred to channel assistance to those towns showing a clear interest in doing so. Progression of a central policy to create a national portfolio of designations to present a French essentialism referred to by the respondent in D1 seemed ambiguous: 'Future designations are evaluated at national level. The Minister is trying to create a balance across the country, filling in gaps' (DRAC official in D1). The policy of infill was explained in spatial terms, and in terms of types of heritage: maritime, industrial, parks and gardens. The priority was to respond to towns showing a will to value heritage, and the matching of portfolio to offers was at best secondary. Asked if towns were asked by DRAC to apply, the official replied, 'No, we wait. Because we want to be sure that a candidate town expresses a genuine will for partnership with the Ministry.' The same was true elsewhere. 'More important is the permanent link between DRAC and the *animateurs*. DRAC's role is to assist the *animateurs* in the implementation of their projects' (DRAC official in D2). Key here is implementation; a focus on service delivery rather than on ongoing product development.

In consequence, locally the administrative role of the DRACs was viewed reactively, rather than proactively. They were seen as advisory and funding bodies, giving advice and access to a professional network. For example, DRACs ran regional seminars that drew together the *animateurs*, reducing their isolation one from the other. These meetings were the basis of informal networking, and building networks of advice and expertise that could be exploited when new challenges were being faced. The DRACs encouraged a common identity between the *animateurs*. Importantly, they helped in the formation and updating of decision networks for the *animateurs*. In effect, the DRACs were seen locally as one more resource, but their advice did not need to be taken:

> We don't have to follow Ministerial policy if it does not fit into local framework. The Ministry has been encouraging us to focus on contemporary architecture. This suits us fine because last year was the tenth anniversary of the new art museum . . . so in this case we could fall in line.
>
> (*animateur* of S4)

and

> I am not responsible to the regional DRAC. If I do not agree with DRAC, I do what I want. I am employed by the municipality. But the regional DRAC provides orientation, advice, assistance and grants.
>
> (*animateur* of N3)

At best, the DRAC officers seemed content to monitor the VPAHs' performance in terms of volume, rather than in terms of demonstrated effectiveness or regional strategy.

Conclusions

This chapter has predicted the emergence of *engaged creative tourism* as a successor wave to the forms of culturally based creative tourism that have been experienced so far in the Western world. Figure 6.2 set out a model for the integrated creation and delivery of CHADs and CHEDs to service this new wave. This model requires the integration of destination positioning, interpretation and promotion, and their integration in turn with attraction interpretation and promotion. The model also emphasizes the role of mediators in *leading* the process. The model is aspirational, and describes a situation yet to be reached.

Few destinations have even the basic structures in place to effect the integration proposed, or the mediators needed to effect the process. The French VPAHs are a partial exception to this situation, and are an example of towns and localities seeking to move in the direction set out in the model proposed by Figure 6.2. The mediation role has been developed in the VPAHs, but essentially as functional leadership rather than as strategic interpretative development. Integration is also largely restricted to built heritage interpreted as UESPs and USSPs. The latter have been developed as local essentialisms, and principally as *traditional* French localist culture. The VPAHs are about being French in a manner consequent of a national identity developed after the hegemony of the Revolution. As a creative tourism product, they proffer collectively an imagined France of small towns and villages as prototype CHADs and CHEDS.

The French VPAHs are a supply-led example of creative tourism development. They develop what is available, commonly converting distressed and obsolete built resources into resources that are valued for multiple use. As they are intended for tourists and residents alike, they also demonstrate the blending of leisure and tourism as creative pursuit. The challenges the VPAHs face is to adapt fully to experiential tourism, and particularly to the engaged creative tourism postulated in this chapter as the likely future for creative tourism offerings. The VPAHs are not alone in this: a recent issue of the journal *Cahiers Espaces* (volume 87) was devoted to the same issue in the context of museums. The VPAHs also need to become much more strategic, and use the central mediator role defined in Figure 6.2 to develop interpretative policies. To fully engage with creative tourism, the VPAHs will also have to more fully anticipate the demands of tourists seeking engagement before and after their visit, and to measure their effectiveness in delivering messages that are wanted and understood. In practice, a greater digital presence will be needed, and one beyond simply listing attractions as information for potential visitors. An even greater challenge is to embrace a wider concept of resource, both in terms of a wider scoping of art and history, and in terms of lived culture, thereby offering greater inclusion to new immigrants.

Irrespective of the limitations of the VPAH system in developing mediation, the system represents a sustained initiative to develop local place-branding. However, this is still only a halfway house. The challenge of many other destinations is greater still, namely putting a mediating structure in place and fully operationalizing it. As such, for all their imperfections in terms of Figure 6.2, the VPAHs represent an example to facilitate creative tourism based on proffered and felt authenticity of product. They represent the beginnings of bridge-building between destination promotion and attraction design and interpretation. They offer a supplier model which combines public and private investment, based firmly in local political realities and local cultures.

7 Tourism quality labels

An incentive for the sustainable development of creative clusters as tourist attractions?

Walter Santagata, Antonio Paolo Russo and Giovanna Segre

Introduction

This chapter aims to introduce and discuss collective intellectual property rights as a juridical framework for economic development that could mark a step forward in tourism planning in the sense of supporting a higher degree of creativity in tourist activity. Our approach is based on the idea that local creative knowledge can be valuable in the production of tourist experiences. Since the main components of creative knowledge are heritage, identity, material culture, lifestyle and *savoir faire*, the local issue is clearly very important. However, we argue that local creative knowledge needs to be embedded into a fully fledged economic production process in order to become a development asset for the community.

The defence and strengthening of creative clusters as pillars of tourist production is subject to two major challenges: the structure of the market to which it is directed (tourism) and the global development of the world economy, which tends to disenfranchise production processes from locally embedded knowledge, dissipating place advantages.

The viability of a process of 'commodification' of creative knowledge can then be viewed in the light of such turbulent environments, focusing on regions where the pressure to market local resources to visitors is higher. It is here in particular that a specific tool for the institution of collective intellectual property rights (CIPR) for culture-based goods and landscape elements may contribute towards the stated aims.

The creative tourism district

Community development strategies, especially in the case of disadvantaged regional economies, can be based on the valorization of local knowledge assets, which face relatively low capital investment barriers. However, communities are sometimes tempted to 'abuse' resources; this is why a tourism development model that preserves and valorizes the physical and symbolic 'genuine' is important. This may be called a 'Creative Tourism District' model, as opposed to a '5-Star Hotel Model' or a 'Traditional Tourism Model', which are based respectively on scale economies in the production of tourist services – a central role for tourist infrastructure in the provision of services – and the thematic (often fake) 'reconstruction' of the cultural identity of a place.

Creative tourism districts (CTDs) could be conceptualized instead as a dynamic tourist production system based on material culture and original lifestyles, linked to cultural heritage and traditions as well as to recent social developments, such as the rise in importance of the 'Creative Class', described by Florida (2002) as the driving force of contemporary urbanism. In the CTD, the focus is on a wide range of activities and attractions: accommodation is important, but not the main reason for visiting a site.

The analysis of CTDs is rooted in the conceptual distinction between the cultural sector and the creative industries. The literature generally distinguishes between these sectors according to their non-profit- and profit-making orientation respectively. This distinction is grounded in the general acceptance that commercial gain should not be the guiding organizational principle of the cultural sector as such, but that there are activities 'at the edge' of cultural or symbolic production that are instead straightforward capitalistic agents. Museums, heritage protection, performing arts and galleries are unquestionably part of the cultural sector, while industries such as architecture, graphic design, film, television, radio, music recording, research, higher education, events and tourist services are not included in the definition and are classified as creative industries.

Within a creative district, these two realities often stand side by side and complement each other. The outcome is the economic production of culture-based goods and services with strong ties to the territory, its history and its knowledge or *savoir faire*.

The CTD as a reference model of tourism development gains importance through increasing interest in 'culturally stimulating' environments, where visitors are likely to be in contact with new ideas, people, products, social rituals, languages and visual expressions. Tourists are particularly attracted by the original expressions of local culture and creativity: cultural heritage (old cities, museums, monuments and archaeological sites), performing arts (festivals, music), gastronomy, adventure and the natural environment (excursions to caves, waterfalls, deserts etc., see also Cloke, Chapter 2 of this volume). The core of a CTD is then creative production *and* consumption, rather than the mere presence of major cultural attractions which tend to support a more 'passive' traditional cultural tourism.

The spatial and social embedding of local creative resources provides the ideal scenario for visitor engagement. Integrating tourist production in such a local system or *district* means taking advantage of good governance, economies of agglomeration, creative entrepreneurship, trust and cooperation which are normally part of the fabric of a place, but also night time animation, street life, unexpected events, cafés and restaurants, ethnicity and diversity which increase the range and intensity of the visitors' experiences.

In fact, one of the main conditions for tourism development is the coordinated provision of a full range of visitor services (accommodation, catering, interpretation, etc.). Any action targeting the realm of creativity and cultural production alone cannot neglect the composite nature of the tourist product, or the fact that visitor services not operationally connected with primary cultural attractions are offered in markets with thoroughly different structures of interest and stakeholdership (Dahles 2000; Orbasli 2001).

Balance in the creativity–tourism relationship

Through developing a CTD, the local community should benefit from the direct impacts of a 'socially embedded' tourist function, as well as the market-size effects from the production and export of culture-based goods. It is important that a balance is struck between the forces leading to the strengthening of a cultural or creative district and its cultural tourism facilities: if not, it is likely that local creativity and tourism will develop in separate ways, or, worse, that tourism will become 'predatory' upon local creativity. Indeed, recognizing the inherent tensions between systems of objectives and market structures within the CTD can be a first step in an analysis of tourism sustainability that focuses on economic rather than physical issues (Russo 2002).

In this light it is useful to compare the elements which are seen as crucial in fostering creative production, and those which stimulate tourism development. On one hand, we argue that the development of a cultural district depends on four '*I*s':

- *Idiosyncrasy*: The leitmotiv is that culture-based goods must be valuable and connected to the place. Otherwise, they could be produced anywhere and place advantages would be dissipated.
- *Innovation*: Culture-based goods must reflect social dynamism, otherwise they are doomed to become disconnected from the 'local' and lose resilience when faced with endogenous social changes.
- *Insider ownership*: If the control over culture-based production processes is expropriated, the 'stewardship' of local culture is lost, along with the economic bases for its conservation.
- *Institutions*: Local knowledge has to be 'defended' and forwarded; market instability has to be harnessed.

On the other hand, the sustainable integration of culture-based goods in tourism relies on:

- *Profitability*: Culture-based goods must guarantee a return to the producer and a foreseeable income to the packager.
- *Empathy between hosts and guests*: Culture-based goods should stimulate curiosity for the place where they are offered and the community that produces them.
- *Coordination*: Destinations need the capacity to offer holistic experiences which integrate cultural and non-cultural elements.
- *Risk minimization for the visitor*: Culture-based goods should be sold at prices that reflect their real value.

Although the conditions for culture and tourism respectively have large areas of overlap, they do not necessarily coincide, and sometimes they even collide. Indeed, tourism commodification may 'kill' the creative process. In such cases, the industrial production of tourism experiences (including primary attractions) affects the individuality of cultural processes. Moreover, the 'imaging' of places by intermediaries – who dominate distribution channels, presenting themselves as the 'gatekeepers of tourism' (Ioannides 1998) – can reduce the chances of host–guest interaction because it produces a counterfeit, conservative landscape, which negates social dynamism.

Hence, in Venice the tourists' 'romantic' is – for its citizens – the 'silent', the 'dead', the negation of a possibility of social and economic evolution. In a similar way, the expropriation of elements of the tourist product (hotels, travel, restaurants, even events) from local ownership or control is likely to bring about a rupture in the process of integration and delivery of cultural products, with global functions developing independently and sometimes at different quality levels from the local. Finally, asymmetric information and lack of empathy between guests and hosts can lead the latter to 'cheating' on quality (for instance in culture-based goods) in order to reap short-term profits, undermining the structure of the cluster (Keane 1997).

The sustainability of tourism could thus be seen in this light as a 'coordination game' played by actors with different strategic horizons. The ground on which the sustainability game in the field of cultural tourism is played depends therefore on two key points:

- The creation of emotive links between local products and visitors, so that cultural empathy is established. This may lead to a higher level of resilience of the tourism development process in the face of external shocks and internal market transformations.
- The maintenance of variety and diversity within the district in the face of the 'standardizing' pressure of the global tourist market.

We focus on the last issue in this chapter, but the first is closely related and in any case deserves further treatment (see Go *et al.* 2004).

Collective intellectual property rights and their role in creativity

In order to maintain diversity, Santagata has emphasized the role of CIPR in fostering the development of culture-based, local knowledge-based, idiosyncratic products (Moreno *et al.* 2005; Santagata and Cuccia 2003). Equivalent to conservation with regard to tangible assets, CIPR can be seen as an instrument to 'preserve' intangible cultural knowledge and boost its value for the community in a context characterized by globalized production processes; loosening connections between local and global, and footloose knowledge.

The main purpose of the introduction of CIPR is then to anchor and stimulate creativity in the location, through an induced process of cooperation between different actors, both within the creative sector and outside. CIPR provide a regulatory instrument that can apply both for the regulation of existing creative tourism districts in mature markets and for the development of new CTDs in markets that are taking off. We argue therefore that a juridical framework is needed to build an economic asset. The reasons are explored in the rest of this section.

Functions of intellectual property rights

In principle, any type of intellectual property right (IPR) serves two main functions. The first is an *information function*, crucial to defend original culture-based production when there are information asymmetries between consumer and producers. This is the case for goods or services for which quality is not easily detected prior to purchase – Nelson's (1970) 'experience goods'[1] – and transactions are not repeated.

Within this context, identifying the ownership of the brand and signalling the producers' reputation contributes to the dissemination of information about the quality of the goods produced or the service provided. This fundamental role serves to reduce the transaction and search costs to consumers (Posner 2003). It also acts as a safeguard against the illegal copying of goods, ideas, tags, labels or logos. By giving assurance of homogeneous and minimum quality, IPR allow consumers to economize on search costs, as well as protect consumers from fraud. These conditions become even more important when the sector is exposed to global low-cost competition.

The second function of intellectual property rights is *organizational*. IPR, especially collective ones, have an organizational role to play which is critically related to the enhancement of the quality of goods and services provided. They entail the introduction of rules, standards, inspection procedures and financial mechanisms for business development into an area, a community or an association of producers. Setting standards on the quality of the products in turn implies maintaining a particular level of cooperation among the local micro and small enterprises. In this sense, CIPR can be assets for sustainable economic development as they enhance the opportunity to develop trust and cooperation.

Three types of collective intellectual property right

Intellectual property rights can be *individual* or *collective*, according to the nature of ownership of the rights. Among collective intellectual property rights, the most important for our study are the Collective Trademark (CT), the Geographical Indication (GI) and the Appellation of Origin (AO).

A *Collective Trademark* is a trademark which the owning collective entity allows its members to use, while excluding others from doing so. The trademark must be indicative of the source of the goods and consumers must be able to distinguish it from other trademarks. The owner of the collective trademark must generally file an application for its registration. As the World Intellectual Property Organization (WIPO)[2] puts it:

> associations of SMEs [Small and Medium sized Enterprises] may register collective marks in order to jointly market the products of a group of SMEs and enhance product recognition. Collective marks may be used together with the individual trademark of the producer of a given good. This allows companies to differentiate their own products from those of competitors, while at the same time benefiting from the confidence of the consumers in products or services offered under the collective mark.
>
> (Rojal 2005: 8)

A *Geographical Indication* 'is a sign used on goods that have specific geographical origin and possess qualities or reputation that are due to their place of origin'. It signals to consumers that 'a good is produced in a certain place and has certain characteristics that are closely associated to that place of production. It may be used by all producers who make their products designated by a geographical indication and whose products share typical qualities' (Rojal 2005: 8).

An *Appellation of Origin* is, according to the WIPO, 'a special kind of geographical indication, used on products that have a specific quality that is exclusively or essentially due to the geographical environment in which the products are produced' (Rojal 2005: 8), including human factors (ITC/WIPO 2003: 10).

GI and AO are thus signs used to guarantee that a product is specific to a given place: they underline the geographical origin of the good or service, whose qualities, reputation and characteristics are rooted in the original place; they guarantee that it is only that place which is capable of supplying those human and natural factors that make the good or service unique.

What do CIPR apply to?

CIPR have various areas of application, which have to be cleverly engineered depending on the geopolitical, geographical and market context. The typical applications are to a production process, a commodity (labelling products at the moment in which they are brought to the market) or an 'idea' or local brand.

The crucial question for this chapter is: how does the institution of CIPR affect the chances of sustainable tourist development? To answer this question, it is useful to distinguish between developing and developed regions. We argue that, while the information role and individual intellectual property rights are particularly fitting in the case of mature markets and in the context of developed countries, in the case of developing countries the managerial role and collective rights are even more relevant.

In fact, in developing countries establishing an environment which provides efficient institutions for the economic start-up of a potential district of micro and small firms producing culture-based goods and services may be the key to local resource-based development. Often, in the case of lagging regions, cultural or creative tourism districts are only a potential solution, which needs to be transformed into an effective structure of territorial industrial relations.

In this light, a Collective Trademark offers some comparative advantages over Geographical Indication or Appellation of Origin, as it fits better the 'composite' but localized nature of the tourist product, made up of many sub-products and services that the collective right helps to reconnect under the same 'local brand'. Collective Trademarks communicate a complex local image, while Geographical Indications are specifically based on the existence of a single production process typical of a region, but a tourist destination needs more to promote its image than the protection of such a process.

In the case of a CTD there is indeed a unique location, but many products and services that are not especially based on peculiar technologies or production processes, and the network between the different elements in the CTD is generally rather loose.

The Appellation of Origin may indeed be used to distinguish a product grounded in the special characteristics of a place, natural and human, such as the wine, olive oil or cheese industry. Yet in certain cases an unfavourable view prevails against it, in terms of a fear of losing discretionary power in managing the right.

There are other possible drawbacks regarding the dynamics of the sector. Establishing a CIPR such as Appellation of Origin tends to stifle innovation and reduce the capacity to adapt to technological and market changes. Moreover, in some cases the individual brand of single producers can be far higher in quality and reputation than the common collective brand, so that they have an incentive to disengage from the trademark agreement and promote themselves individually. Their exit, however, undermines the overall quality and reputation of the trademark, which is made up of the average quality of the associated producers. Finally, it should be considered that, in some cases, markets for high-quality goods may not exist or are stifled by changes in the demand side; in those cases the introduction of a CIPR would be useless or counterproductive to enhancing the quality of the products, and that is especially a problem for traditional tourist destinations with large information asymmetries.

Towards an integral framework for the introduction of CIPR as a development tool for creative tourism

When elaborating a strategy for the development of CTDs, we should carefully consider a number of issues. First of all, the overall objective of a development strategy through the introduction of CIPR should be defined.

The first key aspect to be defended could be *tradition*, intended as the legacy of the identity of a place and its community, which makes that place and that community unique. The institution of collective property rights applied to traditional practices, based on a location's cultural heritage, is likely to foster cooperative behaviour among producers of different goods and services. The success of collective trademarks represents, indeed, a valuable resource in attracting the demand for local production. Within the background created by the CIPR, firms will then, obviously, compete by means of their individual brands for their exclusive market shares. Yet, a conflict could arise between the protection of traditional practices and innovation. As argued before, in the absence of a well designed set of rules, the risk is strong that *innovation* will be stifled as a consequence of the introduction of CIPR and increase of monopolistic rents. Santagata and Cuccia (2003) and Russo and Segre (2005) warn against such possible drawbacks in the institution of property rights.

The second key aspect is the *geography of production*. A production location, for instance traditional industrial heritage such as the glass factories of Murano, can be attributed ethical value and be the object of conservation programmes. However, the risk in this case is that obsolete production models are kept alive artificially: thus, defending the competitive position of culture-based goods may require a modernization and restructuring of production logistics.

In both cases, then, the introduction of CIPR to defend the tourist appeal of culture-based production may collide with the general objective of sustaining the production sectors and guaranteeing the originality of the tourist experience. The counterexample is the transformation of old glass factories in Murano into evocative 'showrooms' where non-original glass items are on display.

A final dilemma is represented by questions of whether Collective Trademarks and Appellations of Origin respond to different objectives and are valid in different contexts.

As already stated, the economic function of trademarks is that, by giving assurance of uniform quality, they allow consumers to reduce search cost. The introduction of collective property rights, in particular, sustaining the quality supplied by the producers involved, may lead to the restructuring of the markets, strengthening the competitive position of a cluster of producers. Our discussion so far has highlighted the fact that among CIPR, Collective Trademarks have a stronger focus on *place*, and regulate an *image*; whereas Appellation of Origin or Geographical Origin focus on the *products*, and regulate a *local market*. To understand exactly how the outcomes of their introduction in different contexts may vary, we will utilize a simple dynamic scheme of the CTD, considering systematically the potential consequences of the application of CIPR in various contexts.

Starting positions

We can describe the starting situation of a destination by referring to three key variables. In the vertical axis the three elements of the creative tourism district are represented in a very simplified way: creativity (primary attractions), hotels (complementary products) and restaurants and other visitor facilities. In the horizontal axis quality is represented: CTD's supply can be of low (L), average (A) or high (H) quality level.

In disadvantaged regions (Figure 7.1) hotels and other visitor facilities are far from the standards that can be found in established tourist destinations. Furthermore, little importance is attached to cultural heritage and local creativity, so local cultural attractions normally exist with below-standard quality levels.

In mature markets (Figure 7.2), the standards in the hotel industry are normally higher and so is the level of cohesion in the cultural and creative production sectors, which are nevertheless unstable and exposed to external shocks.

Quality / Product	L	A	H
1 - creativity	⊗		
2 - hotels	⊗		
3 - other visitor fac.	⊗		

Figure 7.1 Starting situation in disadvantaged regions.

Quality / Product	L	A	H
1 - creativity		⊗	
2 - hotels		⊗	
3 - other visitor fac.		⊗	

Figure 7.2 Starting situation in mature markets.

Unregulated development

Facing an expanding tourist demand, there may be a tendency towards the 'standardization' of tourist products and landscapes. This process may go in two different directions: upward and downward.

The upward trend may bring forward a 'Sheratonization' of tourism. The danger is the diminishment, or the banalization, of local idiosyncrasies, using a term introduced by Muñoz (2006); this also entails the erection of invisible barriers to low-budget travellers, and the creation of 'tourist enclaves'. Equally critical is the downward type of standardization, producing a 'McDonaldization' of the heritage destination. This process recalls Ritzer's argument (1996), but grounds it in economic theory following the seminal works of Shapiro (1983) and other game theorists, which have been applied to the tourist market (Caserta and Russo 2002).

In this case, the danger is loss of competitiveness, which comes from the erosion of cultural capital and its value in the face of economic pressure generated by an increasingly undistinguished and uninformed demand on the local economic system (Figure 7.3).

Introduction of Appellation of Origin or Geographical Indication

As can be seen in the case of Murano (Box 1), just regulating the production of culture-based goods (through an Appellation of Origin, in that case, which sets some sort of minimum quality standard) may not be sufficient to achieve a higher level of quality and integration in the tourist market (Figure 7.4). Actually, the contrary may occur: the profile of the tourist market and its lack of cohesion may ultimately offset the effectiveness of the trademark. It can therefore be argued that the introduction of a minimum quality in culture-based goods may not affect the structure of the tourist market much, if at all.

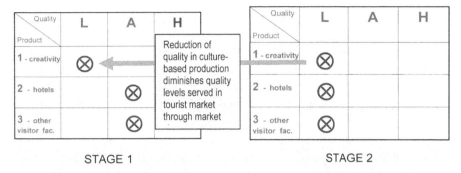

Figure 7.3 Unregulated development downplays level of tourist supply.

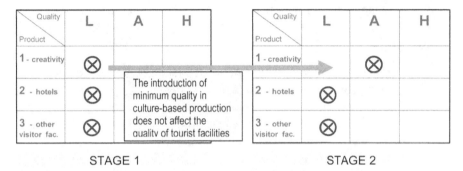

Figure 7.4 Introduction of CIPR does not increase level of tourist supply.

This may or may not be an indication for developing regions, where (i) the priority for tourism is not restructuring but taking off and (ii) intangible cultural elements, impossible to regulate through a GI, are possibly the main attraction for cultural tourism. There is scepticism in the community about 'giving up' the control of traditional production processes to higher levels of government, but there is acceptance of, and in some cases enthusiasm about, the introduction of a 'place brand' to instil cohesion and integration in the potential cultural tourist product.

Box 1   A case of a mature market: art glass production in Venice (based on Russo and Segre 2005)

Today, in the Murano glass industry, 60 per cent of production is sold in the local market – that of Venice – one of the main European tourist destinations with 17 million visits in 2005. In recent years, many traditional glass firms have been converted into showrooms of cheap and imported products for tourists.

To protect the quality and the integrity of the Murano glass production cluster, the Murano Glass trademark (an Appellation of Origin) was introduced in 2001; this was followed by Murano's development pact in 2003, an industrial agreement between the region – the owner of the trademark – and the Italian government funding economic regeneration schemes. All Muranese glassmakers may join, if they respect technical regulations, irrespective of the quality of items produced. A fee is paid to join the trademark and then a per-item 'stamp' (€0.30) to put on the market item (optional). The Promovetro consortium, including a large portion of Murano's glassmakers (but excluding some of the most important ones) promotes the trademark and manages the members' relations.

Problems

In the relatively uninformed traditional tourist markets, the quality of glass production is not easily detected. Tourists, as opposed to local households and international specialized importers, tend to be less informed and 'incidental' buyers who perhaps care less about quality and exhibit lower demand elasticity with respect to price and quality.

Irrespective of the production strategy of the producers and whether the items are labelled or not, the retailer can decide to 'confuse' labelled and non-labelled items (reaping a higher mark-up on non-labelled pieces), or to blackmail the producers into not marking their products, so that the diffusion of the trademark is slowed down and its reputation with buyers is reduced.

There is little technical interdependence between firms, so that sharing the trademark is not necessarily a cooperative gain, and production chains can be easily delocalized. In the face of high costs, some of the most famous glassmakers relocate production facilities in order to reap cost advantages over those who decide to stay and have the right to use the trademark; in this way, the process of dispersion of local knowledge is accelerated rather than contained.

Alternative instruments

'Area labels' for sale of high-quality products. The objective would be to introduce an 'exclusivity' for retailers selling Murano items in the central area (St Mark's square), extending this approach to all commercial activities involved in the sale of culture-based goods.

CIPR on 'ideas' as opposed to GI of products, which might have been a safeguard against the dispersion of the cluster but cannot work in a globalized production environment and with volatile end-markets such as the touristic one. The key concept would be to maintain in Murano the 'intellectual leadership' on design and style, and let local producers do what they want with their production chains, granting renewed dynamism to the sector. The outcome would be an extension of collective property rights (which would be transformed into a hybrid Geographical Origin/ Collective Trademark) to producers outside Murano which adhere to 'high quality design principles'.

In any case, it is suggested that cluster development policies must go hand in hand with a change in the marketing of Venice as a cultural destination: only 3 per cent of visitor expenditure goes on culture and only one visitor in ten gets to visit the main attractions and events (Russo 2006).

Introduction of a Collective Trademark in the creative tourism district

The basic idea is that more diversity in the supply structure induces higher chances of preserving culture-based production, which can be seen as an enhancement of its supply (Figure 7.5). In fact, collective trademarks are meant to grant an 'advantage' to those producers who agree to revise their market strategy according to the genuine atmospheric and historical elements of the place, furthering a tradition, a style, and possibly contributing in this way to the revitalization of creative historical environments.

In such contexts, Collective Trademarks are a more promising instrument for local development than other forms of IPR. The idea is that enhancing the quality, but more importantly the overall 'completeness' of

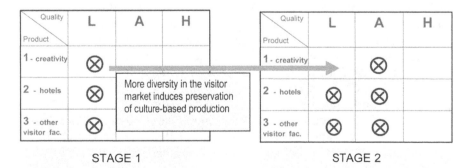

Figure 7.5 Introduction of CIPR increases level of tourist supply.

the creative tourism district (both horizontally, for the range of products involved, and vertically, for the levels of quality), may generate a proper market environment for the restructuring of the cultural production sector, which becomes specialized in the provision of high quality products to curious, aware visitors, and may seek a differentiation in marketing strategies with regard to low-cost competition. This is perfectly consistent with different quality levels, as long as the (planned) 'spirit' of the place is maintained. In addition, it does stimulate change in the demand market as it instills curiosity for diversity and genuine encounters with local culture: encouraging rather than repelling curious visitors whatever their economic capacity.

To sum up, while mature destinations may require cleverly engineered systems of legal protection of production, and above all, commercialization of culture-based goods to sustain high quality in the tourist market, new destinations should instead be oriented towards achieving a larger degree of diversity within their CTD. For instance, the development of a Collective Trademark engineered as a high quality 'area brand' might allow the selection of a tourist market more empathic to high quality culture-based production: a development model which has the potential to be deeply rooted in the social and economic fabric of the resident community.

CIPR as a hierarchy of instruments

As what has to be protected is an ensemble of goods and services which is unique and provided in a specific location, a hierarchical system of collective rights may be imagined. In fact, a tourist system or a CTD can be identified by a general collective sign and by other sub-collective brands that can be assigned to single goods and services. For instance, making reference to the Lebanese heritage city of Saida (Box 2), the introduction of CIPR to promote traditional productions (soap) as a tourist product is associated with the historical architectural heritage of the old city centre. 'Saida is for culture' could be the general Collective Trademark of the tourist destination, and cascading from that we might have 'Saida is for culture/Traditional soap', or 'Saida is for culture/heritage', or 'Saida is for culture/The Suq'. In this sense, a cascading system of Collective Trademarks can be created with reference to different aspects of the tourist product, achieving a wider range of combinations of quality and products in the CTD.

In this way, CIPR on culture-based products would be conceived of as part of a 'brand strategy' in which collective rights reward creativity and local idiosyncrasies, and not necessarily value. The slogan could be: '*It does not have to be expensive: it has to be typical*'. CIPR, like GI and AO, should be applied at the moment of sale and not (necessarily) on production, and sale location is also part of the uniqueness that needs protection (a market, an old square, etc.). CIPR could then be articulated in order to apply at product level and at area level: only original products can be sold in *designated areas*, and the zones of application of a brand may be chosen so as to enhance product integration and innovation in the site and around it, for instance extending to symbolically valuable central areas or to regenerating peripheral districts.

Box 2 A case of disadvantaged regions: Byblos and Saida, Lebanon

Byblos and Saida are among Lebanon's most famous heritage sites – both with a rich tradition in material culture, especially in the production of handicrafts, and dilapidated walled city centres – and are the object of many a hurried visit by people coming from the capital, Beirut, or the nearby traditional resorts. Both resorts are the object of governmental programmes which associate the rehabilitation of architectural heritage with the recuperation of traditional production and markets, hoping to establish an appealing environment for the development of viable tourist districts in the old city centres. In both cases, the WIPO assists national governments in assessing the possible impacts and strategies for the introduction of intellectual property rights as an instrument for tourism development.

Byblos, the ancient capital of Mediterranean writing (the prefix 'biblio-' derives from this place, as does the Bible) receives around 50,000 visitors per year. Primary cultural attractions include the walled city, religious monuments, archaeological sites, pristine maritime and natural fluvial landscapes, cultural events, and commercial activities such as handicraft markets and souks. The old city is clearly a potential creative tourism district and although it suffers from a very poor 'tourist orientation', the prevailing tourism development project is geared more to the '5-star development model' outside the old city. New World Bank-sponsored projects target this area, funding the refurbishing of public spaces, events and animation in the old city.

No existing IPR or CIPR are presently in place, but local stakeholders seem to be willing to take the initiative. The general idea is to establish a strong 'area brand' (e.g. 'Byblos, Phoenician Mother of the Alphabet') encompassing the old city, where local traditional products are promoted, and cultural resources and events evoking Byblos' association with book culture.

Saida or Sidon – a medium-sized city and capital of Southern Lebanon – also has the potential for a CTD, including many museums, cultural foundations and heritage sites, as well as a walled quarter (Old Saida), beautiful natural surroundings and a few remains of Saida's main handicraft tradition: soap. In the picturesque Souk, mostly unoriginal, low quality handicrafts are sold and the legacy of soap production is mostly confined to the soap museum.

Private cultural foundations are very active players in urban management and the revitalization of the old city. New projects sponsored by the World Bank include the rehabilitation of public space and social development initiatives. There is clearly a role for CIPR, which could be differentiated as a CT for the old city (the Saida medina) and AO for traditional products.

Conclusion and policy suggestions

Creativity can be conceived as a resource highly embedded in local knowledge, and closely associated with material culture. It can also be an asset for local development based on tourism. Today, the keyword for sustainable tourism is 'empathy' between host and guest community, hinting at the genuine, transparent, enticing content of the visitor experience. Creative tourism districts – as opposed to other development models – are conducive to empathy: time-optimising, dynamic in nature, inclusive and thriving on quality.

CIPR can – on the one hand – stabilize the structure of a creative tourism district, and – on the other hand – increase its orientation to different visitor profiles through better visibility and increased diversity. However, CIPR alone cannot foster coordination of the different elements of the CTD. Collective Trademarks on 'intangible' qualities of the district are needed as a development tool; standardization, even high quality, should not be sought, but rather cohesion of different production qualities or strategies around a 'project'.

'Area labels' promoting a set of products and a consumption environment could achieve this effect. They could be structured like a CT with a hierarchical structure while providing a stabilizing effect for the district.

Collective Trademarks could provide the appropriate regulatory instrument in the situation of disadvantaged regions, but then the question is how to apply them in the proper way, taking strategic decisions on the start-up, implementation and monitoring of their introduction. As a final contribution, we suggest a practical procedure for establishing CTs that may be articulated in eight steps (Box 3).

Box 3 An eight-step programme for the introduction of Collective Trademarks (after WIPO www.wipo.int)

Step 1: Identifying the symbolic image of the tourist destination. This has to be a clear-cut and appealing image to be stylized and reduced into a CT, which at the same time can be articulated as a logo, design, musical theme, label or slogan.

Step 2: Filing the chosen CT at the IPO. The trademark has to comply with some elementary rules that the Intellectual Property Office will check carefully through an examination process before accepting it.

Step 3: Setting up an association for the ownership of the collective trademark. The collective trademark representing the identity of the CTD in principle concerns all of the local economic agents involved in tourism-related production. As such, the owner of the CT should be an organization such as an association, a government agency or a cooperative. All the accredited members of the association have the exclusive right to use the CT according to the stated regulations. Members can be private companies, foundations, non-profit organizations or local government bodies who agree to be inspected periodically on the quality of their goods and services.

Step 4: Appointing a steering committee. This is a crucial step, because the rationale of the CT requires the establishment of a steering committee charged with controlling access to the trademark and the quality level of its members.

Step 5: Establishing the rule of compromise as the highest level decision tool. The compromise rule should be employed, based on majority rule.

Step 6: Selecting the minimum quality standard. The standards are related to every good produced and service provided in the CTD. The committee will also control to which extent these criteria are being met by the individual firms wishing to use the logo. If firms meet the quality requirements, they are entitled to use the collective trademark.

Step 7: Giving no refusals, but registration and accreditation. If an applicant does not meet the necessary requirements, membership is not refused – rather, they are placed on a waiting list until they meet the required standards. Indeed, there is a softer way to say 'no' – a two-step procedure. The first step, that of registration as a potential member of the Association, is fully open: all the applicants are accepted. The minimal requirement is the active involvement in the tourism activity within the district. The registration then opens a procedural path that begins with the initial assessment of the quality level of the goods and services provided by the candidate. If the standards are met, accreditation follows and, as an official member of the association, the candidate is entitled to use the CT. If the quality is below the minimum, an interactive process will begin to help the candidate achieve the minimum quality. This is done through periodic inspections, advice and institutional support. In this way, accreditation is the result of a cooperative process that helps attain a good average quality for the whole district.

Step 8: Following up and royalties. The management of the CT can be sustainable in economic terms when royalties are collected by selling labels or tags to the members (to be put on the objects sold or given on the provision of the service). If the role of the CT is appreciated and backed by the local community, a positive willingness to pay for it is probable, from both producers and consumers.

(Santagata 2004)

Notes

1 Nelson (1970) divided goods into two varieties – search and experience – according to how they convey information to the consumer. In the case of experience goods, consumers cannot ascertain quality prior to purchase through the standard process of inspection and research applicable to search goods. Consumers have to buy them first and ascertain their quality afterwards.

2 The World Intellectual Property Organization is a UN agency which monitors IPR and advises on their importance as community development tools.

8 Creativity in tourism experiences

The case of Sitges

Esther Binkhorst

Introduction

With access to almost anything they want, people in the developed world are engaged in a quest to satisfy their psychological needs such as inspiration, belonging to a meaningful community and meaning in general (Nijs and Peters 2002; Ter Borg 2003). It is particularly in the experience environment of free time, that leisure and tourism experiences have become a basic contributor to the quality of life (Csikszentmihalyi and Hunter 2003; Richards 1999; Urry 1990). People wonder whether they should spend their next couple of days off in the mountains or in the city, which culture they have not explored yet, what cuisine they would like to try now, whether they should try a boutique hotel, a design hotel or an ice hotel and what would really make their trip memorable.

Selling tourism experiences to this highly experienced and sophisticated clientele is a great challenge. According to Pine and Gilmore (1999), experiences can 'touch' people better than products or services. Experiences are intangible and immaterial and although they tend to be expensive, people attach great value to them because they are memorable. The so-called 'first generation' experiences have been criticized within European contexts for being too staged, commercial and artificial and therefore not always suitable for today's customers (Binkhorst 2002; Boswijk *et al.* 2005; Nijs and Peters 2002). Modern consumers want context-related, authentic experiences. Moreover, the 'new tourist' (Poon 1993) wants to be in charge. The 'second generation' experiences, based on co-creation between company and client, enable this and therefore deserve to be studied (see also Binkhorst 2005).

The concept of creativity is interesting in respect of tourism. It was first introduced by Richards and Raymond (2000: 18) as a form of 'tourism which offers visitors the opportunity to develop their creative potential through active participation in courses and learning experiences which are characteristic of the holiday destination where they are undertaken'. This definition reflects, on the one hand, the growing interest among individuals to creatively construct their own 'narrative of the self' (Giddens 1990). On the other hand, creative tourism explicitly provides tourism destinations with opportunities to engage with their local culture in order to offer a unique experience. Creative tourism could thus be a very welcome alternative to those destinations providing a 'copy and paste' reproduction of culture.

But where and when does creativity in tourism experiences occur? Can the fact that people visit destinations in order to improve their skills or transform themselves be called 'creative'? It surely results in more unique life stories than the experience of traditional tourism when everyone did more or less the same thing. But when diversity is only defined by the way people link their tourism experiences, which in fact are a series of standard and product-driven goods and services, what is so creative about that? Is creativity then focused on a more interactive form of tourism, a *co-creation* between the tourist and other stakeholders in the tourism network, in order to create unique value for both? If so, when does the co-creation take place? At the destination, during some of the cultural tourism activities undertaken? Or also in the home environment before and/or after the actual travel? Are tourism destinations aware of the existence of the co-creation tourism experience yet, and, if so, how can they develop strategies to really embrace creative tourism?

In this chapter, the concepts of experience and creativity are explored in relation to tourism. The first part of the chapter focuses on the relationship between meaningful experiences and creativity. This is followed by a case study of how the Spanish coastal town of Sitges is developing creative alternatives to its traditional '*sol y playa*' ('sun and beach') product.

Tourism experiences and creativity

Experiences and transformations

While 'experiences' are nothing new,[1] they have long been neglected and perceived as similar to everything else supplied in the service sector. However, according to Pine and Gilmore (1999), they have now been (re)discovered by companies as the way to survive ever-increasing competition. Experiences, rather than goods or services, are said to escape the foot-loose character of standardization. But, if one views experiences as modules created by companies and selected by individuals from an overloaded (time) market full of experiences, what protects them from commodification?

In view of how the modernization process has manifested itself so far, the production of goods and services in general is said to be 'McDonaldized', that is, based on efficiency, predictability, calculability and control (Ritzer 1996), while the products of tourism in particular (which generally thrives on intangible experiences) are said to be 'McDisneyified' (Ritzer and Liska 1997). The modularization of daily life (Van der Poel 1993, 1997) shows that experiences, when they are considered to be more or less commodified activities undertaken in a certain period of time, can easily become subject to standardization. Modules are exchangeable in the sense that people today have limited time and unlimited options for time allocation, making the selection of time-modules a rather arbitrary process. Modularization does, however, provide the opportunity to compose a unique experience for every individual consumer, while the basic assortment for everybody is standard.[2] For instance, although packaged as a unique and everlasting memorable experience, every show or time-module chosen by a Disney World guest is a McDisneyified performance conditioned by the guidelines of McDonaldization. While the ultimate composition and experience thereof is in itself unique for each individual, its modular parts are nothing more than a sum of standardized product and service elements. The ultimate consumer experience is the result of these elements together with the contextual time-space and personal conditions that surround the individual. Today's post-tourists, who are already collectors of images and sights, will now also become collectors of memorable experiences. The cumulative experience of every person reflects their story or 'narrative of the self'. But is that what creativity refers to?

Much has been written about why people spend their leisure time the way they do. Authors such as Cohen (1972, 1979, 1988), Lengkeek (1994, 1996), MacCannell (1976) and Urry (1990) all interpret tourism as a consequence of the alienation in everyday life. The interruption of daily temporal and spatial structures is what most obviously distinguishes a holiday from the normal routine. Moreover, during holidays, activities are undertaken for which, usually, no time, space or opportunities are at hand, and which are strenuous but relaxing, risky and/or pleasant and so on, in contrast to daily activity patterns (Lengkeek 1994, 1996). The extent to which counter-structures are different from the ordinary, however, depends on the degree of interest that people have in the 'other' and the extent of alienation in everyday life. Many tourist typologies have been developed in order to categorize tourism behaviour (Cohen 1972, 1979; Elands and Lengkeek 2000; Plog 1972, quoted in Shaw and Williams, 1994: 70; Richards 1996a, 1996c, Richards and Wilson 2003; Stebbins 1997; Thrane 2001).

With respect to cultural tourism in particular, McKercher and Du Cros (2002) found that the major reason for the 'purposeful cultural tourist' to visit a destination is to learn about and to experience the other culture. The 'sightseeing cultural tourist', by contrast, is less concerned with experiencing the other culture than with visiting the cultural highlights. For the 'casual cultural tourist' culture plays a less dominant role in the decision making process for the destination, and being there, the tourist does not get deeply involved. The 'incidental cultural tourist' does not choose a destination based upon culture, and will only be superficially involved with culture during his or her visit. The 'serendipitous cultural tourist' does not seek cultural involvement in the destination choice, but while there gets really involved and has a deep experience. This typology shows that a certain group of cultural tourists is susceptible to being surprised by cultural tourism experiences they encounter along the way. Together with those tourists already interested in culture, this is an interesting and important group for those involved in cultural tourism development. To what extent, however, do these tourists differ in achieving their meaningful experiences? What was the balance for each type of cultural tourist between more passive and more active involvement with their cultural tourism experiences? These questions should not only be investigated regarding tourism experiences at the destination but also in the home environment; when preparing the trip and after returning. Creativity could bloom during all these phases.

It is argued here that experiences can only be unique when people not only play an interactive and participative role in undergoing them, but also in creating, designing, selecting and reflecting upon them. This implies the necessity for a co-creation tourism experience environment, where supply and demand meet and where dialogue between producers and consumers can take place. These trends of future innovation are described by Prahalad and Ramaswamy (2004) but have not yet been specifically studied in the area of leisure and tourism experiences. It is argued here that creative tourism cannot be considered an appropriate alternative to the serial reproduction of culture unless it explicitly refers to a co-creative role of the guest and other stakeholders in the tourism experience network (see also Binkhorst 2005).

Pine and Gilmore (1999) do acknowledge the commodification pitfall for experiences when more and more competitors start offering 'unique' experiences. To avoid this pitfall, they argue that producers need to offer experiences which transform the consumer. People arguably attach more value to such 'transformations' than to services or basic experiences because transformations touch the very source of all other needs, including the origin of the desire for commodities, goods, services and experiences. Memories can be seen as souvenirs of experiences, but they slowly fade away. In the case of transformations, however, the client *is* the 'product'. No transformation can be copied or simulated exactly, as the unique relationship between the transformer and the transformed can never be made common property. Therefore, transformations cannot degrade to commodities and, at least when adequate aftercare is provided, will endure until the next transformation presents itself (Pine and Gilmore 1999).

Transformations can thus be considered to be co-creations between the transformer and the transformed. The involvement and engagement of the transformed is exactly what explains the added value of transformations, as opposed to memories of experiences that fade away. But does this mean that all tourism experiences undertaken to develop creative potential at the tourism destination could be considered transformations? No. Crucial here is the concept of co-creation; the involvement of the individual in designing or co-creating, undergoing and evaluating their own experiences. These phases are connected in the tourism experience environment in which creativity could bloom.

Therefore, the focus should shift from discussing whether something should be called an 'experience' or a 'transformation' to studying the process of how the products come alive. Who initiates and designs experiences? Who participates in experiencing them? Who participates in optimizing them? How can the process of co-creation be characterized for tourism experiences?

Meaningful experiences during travel

Csikszentmihalyi (1990) characterized the optimal experience as containing a feeling of play, a feeling of control, maximum concentration, enjoying the activity in itself, losing time awareness, a good balance between challenge and own capacities and having a clear objective. Based on an extensive literature review, Boswijk *et al.* (2005: 27) added three other characteristics to the seven already defined by Csikszentmihalyi, namely a unique process for the individual that has intrinsic value, contact with the 'raw stuff' or the 'real thing' and finally, active involvement.

Considering these characteristics, not everything one does can be considered a meaningful experience. Doing routine daily activities such as waking up to the alarm clock in the morning, preparing for school or work or rushing to the bus or train, do not help you to lose time awareness or to have a sense of playfulness. Neither do these types of activity provide us with an increased level of concentration or get us emotionally involved. Consequently, meaningful experiences are generally to be found outside the range of usual, routine, repetitive daily activities. Rather, they are 'first time' experiences or they are 'once in a lifetime' experiences, of which many occur when discovering new places during travel. The area of free time in which leisure and tourism activities are usually undertaken can therefore be considered to be the main generator for meaningful experiences (see also Binkhorst 2005). Tourism experiences are generally undertaken as part of 'paid for' experiences. Consequently, those who buy and those who sell experiences interact in what is called the 'co-creation tourism experience environment'. This experience environment where demand and supply meet is still underestimated as a useful tool to stimulate creative tourism.

It is argued here that both in the home environment when preparing the trip or reflecting upon it, and during the actual travel, virtual and real experiences continuously shape the individual's tourism network and determine the final tourism experience (see also Binkhorst 2005). The current setting in which tourism experiences are designed and offered does not stimulate but rather limits creativity.

The concept of creativity

Traditionally, creativity is associated with 'doing something manually', or with 'the creation of things' such as painting, making music, making handicrafts and so on. It also refers to being inventive, imaginative and original as Van Dale (2007) puts it. Creativity in the concept development process refers to 'finding solutions for problems that others have not found yet and applying combinations of knowledge to new problem areas' (Walravens, cited in Nijs and Peters 2002). Although know-how and an analytical approach are absolutely necessary in any concept development process, creativity is an indispensable ingredient. Nijs and Peters (2002) point to several key aspects of creativity as part of the concept development process: creativity is about solving problems; creativity is about innovation, that is, finding solutions that others have not found yet; creativity is about crossing borders and looking into other fields; and creativity is about combining knowledge from different fields, not necessarily about the development of completely new ideas. The descriptions in the literature of the process of creativity resulting in new developments has varied a lot over the years. It is summarized by Nijs and Peters (2002) as: problem definition, preparation, breeding, 'aha-erlebnis' (moment of finding the solution), evaluation and selection, elaboration and implementation. These phases do not necessarily have to progress in a linear fashion during the concept development process; one can easily start breeding again after the aha-erlebnis, for instance. Csikszentmihalyi argues that creativity results from the interaction between three elements in a system: an individual launches something new into a symbolic field with its own culture of norms and rules and a market of experts recognizes and then evaluates the innovations (Csikszentmihalyi 1999, cited in Nijs and Peters 2002).

Based on this model, creativity is a pervasive and ongoing process that continually shapes everyday life. Consumers will play an ever bigger role in this, as they increasingly gain more power and control in future experience environments in which dialogues can take place between them and companies (Prahalad and Ramaswamy 2004). Florida (2002) argues that the rise of human creativity is the key factor in our economy and society and identifies the 'creative class' as the driving force behind the transformation of society.

Creative tourism

Based on a study of tourism experiences of US tourists in the Netherlands, Binkhorst (2002) warns against the creation of too many 'fun-seeking sights' there. These 'footloose' reproductions of reproductions may occur in any other country and will therefore not contribute to the uniqueness of the Netherlands as a tourism destination. Several of its popular and unique cultural tourism attractions, such as the red light district of Amsterdam, the coffee shops and the Delta Works, which are limited to passive sightseeing so far, could be transformed creatively into more interactive tourism experiences. Moreover, the intangible aspects of the Netherlands very much appreciated by American travellers, such as the atmosphere, culture, food and people, are invaluable resources for creative tourism experiences. A recent initiative, 'Dine with the Dutch', invites tourists to have dinner in real, private, local Dutch settings at people's homes:

> Get to know Dutch culture in a unique way and join an Amsterdammer for dinner. Book an at-home dinner with one of our specially selected hosts. Listen to their stories about *Dutch culture*, find out more about Amsterdam and sample our kitchen!
>
> (www.dinewiththedutch.nl)

According to the ten characteristics detailed above (Boswijk *et al.* 2005), this can be considered a meaningful experience. An American couple reflects, after having returned home, that they had a wonderful time visiting Amsterdam and the Netherlands and that their dinner with 'Hans and Marreke' was one of the highlights of their trip. They report that this unique experience does indeed give the visitor an entirely different perspective of the country (ibid.).

Creativity in tourism has so far basically been referred to in two ways. First, people today are more decisive in the process of shaping their own narratives, a phenomenon that pre-eminently develops during leisure and tourism. This results in numerous stories full of ever-more original tourism experiences. Second, it involves 'the creation of things' at the destination – the painting, cooking, making handicrafts and so on mentioned above. Indeed, there is a growing interest in having such creative experiences during holidays. Richards and Wilson (2006) point out various ways in which the application of creativity can help develop cultural tourism into creative tourism. These are:

- creative spectacles: the production of creative experiences for passive consumption by tourists;
- creative spaces: the development of a spatially demarcated creative 'enclave' populated by creatives to attract visitors;
- creative tourism: a more active involvement of tourists, not just spectating or 'being there' but reflexive interaction.

Having dinner in a real Dutch home can be considered creative tourism. The tourist interacts with the local people at the destination, enters places that normally would not easily be accessible, learns about how they live, eat, cook and so on, and both visitor and visited are able to share their cultures. This example illustrates as well that creativity in itself cannot be considered an appropriate alternative to the serial reproduction of culture unless it refers to a participative role of the individual and other stakeholders in the tourism network, in the co-creation of the tourism experience before, during and after travel. The following section focuses on how the coastal Spanish town of Sitges deals with creativity in tourism.

Creativity and tourism in Sitges

Plan of Excellence *2004*

In Sitges, a Spanish coastal town just south of Barcelona, attempts are being made to fight the traditional tourism developments of the 1960s that made the destination suffer from typical problems such as overcrowding. Sitges is basically known for its beaches and palm tree boulevard, its picturesque church, its art and galleries, the traditional celebrations, the parties, its numerous events, its gay scene, the boutiques, nineteenth- and twentieth-century modernist architecture, and its international and cosmopolitan atmosphere. In 2004, the *Plan of Excellence* was launched by the local tourism authorities to relocate services and amenities, to create new products adapted to current trends in demand, and to work towards enhancing natural resources. The plan consists of four programmes focusing on the creation and consolidation of tourist products (product portfolios of cultural tourism, gastronomic events, active tourism, company products); tourist promotion and publicity (the Sitges brand, infrastructure and services, a marketing plan); strengthening tourist commercialization (new promotional DVD, new website); and improvements of beach services (services, accessibility and sustainability).

Some recent initiatives show how tourists are being offered cultural- and natural-related experiences by inviting them inland to experience the 'other side of the beach'. Current supply ranges from standard tours to innovative and context-related activities. They can basically be organized around the following topics:

- Cultural heritage (passive sightseeing to museums, art galleries, monasteries, markets and so on, and routes passing modernist buildings);
- Festivals and events (observing, spectating activities);
- Gastronomy (active participation in wineries and other food- and drink-related settings);
- Nature (passive and/or active activities in the Garraf National Park);
- Adventure (active sightseeing such as hot-air ballooning, helicopter tours, sailing trips, 4x4 excursions or activities such as carting and skydiving).

Each of these will be discussed in the following sections.

Sitges' cultural heritage experiences

The cultural heritage highlights of the town are the three museums. The museum of Cau Ferrat is the former house and studio of artist and painter Santiago Rusiñol. He bought this old fisherman's house in 1894 and transformed it into a meeting place for artists and intellectuals of the time. Later it was transformed into a museum where works by El Greco, Picasso, Miró,

Rusiñol, Utrillo and others are exhibited. Today it is one of the icons of Catalan Modernism. Visitors may enjoy the cultural spirit of the era during concerts organized in the museum Cau Ferrat. The Romantic Museum shows how a Sitgetan family lived at the end of the nineteenth century and the museum also houses an interesting collection of dolls, a donation of the writer and illustrator Lola Anglada. In summer, outdoor opera concerts can be enjoyed. The Museum Maricel houses three art collections and forms, together with the Maricel Palace (a building constructed at the beginning of the twentieth century by the American collector Charles Deering with the help of the engineer and art critic Miquel Utrillo), an impressive entity of modernist architecture right by the sea in the old town centre. In summer, the music of composers such as Bizet and Albéniz can be enjoyed accompanied by castanets and a glass of cava (www.sitgestour.com/sitgestiucultura-13.html). Other cultural heritage sights in Sitges include the churches of Vinyet and Sant Bartomeu and Santa Tecla and the modernist architectural route along the villas of the Catalans who made their fortune in America.

The year 2006 was dedicated to Santiago Rusiñol, the artist and painter who is generally held responsible for the transformation of Sitges from a fishing village into a meeting place for artists, building the foundations of the magical appeal of Sitges today. Seventy-five years after his death, the municipality of Sitges celebrates the spirit of modernism and wants to strengthen the image of Sitges as Santiago Rusiñol created it: an image of art and culture, of civilization, of modernity and, moreover, a European image. To this end, the municipality organized a range of activities from June 2006 to June 2007, including conferences, courses, symposia, itineraries, web pages, book (re)editions, documents and the restoration and improvement of the architectural and cultural heritage of the town (Municipality of Sitges 2006).

In the area around Sitges, markets, castles, monuments and monasteries can be visited, as well as, obviously, the nearby city of Barcelona with its numerous cultural sights and attractions. Sitges' cultural heritage can basically be experienced by passing by and admiring the buildings from the outside, having a look at the expositions inside, joining a guided sightseeing tour or joining one of the events organized by the museums. Through seeing, listening and a bit of tasting, tourists can familiarize themselves with the town's cultural heritage.

Sitges' festivals and events

Sitges es una fiesta permanente ('Sitges is a permanent festival') according to the Patronato Municipal de Turismo de Sitges (2005b), and an expansive calendar of local, national and international events is held in Sitges. Whether it is a traditional celebration, cinema, theatre, music, dance or gastronomy, Sitges hosts events of all kinds and sizes. Some of the principal events are the Carnival, the vintage car rally, Sant Jordi book and rose day, the celebration of the 'Corpus Christi' flower carpet competition, the annual festival (*Fiesta Mayor*) in honour of Sant Bartomeu (the town's Patron Saint), the 'Fiestas de la Vendimia' (the grape harvest festival), Santa Tecla and the International Film Festival of Catalunya (Patronato Municipal de Turismo de Sitges 2006). The events guarantee a continuous flow of visitors to the town. Copious events announced weekly in the local newspaper 'L'Eco Sitges' reflect the nature of Sitges as a creative space.

Lately, Sitges has also increasingly developed as a conference and meeting destination. The recent opening of the Dolce Sitges Conference Centre and the establishment of the Sitges Convention Bureau has professionalized this segment even more. Companies such as Porsche, Nike and others come to Sitges to celebrate all kinds of event such as product launches, conferences, team building activities and so on. Located almost next door to Barcelona, Sitges benefits from the increase in popularity of Barcelona as an event destination.

Sitges' gastronomic experiences

Boosted by the year of gastronomy in Catalunya in 2005, the area of gastronomy is being exploited more than ever before in the creation of cultural tourism experiences. As the senses of taste and smell are explicitly involved in gastronomy, it is a very valuable resource for creating meaningful experiences.

In a recent interview with the Gremio de Hostelería de Sitges (Arenas 2006), their President argues that the level of gastronomy in Sitges has increased tremendously during the last 20 years. Gastronomy seems to be a year-round reason for people to visit the town. 'Sometimes clients come to Sitges just to have a good lunch', he says. Therefore, some of the restaurants in Sitges have more work during winter than during the tourist season. 'Most of the clients are of Spanish origin who tend to spend more on food generally' (Arenas 2006: 14).

El parc a taula ('Park at your Table') for instance, is part of the project *Viu el Parc* ('Enjoy the park'). It is aimed at sharing significant aspects of Catalan culture, for example gastronomy, with visitors to the Garraf National Park. Twelve restaurants offer specially prepared dishes with products that are typical of the national park and cooked by experts. In 2005, the wine and cava producers of the municipalities in the park also joined the project to complement the Garraf gastronomic experience.

On winery excursions, people come to see, smell and taste diverse wines besides learning about the thousand-year-old tradition of winemaking and viticulture. A winery experience in the surroundings of Sitges usually includes a walking or 4x4 tour of the vineyard with its chapel, tower and cellar, where tourists can learn about viticulture from a qualified oenologist who guides the wine tasting. To get a taste of the typical local cuisine, a Catalan lunch is often served.

At the famous Torres winery (www.torres.es) the one-hour tour begins inside the 1300 square metre visitors' centre, with the screening of a video which explains Torres' history. The estate where the mythical Mas la Plana wine was born can be visited. The secrets of the Torres way of planting and picking grapes are shown, visitors can see the arrival of the grapes, the fermentation process and visit the ample underground cellars. Then they will be able to walk among the casks, smell the wine while it ages and understand the secrets of this work. After a wine tasting session at the end of the tour in the vaulted-roofed tasting hall, people will leave the Penedès region with the Torres taste. The Torres family makes it clear that their vocation does not lie in the restaurant industry but in the production of wine. However, the customer has the opportunity to accompany his or her wine with some gourmet dishes, essentially based on selected products of the *Torres Real* brand, also owned by the family, including asparagus, tuna, olives and so on. *La Vinoteca Torres* is a space of modern design where one can taste any of the more than 50 wines and brandies produced by the Torres family in Catalonia, Penedès, Conca de Barberà, Ribera del Duero, Chile and California. Wines of the PFV (Primum Familiae Vini) association, to which the Torres Family belongs, may also be found. Also at Jean Leon (www.jeanleon.com) another winery in the Penedès region owned by Torres, they have succeeded very well in creating an experience environment in which visitors are being exposed to the legend behind the brand by guiding them through different experience spaces. Visitors arrive at a brand new modern visitor centre at the heart of the Penedès, perched on top of a hill with views of the entire valley. A film shows visitors the fascinating life of Jean Leon, a Spanish immigrant who, with his extraordinary and unbreakable willpower, made it through a relentless string of adventures until he saw his most desired dreams become reality. The museum contains a mine of information visualizing Jean Leon's life story on the one hand, and about the viticulture in different areas of Leon's vineyards on the other. Visitors are then taken on a guided tour through the vineyard and bodega with explanations of viticulture and the different winemaking techniques. Moreover, there are opportunities to taste several wines, and lunches can also be served.

What is remarkable about all of these gastronomic experiences, however, is that the participation of the visitor is still limited to seeing, listening, smelling and tasting. There are no options yet to be completely engaged in, for instance, the Torres brand through the principle of co-creation.

Sitges' nature experiences

Sitges is located between the Mediterranean and the Garraf National Park. The latter is located between the regions Baix Llobregat, l'Alt Penedès and El Garraf. It borders the inland area of Llobregat, the Mediterranean and the Penedès. The Park occupies 12,820 hectares and its two highest mountains peaks are 'la Morella' (595 metres) and 'el Rascler' (572 metres). Some interesting examples of creative tourism environments can be found in the Park.

Jafra Natura (www.jafranatura.com), a governmental body of the *Diputació de Barcelona*, organizes guided walks along the coast to observe transformations of the landscape. They also organize specialized guided tours in the Garraf National Park to discover its flora and fauna, its history, its inhabitants, its fruits and vegetables and so on, with a link to Catalan cuisine. For the more adventurous tourist they offer guided sailing and cata-maran tours to experience the Park's coastline from the sea, as well as excursions by 4x4, mountain biking, climbing and so on. Experiencing the park is creatively linked with gastronomy in *El Parc a Taula* (see above).

Besides various educational centres, the Garraf National Park houses *Vallgrasa*, an experimental centre for the arts that was established some 20 years ago. In 2001 it was transformed into an innovative project based on both the natural surroundings of the Garraf National Park and the Mediterranean cultural identity. It is a meeting point for artists and visitors; a gathering place for creativity and observation.

> It is a place where the universal, open language of today's art allows for the discovery and integration of the soul of the Park of Garraf. . . . It is a meeting point and a departure point too, where bridges are created to connect the Park and its surrounding territory through workshops, seminars, exhibits, poetry recitals, music concerts.
>
> (Vallgrasa 2005)

One of the workshops that Vallgrasa organizes is set up around the sense of smell, giving form to emotions and sensations arising from the experience of nature, through different smells that can be found in the Park such as rosemary, pine sap, sea and so on.

The examples described above show that the natural heritage of Sitges and its surroundings can be experienced in a more participative way. A more in-depth analysis of the supply side will tell us who the visitors are that actually consume these experiences. The initial results of fieldwork under-taken in Sitges during the summer of 2006 indicate that creativity is not yet a well-known part of the tourism product.

Sitges' adventure experiences

The more adventurous experiences that are currently being offered in Sitges are hot-air ballooning, helicopter tours, carting, sailing, surfing, 4x4 and skydiving. The bicycle has also become a visible element in town as cycle tracks are becoming part of Sitges' street scene. 'Sitges Bike' (www.sitges-bike.com) now makes it possible for tourists to experience the town and its surroundings on two wheels.

Garrafactiu (www.garrafactiu.com) is a company specialising in sport activities in the Garraf National Park. Through activities such as quad rides, archery, kayaking, orientation games on mountain bikes, horseriding, discovering all the secrets of the Park while trekking and other personalized activities, emotions are touched while people are in contact with the natural environment. The experiences they offer are fun and educational at the same time.

As the examples described above are sports and adventurous experiences in which the participant is personally and actively involved, the level of participation is obviously higher than in some of the other tourism experiences mentioned before. A more in-depth analysis of the suppliers in this category could shed light on the question 'who are the people that sign up for these adventurous experiences?' Again, from the results of fieldwork undertaken during the summer of 2006 among tourists in the centre of town, it is not currently an experience that tourists have heard of or actually participate in during their stay in Sitges.

Tourists in the summer of 2006 in Sitges

Methodology

No recent and reliable data were available on visitor profiles or travel behaviour in Sitges. In order to get an insight into the characteristics, motivations and behaviour of cultural tourists in the town, it was decided to use the ATLAS[3] cultural tourism questionnaire. The advantage of using this questionnaire is the ability to compare the data collected locally with the worldwide ATLAS database. A disadvantage, however, is its more or less fixed structure and content. Some specific questions about the tourism supply in Sitges were added but, to be able to compare the data with the rest of the ATLAS database and to prevent the questionnaire from being too long, the basic content of the ATLAS questionnaire was followed. In future, tailor-made questionnaires and in-depth interviews with a number of tourists and suppliers would be very useful for getting an insight into more specific topics, especially when it comes to the impacts of the various tourism experiences that people undergo.

The survey was conducted in Spanish and English among 350 international tourists in Sitges from mid July to mid August 2006. Two students from ESADE/St. Ignasi[4] were instructed to approach tourists throughout town, along the boulevard and on the beach, either in the morning from 10 am to 2 pm or from 5 pm to 9 pm in the afternoon. Due to very high temperatures during the month of July, an exception was sometimes made, and fieldwork was conducted later at night in cooler temperatures. Generally, tourists were quite willing to participate and to talk about their holidays. Nevertheless, as the survey was conducted among all tourists and not directed at those who had visited cultural attractions at the moment of interviewing, some of the questions were not applicable or did not fit into the tourist's mindset at that time.

Obviously, this research has some limitations. Most importantly, as fieldwork was only done during the summer, it is not surprising beach tourists dominated the sample. The research was also limited because the questionnaires were conducted only in the centre of Sitges. Consequently, none of the specific locations for cultural, natural, gastronomic, sports and adventurous tourism experiences were included. This increases the chance of having a sample of tourists who are only walking around in town or lying on the beach. This problem was reduced, however, by including questions about the cultural, natural, gastronomic or adventurous sights and attractions tourists had heard about or planned to visit during their stay.

Visitor profile

The total sample consisted of 350 people who spent their holidays in Sitges during the survey period. Two-thirds of them were male and one-third were female.

Ten per cent came from the local area (basically from Barcelona; those residing in Sitges were excluded from the survey), 41 per cent from the rest of Spain and 49 per cent from abroad. The sample represents a very international visitor profile. The majority of those who came from abroad (81 per cent) came from other countries within the European Union, most of them from France (23 per cent), followed by the United Kingdom (18 per cent), Germany (11 per cent) and the Netherlands (10 per cent). Other European countries that were less well represented were Andorra, Finland, Greece, Ireland, Italy, Norway, Portugal, Switzerland and Sweden. Of the visitors originating from non-European countries, 11 per cent came from the USA and others (only 1 to 2 per cent) came from Argentina, Australia, Bulgaria, Canada, Colombia, Japan, Kuwait, Russia and Uruguay.

Almost one-third of the total sample of tourists was 30–39 years old, closely followed by the categories 40–49 and then 20–29. Only a few of them were younger than 20 years or 50 years and older. There are generally no significant differences in age regarding the country of origin. However, tourists from the local area tended to be younger. The sample was quite highly educated: a quarter having completed vocational education and another quarter a masters or doctoral degree. Remarkable is the high percentage of visitors from outside Spain with a masters or doctoral degree (45 per cent), while only 7 per cent of people from the local area had a similar qualification. Of those in paid employment, over half were professionals, and this meant that the tourists had relatively high incomes. Over a quarter of the sample had an annual household gross income level of €40,000–50,000.

In terms of travel style, most people were travelling with their family (36 per cent) or their partner (31 per cent). Most tourists stayed in a hotel (57 per cent), although a large proportion of the tourists from Barcelona stayed in a second residence and 11 per cent of them stayed with family and friends. When looking at the people coming from other places in Spain, self catering accommodation such as apartments (19 per cent), caravan or tent (10 per cent) and staying with family and friends (10 per cent) were also popular. For those who originated from outside Spain, most were staying in a hotel (64 per cent) or in a caravan or tent (15 per cent).

The majority of tourists who booked their transport separately indicated that they did not book it in advance (61 per cent). Many of them probably came with their own car from other places in Spain or Europe. Twenty-one per cent arranged their transport via the Internet. When looking at how they booked their accommodation, 30 per cent indicated that they made their own travel arrangements directly (by phone or fax), 27 per cent booked accommodation via the Internet, 26 per cent did not book their accommodation in advance and 17 per cent booked through a travel agent or tour operator. The average length of stay for the total sample was six nights.

Purpose of visit

When asking tourists about the main purpose of their trip to Sitges, the vast majority (63 per cent) answered that they came for a holiday. Other options they could choose from were 'for a cultural event' (12 per cent), 'visiting relatives and friends' (5 per cent), 'for business' (5 per cent), 'for a sports event' (6 per cent), 'for shopping' (3 per cent) or for 'other', unspecified, reasons (6 per cent). The fact that 12 per cent visited Sitges for 'a cultural event' reflects the specific interest in cultural tourism that many people had when choosing Sitges as a holiday destination.

When reflecting upon their holiday in Sitges, most tourists (61 per cent) described it as a sun and beach holiday. For the tourists originating from the local area this was the most important category (74 per cent), while among tourists from the rest of Spain and foreigners it was mentioned by 61 per cent and 60 per cent respectively. Only 15 per cent of all tourists described their trip to Sitges as a cultural holiday. Even fewer local people characterized their trip as 'cultural' (3 per cent), compared with 19 per cent of the foreigners.

From these results it can be concluded that, according to the tourists themselves, Sitges is still basically seen as a sun and beach holiday destination.

Cultural tourism experiences in Sitges

Tourists were asked about their visits to cultural attractions such as museums, monuments, art galleries, religious or historic sites, theatres, cinema, pop concerts, traditional festivals and gastronomic events. More than one-third of the sample did not visit any of the 14 cultural attractions mentioned in the questionnaire. Twenty per cent planned to visit one cultural attraction, another 19 per cent planned to visit two cultural attractions and 11 per cent said they would visit three cultural attractions. There seemed to be little enthusiasm for cultural tourism experiences among the respondents.

International tourists were also asked about some specific cultural attractions in and around Sitges, such as the Garraf National Park, *El Parc a Taula*, wineries, the three museums of Sitges, activities linked to the year of Rusiñol, the annual *Fiesta Mayor*, having lunch or dinner with local people, or other sights or events. Figure 8.1 shows the results for this question.

Over a quarter of the total sample had not heard about any of the cultural attractions in Sitges at all. Over half of the sample had heard of between one and five cultural attractions and another 19 per cent indicated being aware of between six and ten of the attractions. Figure 8.1 indicates that even when tourists were aware of cultural attractions, they did not necessarily visit them. The three museums, wineries, the *Fiesta Mayor* and to a lesser extent the Garraf National Park are the most well-known cultural sights among tourists. Even so, the numbers of tourists who visited, for instance, the three museums are low (between a quarter and a fifth of respondents). Tourists were even less familiar with intangible cultural products, such as the gastronomic project *El Parc a Taula*, the year of Rusiñol and having lunch or dinner with locals. 'Having lunch or dinner with locals' is obviously not very well known, due to the fact that it does not exist (yet) as a cultural tourism experience. It seems that the *Fiesta Mayor* is visited by many tourists. This is remarkable because most of the tourists were questioned in the days leading up to the *Fiesta Mayor*. This percentage is probably boosted by repeat visitors who had visited the *Fiesta Mayor* during previous years. Gastronomic

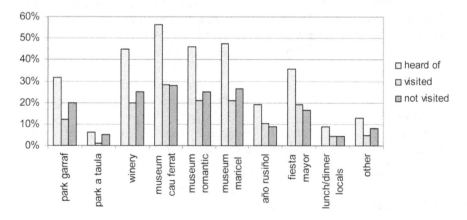

Figure 8.1 Awareness and visiting behaviour of cultural sights in Sitges for those who indicated having heard about at least one of the cultural attractions (N=259).

experiences in the Garraf National Park and activities organized around the year of Rusiñol do not seem to be reaching the tourism market as almost no one had heard about them.

In general, it seems that not too many people are aware of the range of cultural tourism experiences in and around Sitges. More remarkable, however, is that a lot of people who are familiar with them do not visit. We will now look at how tourists evaluate their trip to Sitges.

How do tourists evaluate their Sitges tourism experience?

The fact that more than half of the sample (55 per cent) were repeat visitors to Sitges tells us something about the large number of people that must be highly satisfied with their tourism experience in Sitges. Ninety-three per cent of the tourists coming from the local area (basically Barcelona), 55 per cent of the Spanish tourists and 49 per cent of the foreign tourists indicated that they had been on a holiday in Sitges before.

Figure 8.2 shows to what extent tourists agreed or disagreed with some statements about Sitges as their tourism destination. Again, tourists are extremely positive about Sitges. The average rating of the total sample for their overall satisfaction of their visit to Sitges is 8.5 on a scale from one to ten. Local, Spanish or foreign tourists do not differ in their satisfaction with Sitges.

Most of all, tourists indicated that it was very pleasant being in Sitges (average rating 4.55 on a scale from one to five), that Sitges is a very attractive place (4.49) and that the people in Sitges are fun to be with (4.38). 'My visit to Sitges has stimulated my curiosity' was rated 4.35. Although rated a little lower, 'this place feels very different' still had a score of 4.22 out of five. Tourists tended to agree least with statements such as 'my visit to Sitges has increased my knowledge' (4.07), that 'there are lots of interesting things

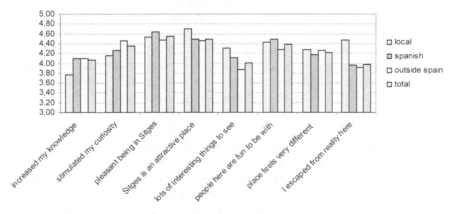

Figure 8.2 Extent to which tourists agreed with statements about Sitges (N=350).

to see in Sitges' (4.0) or that they had 'completely escaped from reality here' (3.97).

These scores seem to indicate that the visitors to Sitges still see it basically as a place for relaxation and fun, and not so strongly related to 'classic' cultural tourism motivations, such as learning new things.

From *sol y playa* to creative tourism?

Current tourism experiences in Sitges

According to the tourists themselves, Sitges is basically seen as a sun and beach holiday destination and relatively few tourists visit cultural sights and attractions. From these results, Sitges cannot be considered a typical cultural tourism destination. Nor would the activities undertaken by tourists identify Sitges as a creative tourism destination where people go to participate in courses or join other learning experiences. Looking at the supply side, there are many cultural tourism experiences on offer. They are not very well known yet, at least not among summer tourists. Generally, creativity is seen as a supply-driven affair during the product development process. There are a number of new 'creative tourism' experiences being developed in and around the town, but to develop Sitges as a creative destination, cultural and natural heritage should be utilized and communicated to visitors in a more engaging way.

An innovative perspective on creative tourism could help Sitges to develop as a creative tourism destination, with both real and virtual tourism experiences. First, a study of the supply and the actual use thereof by different types of tourist could shed light on the current state of creative tourism. Second, the challenge is to explore whether and how tourists could actually be involved in designing and co-creating their own tourism experiences and reflecting upon them afterwards; not just as participative co-creators in the experience setting, but also in the process of designing such settings. As in other businesses where customers co-create innovations, Sitges should seriously consider including knowledgeable tourists in the tourism development process so as to co-create relevant and creative innovative tourism experiences. Consumers are, together with the local entrepreneurs and inhabitants, the carriers of knowledge and experience needed to innovate and develop the products (See Binkhorst 2005 on the 'cocreation tourism experience'). Or, as Nijs (2003: 17) argues, 'People, and particularly their creativity and passion, eventually will appear to be the only possible and most sustainable source for competitive advantage'.

Regarding the gastronomic experiences in Sitges, for instance, a range of more participative and engaging – and therefore more memorable – tourism experiences could be implemented, such as preparing a typical Catalan dish in an authentic local setting, helping with the grape harvest, making wine or staying on the authentic Torres vinyard to live the life of a Torres wine grower. In addition, one can think of an immense range of virtual winery or gourmet experiences that could enhance gastronomic tourism experiences, even before arriving at the destination and again after returning home. The natural heritage of Sitges and its surroundings could also be experienced in a more participative way.

However, from the results of the fieldwork undertaken for this study, it is not yet a well-known or very common experience to undergo during a (summer) stay in Sitges. A more in-depth analysis of the supply side could tell us something about the demand – who the visitors are that actually participate in these experiences and why they participate.

A creative future for cultural tourism in Sitges?

'Sitges, the art of living' is the new slogan to brand the town of Sitges as a tourism destination. It refers to the Mediterranean lifestyle, captures its arty image and even expresses it as a way of living. At the same time, it expresses the magic that surrounds the town of Sitges.

Sitges is a highly popular destination, as the growing number of inhabitants, flows of immigrants, tourists and visitors reflect. Creative tourism in Sitges, however, can basically be found in the first two spheres of 'creative spectacles' and 'creative space', while the more active form of 'creative tourism' is not yet very developed. Although there is an increase in alternatives to the core of *sol y playa*, very few of these aim to have tourists actively co-creating their stay in Sitges, let alone during the preparation of their visit or after having returned home. Rather, tourists are restricted to passive consumption through sightseeing or spectating.

Recommendations for further research

Much research needs to be done here. To begin with, few data are available on the visitor profiles, their motivations, their time-spatial patterns and so on, and the demand for, use of and satisfaction with the traditional tourism supply and recently developed alternatives in Sitges. What strategies could be developed to achieve more creative tourism? It would be interesting to conduct, after a basic and year-round study of the tourist profile, a time-space analysis of tourism behaviour in the town, and to have a closer look at peoples' interest in and evaluation of different experiences in the area. This study and preliminary findings through participant observation show that most tourists are hardly aware of the alternatives off the beach. On the other hand, it is doubtful whether creators of tourism experiences realize that a significant market is now passing by. Often, 'not knowing about' or not being able to understand promotional material due to language barriers keeps people on the beach, or in the shops, bars and restaurants. This leaves the 'other side of the beach' unknown, while Sitges in fact has a rich cultural and natural heritage that provides the town with an indefinite resource for creative tourism experiences.

An interesting question for further research related to the world of experiences and creative tourism would be 'what are the impacts on future tourism behaviour of the various tourism experiences?' The hypothesis is that the higher the level of active participation or even co-creation in the design, undertaking and evaluation of the tourism experience, the more engaged tourists are, the more memorable the tourism experience will be and the more attached people become to the Sitges brand. If this is the case, a coherent portfolio of real and virtual Sitges tourism experiences, targeted at different markets with varying interests, could really brand Sitges as an innovative and creative tourism destination.

Notes

1 The founder of the experience expansion is Walt Disney who, after he made a name for himself with cartoons, opened the theme park Disneyland in 1955 and designed Disney World, which opened in 1971. In Disneyland, guests are engaged in stories told with images, sounds, tastes, smells and sensations, together inducing a unique experience (Pine and Gilmore 1999). More about the cultural construct of the Disney phenomenon can be read in, for instance, *Disney and his Worlds* (Bryman 1995).

2 Pine and Gilmore (1999) also point to the phenomenon of modularization as a means to make mass customization possible. Besides environmental architecture (consisting of a 'design instrument' to couple the needs of consumers to the capacities of producers, and a 'designed interaction' in which the producer constructs an example to help the client decide exactly what experience he or she wants), modular architecture (a supply of modules and a coupling system to dynamically connect the independent modules, cf. 'modular system', Van der Poel 1993) is also required to make modularization possible (Pine and Gilmore 1999: 107).

3 ATLAS is the Association for Tourism and Leisure Education. The Cultural Tourism Research Programme of ATLAS has conducted visitor surveys and studies of cultural tourism policies and suppliers over the past 15 years (see www. tram-research.com/atlas).

4 I would like to thank ESADE/St. Ignasi for supporting this research. Many thanks to Vinyet Gonzalez and Carlos Martín who, despite the demands of their summer jobs, conducted this survey during their scarce free time!

9   Creative Tourism New Zealand

The practical challenges of developing creative tourism

Crispin Raymond

This chapter presents a concrete example of the development of creative tourism and outlines some of the challenges involved in turning creative experiences into saleable products. Over the last few years, the Creative Tourism New Zealand (CTNZ) initiative has been attempting to develop creative tourism into a sustainable business. Here we detail progress to date and draw some tentative conclusions that may help others who wish to establish creative tourism in their own communities.

The creative tourism concept

As outlined in earlier chapters, the concept of 'creative tourism' is relatively recent. As Binkhorst has noted in Chapter 8, creative tourism was originally defined by Richards and Raymond (2000: 18) as a tourism experience which allows tourists to develop their creative potential. In New Zealand, the practical experience of developing creative tourism has led us to re-define the concept in terms of the specific model which has been developed here. In the light of our experience, CTNZ now defines creative tourism as:

> A more sustainable form of tourism that provides an authentic feel for a local culture through informal, hands-on workshops and creative experiences. Workshops take place in small groups at tutors' homes and places of work; they allow visitors to explore their creativity while getting closer to local people.
>
> (CTNZ 2007)

This definition has a number of concrete implications for the way in which we produce creative tourism experiences, and the way in which these are consumed by visitors:

* *Local culture.* We interpret 'local' rather narrowly in the sense that we look for workshop topics that differ in some distinctly New Zealand way from those offered in other countries. At the same time, we interpret 'culture' rather broadly to include topics in food and nature as well as Maori traditions and the arts.

- *Informal* – in contrast to formal learning. Our tutors are knowledgeable enthusiasts who share their experience with participants in a relaxed manner; few are trained teachers. Workshop participants are not offered a certificate, nor do they earn letters after their names!
- *Hands-on*. Our workshops are as interactive as possible because most visitors enjoy themselves more if they 'get their hands dirty'. In practice, some workshop topics lend themselves more to interactivity than others.
- *Small groups*. Our tutors decide their own group size up to a maximum of 12 people. Some like to work on a one-to-one basis; most prefer groups of around half a dozen, partly because they then earn more and partly because both tutors and participants enjoy the greater variety of interaction that then takes place.
- *Tutors' homes and places of work*. These locations add to the informality and authenticity of workshops and are cheaper than renting purpose-built classrooms. Many participants particularly enjoy workshops in tutors' homes because it gives them the opportunity to observe how tutors live.
- *Allow visitors to explore their creativity*. Our workshops offer a taste of a topic rather than in-depth learning: most visitors do not have time for more. That said, visitors sometimes enjoy a workshop so much that they come back and do a repeat the next day.
- *Getting closer to local people*. Although this is never a participant's main motivation for choosing a workshop, when questioned afterwards, participants frequently emphasize that their conversations with the tutor, often ranging well beyond the workshop topic, were a key factor in their enjoyment.
- *More sustainable*. Whether creative tourism is a sustainable activity or not is debatable. However, because it adds to the market for the traditional skills of a community – a market that is sometimes small and may even be diminishing – and because it takes place in existing buildings, creative tourism is more sustainable than many other forms of tourism.

Behind this definition lie a number of practical propositions that can be summarized as follows:

- Many visitors want to learn about the particular culture of the places they visit and this motivation is increasing as a proportion of the various motivations for travel (Richards and Wilson 2006).
- To a greater or lesser extent, all communities develop forms of creativity that distinguish them from their neighbours.
- Therefore, in many places, there is an increasing tourist demand for learning experiences that are based on a destination's distinctive forms of creativity.

- This demand is likely to be greater if visitors are offered a range of learning experiences to choose from, rather than one-off workshop topics.
- This demand is also likely to be greater if workshops are marketed under a name, 'creative tourism', that has the potential to become a generic expression (like, for example, 'adventure tourism').
- This demand can be stimulated more efficiently if these learning experiences are promoted together as a cluster, rather than on a one-off basis.

While each of these relatively simplistic propositions can be disputed, taken together they have underpinned the development of creative tourism in New Zealand so far.

Beginnings

The New Zealand initiative started in May 2002 with the publication of an article in a regional newspaper, outlining the creative tourism concept and arguing that it had potential for the 'Top of the South' region around the city of Nelson (Raymond 2002).

Nelson is on the north coast of the South Island (Figure 9.1) and at the centre of a region with a population approaching 100,000 (of New Zealand's four million inhabitants). The region attracts around 400,000 international overnight visitors a year (of New Zealand's 2.4 million annual international visitors), plus a further 800,000 domestic overnight visitors (Tourism Research Council New Zealand 2006). Nelson prides itself on its arts and creativity, and is home to many New Zealand artists. Nelson seemed a promising place to start creative tourism in New Zealand.

Although the newspaper article attracted little apparent interest it provided some initial credibility and helped to secure the help of Nelson Bays Arts Marketing, a small not-for-profit trust that promotes and develops arts in the region. Its Chief Executive liked the idea of creative tourism because of its potential to provide a new income stream for artists and crafts people. She offered support in the form of contacts, administrative back-up and, always important at the start of any new enterprise, enthusiasm. Arts Marketing's backing made the creative tourism idea realistic.

We decided to seek 20 local tutors, our 'Founder Members', to join an embryonic network that would offer a mix of workshops for a 12 month 'pilot' period. We called the initiative Creative Tourism New Zealand (CTNZ) in anticipation of becoming a national network in due course. Partly through Arts Marketing's existing contacts with artists, partly with the help of Latitude Nelson, the regional tourist office, partly by chance (one eventual tutor made the mistake of asking me what I did while sitting next to me at the hairdressers!), our Members came together quite easily. By January 2003, we had 23 tutors willing to offer 29 different workshop topics.

Figure 9.1 The location of the Nelson region.

These tutors were all Kiwis living in the region. They ranged from Mike, who could trace his Maori ancestry back for 47 generations, to Stephan who had migrated a decade before. Since New Zealand had always been a land of migrants, we decided that any permanent resident with appropriate knowledge could potentially be a tutor.

The topics they offered covered a range of activities that reflected local culture. Participants could learn bone carving or bronze casting, *harakeke* (flax) weaving or seafood cookery. They could propagate native plants or appreciate the local olive oil; take a Maori Journey or make a woolly scarf.

Workshops lasted from four hours (wine appreciation) to four days (bronze casting). Tutors decided the frequency with which they presented their workshops, from as many as six per week to flexible provision to meet demand. Tutors also set their own workshop prices which ranged from NZ$55 to NZ$650 (€27–325).

The main role of CTNZ, working through Arts Marketing, was to promote the concept of creative tourism and generate demand for our members' workshops. The core of our promotional efforts was the website, www.creativetourism.co.nz, and awareness of the site was encouraged through a variety of means. Tourism New Zealand, the government funded agency that promotes New Zealand tourism internationally, was very helpful and offered to encourage visiting journalists to try out our workshops. Maureen Wheeler, the co-founder of *Lonely Planet*, endorsed the creative tourism idea and provided a link to our website from *Lonely Planet*'s. Web-links from New Zealand based tourism-related businesses were developed too.

We imagined that our workshops would appeal most to international visitors wanting to find out more about New Zealand's way of life. Some domestic tourists might also be interested. However, local residents seemed less likely either because they would feel that they already knew the workshop topics or because they would be able to learn the topics more thoroughly, and at a lower price, through workshops offered by local arts councils or as evening classes.

Within the international market, we reasoned that our most probable customers would be backpackers and baby-boomers: the first, younger group, seemed likely to have the time and the interest to want to find out more about how New Zealanders lived; the second, older group, while having less time available, seemed likely to be looking for intellectual challenges while on holiday. We also decided to concentrate on attracting independent travellers and not to approach tour operators until we were confident of our product.

CTNZ was publicly launched, with mini-workshop presentations by some of the tutors, on 1 May 2003. Dame Cheryll Southeran, who set up and ran Wellington's impressive Te Papa museum, did the honours. The local mayors attended. After much work, our website was 'live' and looked great. Energy levels were high. The launch generated considerable media interest and our tutors waited keenly for their phones to ring. Unfortunately they didn't.

Initial disappointments

We had deliberately launched in the New Zealand winter thinking that workshops could be fine-tuned during the low season prior to the pressures of the summer tourist rush which starts at Christmas. In practice, few of the tutors had any contacts during the first few months and things did not improve much during the summer either. By the end of the 12 month pilot in June 2004, total paid attendances for our 29 workshops had been just 23 participants; half the tutors had had no bookings at all.

What had gone wrong and what should we do next?

Some mistakes had become obvious during the year. Many of the workshop descriptions on our website were over-complicated. There were not enough visitors staying in the region for long enough to do our longer workshops and it was these that had attracted least interest. The dates on which workshops were available changed from week to week depending on tutors' availability. Generally, our product was not as easy to understand as it should have been.

Making bookings was also not straightforward. Tutors were working on other projects and did not always answer their phones. Some struggled to reply to their emails. We were not professional enough.

Our meagre finances during the pilot had also restricted us. Each tutor had paid a modest fee of NZ$100 (€50) for a year's membership. CTNZ had also charged NZ$50 (€25) to accommodation providers, who were recommended by Members and given a web-link from the CTNZ website. Other income had come from grants raised by Arts Marketing on the basis that it was acting as an incubator for an initiative that could benefit the arts. Two sponsors, the local technology college and the local airline, had also contributed. However, total income for the year had been only NZ$18,000 (€9,000), most of which had been spent on website development, and leaflet printing and distribution. There had been no paid advertising and no one involved had earned anything.

Perhaps most of all, we realized that we were still a largely unseen pinpoint of light in the vast universe that makes up the global tourism industry today. Our promotional efforts may have been imaginative and visible to us but they had been limited by our tiny budget and had made limited impact outside the Nelson region and negligible impact internationally.

Encouraging signs

However, in spite of the disappointments, there had been positive developments during the pilot too.

First, the Intellectual Property Office of New Zealand had allowed us to trade-mark the expression 'creative tourism', thereby underpinning our efforts to establish the meaning of the term and giving us ownership of an asset that could eventually have value.

Second, it had become clear that the concept of creative tourism appealed to journalists and travel writers. It was both sufficiently new and sufficiently simple to attract attention and media coverage. This seemed likely to continue and, in time, lead to CTNZ becoming better known.

Third, half way through the pilot, Tourism New Zealand had announced that it was shifting its marketing focus to attracting a target group it called 'interactive travellers'. It defined these people as, 'regular international travellers who consume a wide range of tourism products and services, seek out new experiences that involve engagement and interaction, and demonstrate respect for natural, social and cultural environments' (Tourism New Zealand 2003). Interactive travellers could be of any age and from any income bracket. It seemed probable that our workshops would be a perfect product for interactive travellers and that CTNZ would therefore benefit from Tourism New Zealand's new marketing direction.

Fourth, we had entered a competition being run in Auckland to stimulate new creative ideas and were through to the group of ten finalists out of more than 400 entries. There were two prizes of NZ$50,000 (€25,000) available and we felt we had a good chance of success.

And, fifth, we had been approached by a talented local person with a strong marketing background in New Zealand who was keen to be involved in CTNZ's development.

We concluded that a year's pilot was not a fair test of creative tourism's potential in New Zealand and that we should persevere if we possibly could. And so we adjusted our expectations and took the next plunge.

Becoming a company

We decided to continue CTNZ and to set it up as a separate organization. We debated whether to become a not-for-profit trust or a company and chose the latter for three main reasons. First, there seemed little prospect of obtaining grants even if CTNZ was a trust. Although Arts Marketing had been able to attract some grants during the pilot, we did not think this would continue once we were an independent entity: tourism in New Zealand is regarded entirely as a commercial activity. Second, by becoming a company we could give a shareholding to our new Chief Executive, partly in lieu of salary. And, third, it was simpler to start a company than go through the

Figure 9.2 John Fraser, 'the Bone Dude', teaches the art of bone carving at his workshop in Christchurch, New Zealand.

Figure 9.3 A small group learns to weave baskets from *harakeke* (New Zealand native flax) in the Maori tradition with Arohanui Ropata at Ngatimoti, New Zealand.

procedures of establishing trust status and finding a suitable Board of Trustees. Accordingly, CTNZ Ltd began trading on 1 June 2004 with a capital of NZ$50,000 (€25,000), injected by the shareholders in anticipation of a prize in Auckland.

We reduced the number of workshop topics on offer in the Nelson region to 15 and retained only workshops that lasted for a day or less. Each workshop had to be available on a certain day (or days) each week, as many or as few as the tutor wished, but could not change day(s) from week to week. We took over the bookings ourselves and introduced a free-phone number on the website. We produced a more professional brochure and paid for it to be displayed in a number of key visitor locations. We paid for a few carefully targeted adverts in journals likely to be read by backpackers. We increased our charge to tutors to NZ$500 (€250) a year, but were now offering them much more for their money. Energy levels were once again high.

Although the company got off to a disappointing start when we learnt that we had been runner-up in the Auckland competition and had just missed out on 'our' NZ$50,000 prize, having a full-time Chief Executive made a big difference to what got done and CTNZ now had a much stronger sense of purpose. Taking bookings ourselves also gave us a closer feel for our market. Feedback from workshop participants was invariably excellent and journalists, usually paid for by Tourism New Zealand, continued to do workshops and write about, and televise, them internationally.

Nevertheless, as the year progressed, it became obvious that creative tourism was still not ready to take off in Nelson. Bookings had increased, but after the first 12 months of trading as a company, were still less than 200 in total – impressive if expressed as a percentage of the bookings made during the pilot perhaps, but still a very long way from providing a sufficient income for tutors to justify their NZ$500 (€250) fee to us. Our tutors were losing faith in CTNZ and our capital was almost used up. We could no longer afford to pay our Chief Executive. Once again, we faced the hard decision: should we try to continue the creative tourism initiative and, if so, how?

New focus

Whether through admirable persistence or foolish obstinacy we decided to continue, but to take a rather different approach. Up to that point (June 2005) we had viewed workshop participants as our primary customers and, by and large, had looked for tutors to provide workshop topics that we thought would be attractive to these customers. We now decided to shift our focus and see our tutors as our primary customers and to concentrate on providing them with a value-for-money promotional service for their workshops. Of course, we would still need to attract more participants to workshops if we were to retain our tutor-customers but, provided we kept our charges to tutors at a modest level, we could at least offer them a way of promoting their workshops that would cost them less than trying to do this by themselves. This approach would allow CTNZ to continue, albeit in a modest way, and, even if there was no immediate prospect of making money, at least the losses could be contained at a level acceptable to the shareholders.

For the next 12 months, we decided not to charge our loyal Nelson tutors anything and to charge new tutors only NZ$250 (€125) a year. We started to seek tutors in other parts of New Zealand and, with the help of the local arts centre, attracted a cluster around Christchurch. A new brochure was produced with 23 Nelson and Christchurch workshops and displayed in the i-sites (information centres) and key accommodation providers in Nelson and Christchurch. The CTNZ website was extended to include these new workshops. Further web-links were steadily added and Michael Palin, famous for his many roles in the *Monty Python* series and now the writer and presenter of a range of international travel programmes, added his support.

The shareholders were only prepared to put in a further NZ$5,000 (€2,500) for the year so our annual budget was lower than at any point since the pilot began. All work was now being done on a voluntary basis by the shareholders. We no longer took bookings centrally ourselves but encouraged tutors to make use of the booking services at their local i-sites. We also offered tutors a handbook we had written entitled, *Making the Most of your Workshops* which drew on what we had learnt to date and advised them about some of the practical issues in running workshops.

By June 2006, although our tutors had had no major upward shift in their bookings, enough of them were hoping that CTNZ continued operating to suggest that we had found a semi-sustainable existence – for as long as the shareholders were willing to cover the modest annual losses and work without pay. Furthermore, the main guidebooks to New Zealand, which are mostly updated biannually, seemed likely to include mention of CTNZ in their next editions due out prior to Christmas 2006.

The current situation

CTNZ's next aim is to sign up tutors throughout New Zealand so that our website becomes THE place where visitors can find out about short New Zealand workshops based on local culture. We are going to charge tutors a modest NZ$120 (€60) for their first year and NZ$100 (€50) thereafter. We believe that if we can keep going for the next few years and expand in this way, we will become more visible and start to attract more business for our tutors. We should then be able to justify a price rise to them and eventually become financially sustainable.

Underpinning this determination to continue is our belief that creative tourism is a good idea with potential, both in New Zealand and elsewhere, to make a worthwhile contribution to tourism development. We are also absorbed – though frequently frustrated – by the challenge of trying to find a way to develop creative tourism sustainably.

Implications

So what has been learnt about our attempts to develop creative tourism in New Zealand that may be useful to others? Given the cautionary tale above, all suggestions should be treated with the utmost care! Nevertheless, here are several issues that are worth considering carefully when developing creative tourism elsewhere.

Language

Early on we decided that we needed a vocabulary to describe the three key elements of what creative tourism was about; we settled on the expressions, 'tutors', 'workshops' and 'participants'. However, none of these words is ideal and to different ears, and in different English-speaking countries, can sound too formal or even ambiguous. They are the best that we have found and have only been chosen after rejecting others that seemed less appropriate. *Are there better alternatives?*

Workshop topics

We have had considerable discussion on what workshop topics should be allowed within our definition of creative tourism. Since our New Zealand culture is derived from over 700 years of Maori traditions and just 200 years of European traditions, the former offer plenty of possible topics that are distinctly New Zealand, but the latter are often difficult to differentiate from current European traditions.

So, for example, should a workshop on rugby be included: a game that is at the heart of the New Zealand way of life today but which originated in England and is played internationally? Our compromise is to have a workshop that teaches the *haka*, the traditional challenge that the All Blacks (New Zealand's rugby team), and only the All Blacks, perform at the start of each international rugby game.

Generally we ask three questions of a potential workshop topic: 'Is this topic part of New Zealand's way of life?'; 'Can it be taught in a way that is distinctly New Zealand?'; and 'Will visitors be interested?' *Is this the best way to decide what workshop topics to include?*

Supply-led

CTNZ is largely a supply-driven initiative. Our starting point has been our tutors' workshops for which we try to find a market. We are thus like much of the arts: seeking a market for a given product rather than trying to match our product to an established demand. From a business perspective, this is not ideal. For example, we have not carried out market research (other than with participants after they had completed a workshop). *Are there ways in which creative tourism can piggy-back onto existing demand and thus improve its market focus?*

Demand

While our initial assumption that workshop participants were most likely to be backpackers and baby-boomers has been largely confirmed in practice, these two market segments differ in an important way. The former are usually in New Zealand for an extended period but are being careful with their money: time-rich but cash-poor. The latter are here for shorter periods but have more to spend: cash-rich but time-poor. Although both groups can be included within Tourism New Zealand's definition of the interactive traveller, trying to appeal to both segments through one set of marketing initiatives may be flawed. We are particularly concerned about the implications of promoting workshops with widely differing price tags alongside each other. *Does this matter?*

New Zealand

We believe that creative tourism will eventually succeed in New Zealand for three main reasons: there is a growing interest in cultural tourism everywhere (Richards 2007); New Zealand has less historic infrastructure around which to base cultural tourism experiences than many countries, making creative tourism relatively more attractive; and Tourism New Zealand's marketing focus on the interactive traveller helps us.

However, it is also possible to argue the opposite, namely that New Zealand will always be a tourist destination that appeals primarily to those looking for beautiful landscape and outdoor experiences: cultural tourism in any form will remain a peripheral activity. *Will creative tourism become more successful in other countries?*

Finances and business models

CTNZ has been under-capitalized from the start and we have always worked within severe financial constraints. On the one hand, this has been reflected in our inability to pay staff consistently and to promote workshops aggressively. But on the other hand, and perhaps more importantly, it has also been reflected in the sort of 'business models' we have pursued. In particular, we have always asked our tutors to pay us something.

If more substantial investment had been available a different approach could have been taken, with CTNZ paying its tutors to teach the workshop topics we wanted to offer. Our tutors would then have been our suppliers rather than our customers. This approach would have allowed us to develop a 'commissionable product', a fundamental step for involving tour operators. As it is, tour operators, while often interested in the creative tourism idea, are not willing to sell our workshops because they want to deal with an established company offering an established product from which they can earn a commission. *Are there more entrepreneurial ways to develop creative tourism?*

Commercial company or trust?

While we have adopted a commercial structure for developing creative tourism for the reasons explained above, this is not necessarily the best approach for all countries. The development of creative tourism stands to benefit a community by showcasing its creativity and creating employment. In countries where there is a stronger tradition of providing public funding for such purposes, a not-for-profit structure might therefore be more appropriate. *Which structure is the most suitable for developing creative tourism?*

Conclusion

Creative tourism is not new. People have always wanted to learn on holiday and have been doing workshops at least since the nineteenth century when the English gentry went on a 'Grand Tour' to learn about Renaissance art in Italy and have painting lessons there. Today, workshops are on offer to tourists all over the world. But CTNZ may be the first organization that seeks to promote a range of hands-on creative activities that reflect a country's culture.

Although we have trade-marked the term in New Zealand, we do not see 'creative tourism' as a brand to be rolled out in different places like another Guggenheim museum. Rather we hope that it will become a recognized term that different communities can use and benefit from. Just as we are helping the promotional efforts of New Zealand tutors go further by developing a tutor network under the creative tourism name, so the network will become even stronger if creative tourism is developed in other countries too.

Creative Tourism New Zealand is always happy to share what it has learnt with others who also believe in the potential for creative tourism in their communities.

Part 3

Consuming lifestyles

10 Student communities as creative landscapes

Evidence from Venice

Antonio Paolo Russo and Albert Arias Sans

Introduction

In the race to restructure and market themselves for the consumer class
(Eisinger 2000; Zukin 1995), many cities run the risk of depleting their original character. A trend towards the standardization of cityscapes has been
termed 'McDonaldization' (Ritzer 1996) and 'urbanalization' (Muñoz 2006),
implying that beyond the corporate-driven reconstruction of city images,
there is also a (dangerous) simplification in functions and habitats which
may make cities less resilient to global changes and new cultural stances.
Paraphrasing Martinotti (1993), the 'fourth generation metropolis' today
has less and less space for differentiation, in spite of the driving force of the
diversity discourse which is pushing urban change, and of the rhetoric of the
knowledge-intensive urban society in the last decade.

This could be seen as the natural outcome of the partnership model of
regeneration that brings together cash-strapped city governments and
private developers who favour an 'industrial model' of redevelopment-by-
spectacularization, consisting of luxury housing, multifunctional leisure
infrastructure, iconic cultural landmarks and 'pacified' public space under
private control (Zukin 1995). This process is fuelled by the emulation of a
few successful cases (e.g. Barcelona, Rotterdam, Berlin, Lille), and often
has little regard for local idiosyncrasies or long-term impacts.

Cities have therefore become increasingly similar in terms of their physical landscapes, their atmosphere, their lifestyles, their imaging and their
soundscapes. Richards and Wilson (2006) talk about 'serial reproduction of
culture'; their argument could be extended to identify a *serial reproduction
of landscapes* in the postmodern city. At the same time, however, there are
persisting conflicts of identity, for instance as far as class or gender spaces
are concerned (Arias Sans 2003; Boyle and Hughes 1991; Eisinger 2000;
Griffiths 1993; Marrero Guillamón 2005).

Tourism development strategies are deeply embedded in such processes
of urban change, to the point that it could be argued that there is confusion
or even convergence in the visitors' and residents' gaze on the urban

landscape. A paradox persists, which may be typical of cyclical tourism development models (Butler 2006). As cities become more spectacular and better equipped to attract global travellers and short-break visitors, more culturally aware and experienced visitors may turn away from them in search of more original experiences, driving a trend towards diversification in the tourist market. At a finer scale, tourism-ridden neighbourhoods are abandoned by 'pioneers' in search of more appealing areas of the city where they can establish a more vivid empathy with local communities. The impact of trendsetters in leading the way to tourist restructuring is illustrated by Maitland's analysis of the success of 'unplanned' new tourist areas which, to some extent, occur at the expense of the tourist performance of more mature tourist clusters (Chapter 5 of this volume). Although many cities waste their competitive advantage because of excessive reliance on infrastructure development and the creation of artificial tourist landscapes, the 'migratory' aspect of tourist development is particularly interesting, because capital investments, space and images are continuously 'burned out' and have to be replaced.

In this problematic context, distinct groups keep the urban experience alive: collectives recycling obsolete infrastructure for creative consumption (exemplary cases in Amsterdam, Helsinki and Milan are quoted in Bonink and Hitters 2001; Cantell 2005; Lavanga 2004); street artists introducing a conflictive discourse in the 'pacified' corporate-driven restructuring of public spaces (as in the case of Plaça dels Angels, Barcelona, where the space in front of Meier's MACBA building has become the permanent playscape of skaters – European Travel Commission/World Tourism Organization 2005; Project for Public Spaces Association 2003); and higher education communities (academics, researchers, students) generating spaces of conviviality and creative expression around their habitats.

Students are the focus of the present chapter. Following a theoretical consideration of landscapes of cultural consumption, we highlight the increasing importance of student life in the tourist competitiveness of cities, through the generation of a creative landscape (through skilled consumption, flexible production and lifestyles 'encoding' city images: Jannson 2003) embedded in new centralities and new images of the city. This represents an appealing alternative for sophisticated visitors who want to escape clichéd city images and serialized cultural consumption. The case study of Venice as a student city is then used to illustrate this process and to identify the critical points which may hamper it and possibly turn it into a further element of 'standardization' of urban landscapes. In particular, we focus on the weak position of students in urban property regimes, with the risk that student communities become 'eternal migrants' in the city, producing an 'unstable' pattern in the creative landscape. In the concluding section we identify some basic principles for policy when dealing with the 'touristification' of creative student habitats, proposing urban management actions to balance the ambitions of different stakeholders, including the local population, city developers, student communities and visitors.

The chapter is based on qualitative research conducted by the European Institute for Comparative Urban Research (EURICUR) in a number of European 'student cities' (Van den Berg and Russo 2004).

Cultural consumption landscapes and urban tourism

The structural economic dynamics of globalization (Bell 1973; Harvey 1989; Knox and Taylor 1995) and the political and monetary unification of the European Union (EU), which facilitated the circulation of capital, goods and people, are defining new roles for cities. Cities are becoming global nodes of decision-making and financial transactions (Sassen 1991), value-adding producers of information, symbols and experiences engaged in global processes of exchange (Castells 1989; Lash and Urry 1994; Zukin 1995) and poles of human, social and cultural capital (Logan and Molotch 1987).

In European cities, such multiplication of roles has clear repercussions on the urban landscape. Global and local processes are mixed and hybridized in the urban space, configuring new scenarios of interaction for a plethora of stakeholders moved by different interests and behavioural paradigms. New global functions and new actors are juxtaposed with the local and vernacular urban structure, generating new conflicts and new opportunities. The emergence of such new 'glocal' landscapes can be understood as the result of 'an ensemble of material and social practices and their symbolical representation' (Zukin 1991: 16). These are changing the 'original' character of places; re-mapping their centrality and accessibility; changing their functions, uses and meanings. This process produces a diversification of the patterns of 'living' the city, associated with the appearance of new metropolitan city users (Martinotti 1993) and 'transient' populations.

This chapter focuses on a specific type of 'glocal' landscape, the *cultural consumption landscape*, reflecting a renewed interest in culture as an explicit part of urban regeneration strategies, which aims to increase 'the city's symbolic capital and to catalyze other unsubsidized commercial activities' (Storm 2002: 6). Cultural consumption landscapes are configured by new urban consumption spaces, embodying emerging postmodern patterns of leisure, travel and culture for the pursuit of cultural capital for the residents, users and tourists. They are strictly related to both the commodification and aesthetization of everyday life, and with the global homogenization in patterns of consumption (Featherstone 1991; O'Connor and Wynne 1996). The reflexivity between producer and consumer is also a key factor in understanding the cultural consumption landscape, given that 'urban lifestyles are not only the result, but also the raw materials of the symbolic economy's growth' (Zukin 1995: 826). Such landscapes are based on the interrelated global and local production and consumption of commodified

cultural goods and services, set up and regulated either by public administrations or by market forces through selective consumption of time and
space (Zukin 1991). Cultural consumption landscapes are thus the material
product and the symbolic representation of the dominance of cultural
industry outputs in urban economies (Scott 2000), embedded in specific
places of the city where they are available for consumption (Urry 1995).

In our view, cultural consumption landscapes are defined by two dimensions: a hardware–software dimension and a formal–informal dimension. In
the first dimension, *hardware* is understood as the physical infrastructure,
such as buildings, transport networks and public space. *Software* relates to
the non-fixed supply of intangible experiences such as events, activities,
narratives and anything else related to the symbolic representation of place.
In the second dimension, the *formal* is understood as the institutional supply
of culture and folklore, while the *informal* is conceived as individual or
collective non-institutionalized expressions of creativity: artistic avant-
gardes, lifestyles and everyday practices, such as cosmopolitanism, multiculturalism and sexuality. In spite of the fact that in some cases these
dimensions might be blurred or overlap, they can help us to understand the
configuration of *liminal* landscapes as an integral object of consumption
which 'situates buyers and sellers in a brief, socially recognized transition or
"transaction"' (Zukin 1991: 28).

In recent decades, tourism has emerged as one of the most important
sectors of European urban economies (Begg 2002; Van den Berg *et al.*
1995). The democratization of tourism as a social practice caused by the
abatement of economic, social and geopolitical barriers, and the institution-
alization and consolidation of the tourist industry, are keys to understanding
the changing patterns of demand which have induced a rise in visitors to
European cities. Cities compete to attract visitors in order to generate
economic returns, and make significant financial efforts to achieve this
objective (Eisinger 2000).

Urban tourism is based on culture and its materialization in the landscape. Involved in a competitive struggle to attract visitors, cities are diver-
sifying their tourism products as far as possible, looking for differentiation
and exclusivity (Harvey 2002). But reality differs from expectations. The
emergence of global social practices and lifestyles (Beck 1992; O'Connor
and Wynne 1996; Verdú 2003); the increasing influence of global brands on
the image of the city (Klein 1999); and the ease of using simulacra as a local
collective representation, simplifying complexity and reproducing clichés as
themes (Amendola 1997; Hannigan 1998) has created a 'multiplicity of standardized attractions that reduce the uniqueness of urban identities even
while claims of uniqueness grow more intense' (Zukin 1998: 837). As Muñoz
(2006) points out, the use of the same strategies and patterns for the creation
of global sites produces common urban landscapes, standard habitats which
produce standardized behaviour due to converging organizational and
consumption patterns (Ritzer 1996), although conserving local idiosyncra-

sies through the commodification of the 'socially embedded aesthetics' (Degen 2003) such as localism, multiculturalism, cosmopolitanism, sexuality or creativity.

Yet in the new globalized cultural landscape tourists are also changing their *gaze*, looking for a more 'authentic' and 'original' experience that classic and formal supply cannot offer. In their quest to move away from the postmodern simulacra and the branded and guided experience, informality, exclusivity, diversity and creativity have been converted into the flagships of the new experienced tourist. However, this places a strong emphasis on the quest for new producers who can develop creative consumption landscapes. The very concept of the tourist gaze, which according to Urry (1990) reflects negatively the 'everyday life' of the gazers, loses significance when there is identification between visitors and the visited society. Such dynamics are not exempt from negative spin-offs associated with tourist thematization or the *touristification* of public spaces, the inflation of prices, and the specialization of tourism labour and infrastructure. Yet tourist revenues remain a priority for public administrations even in the light of such dangers, and the tourism industry is now a key aspect of the urban landscape. 'Building the city for the visitor class' (Eisinger 2000) is not substantially different, today, to building it for the global community.

Student life and creative landscapes

Among the emerging producers of the 'glocal' landscape in European cities, students are acquiring increased importance, as a result of various factors.

First, increased student mobility in European Union (EU)-funded exchange programmes. More than 2,200 higher education institutions in 31 countries are currently participating in the largest student mobility funding scheme of the EU: the Erasmus programme. In 2004/5, 144,000 students spent short study periods in foreign host universities; there have been 1.2 million in total since the creation of Erasmus in 1987. The Erasmus budget for the year 2004 is more than €187.5 million. Figure 10.1 illustrates the steady growth of student mobility in the last two decades.

Among long-term foreign students enrolled in European universities, three European countries educate a considerable number of foreign students: France, Germany and the United Kingdom, in order of the number of foreign students enrolled. Foreign students represented between 5 and 7 per cent of the total enrolments in higher education in these three countries in 1990–91, and this percentage has been growing ever since. More recent data from the OECD (2005) indicate that there are now over 2 million foreign students enrolled in universities in OECD member states, and that the numbers are rising at over 11 per cent a year.

Second, there is recognition of the failure of the 'campus' model of higher education settlement in some countries (Various authors 2001). In recent decades, public administrations preferred to isolate and concentrate the

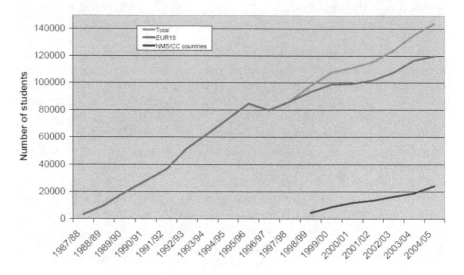

Figure 10.1 Student mobility in Europe. Student participation in Erasmus
programme, 1978–2005.

Source: http://ec.europa.eu/education/programmes/socrates/erasmus

university community, instead of mixing it with the other populations of the
city and especially the residents. This strategy was directed at achieving the
minimum stress for the local community – low land prices, cost-saving
concentration of the university facilities, capability to host residences for
the students – and exercising, at the same time, social control. This eventually led to the relocation of historical universities to suburban locations
(Hall 1997: 301 quotes several examples), with mixed results (Vassal 1987).[1]
However, in recent years, we have seen a renewed interest by the public
administration in bringing back the university and the students to the city
centre as part of integral programmes of urban regeneration based on
cultural institutions.

Third, the increased importance of medium and large cities as higher
education locations. Large cities tend not only to attract larger student numbers, but are also more attractive as destinations for international students
(see student testimonies on http://ec.europa.eu/education/programmes/
socrates/erasmus). According to district theorists such as Jane Jacobs, the
generalist and wide-ranging knowledge generated in state-subsidized higher
education institutes has a wider impact on local growth than specialized
knowledge produced in private-sector R&D departments. In turn, an innovative and dynamic working environment is supposed to stimulate those
urban amenities that generate further attractiveness, in a 'virtuous cycle' of
knowledge-driven development. In particular, educated young people are
presently seen as the backbone of the 'creative city' (Chatterton 1999;

Florida 2002) where lifestyles, social networks and information assets blend in a unique environment supporting the growth of competitive and sustainable economies (Youl Lee *et al.* 2004). This notion is today very popular with public managers and economic stakeholders, triggering a real 'global competition for talent' (Florida 2005) between cities, and pushing national and local government to adopt a strategic approach in higher education policy based on strengthening urban economies.

On account of these factors, higher education students can currently be considered an element of the 'critical infrastructure' (Zukin, 1991) required for the development of creative clusters, the re-signification of the city landscape and the production of creative landscapes. Creative landscapes generated by the presence of students can also be analyzed with reference to their hardware–software and formal–informal dimensions. On the one hand, the *formal hardware* corresponds to the fixed assets of student life: university infrastructure, faculties, libraries, residences, campuses and so on. The *formal software* concerns academic life: didactic activity beyond university walls but also events, congresses and festivals promoted by the university. On the other hand, the *informal hardware* identifies the space (public and private) used by students such as: open university premises and public spaces, bars, cinemas and other leisure or cultural infrastructure. Finally, the *informal software* defines the patterns of student activity, lifestyle and behaviour, such as university gig circuits (Figure 10.2).

Students play a triple role in the configuration of the creative landscape of a city. First, students are among the *producers* of such landscapes, not only through their physical presence, but also in terms of the direct relationships which university institutions (formal hardware and formal software)

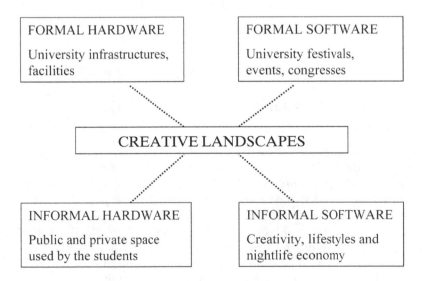

Figure 10.2 Creative landscapes and student life.

build with the city. Second, students are the *mediators* of the landscape through the interaction with the local community, generating new centralities through their use of informal software (public and private space), and new cultural codes in the use they make of that space. Thus, their social habits, their geographical patterns and their symbolic values are a major element behind the development of creative clusters. Third, students are one of the main *consumers* of such landscapes.

The re-centralization of academic and student life is a sign of the widespread acceptance of the central argument of the 'knowledge society', that is that knowledge-related functions and the experience economy generate clear benefits for the inner cities. High-quality cultural institutions, combined with vibrant student lifestyles, supposedly enhance city image and attract global capital, skilled labour and visitors. Public investment in universities also fosters the development of non-subsidized clusters of culture-related production. As Storm (2002: 7) points out, 'cultural institutions are drawn by their own economic needs and by the imperatives of their funding sources to seek broader audiences and exploit more commercial, income-generating strategies'.

Creative landscapes generated by student communities also represent a key asset for tourist development in an era in which Feifer's (1985) 'post-tourist' seeks playful interaction with the local social fabric rather than idyllic, shallow representations of place.

Creative landscapes, however, are not independent of other *glocal* dynamics. An active role for tourists in such landscapes means that visitors and students will come to co-determine spaces of creativity. The result of this encounter is, however, unpredictable. The interrelation between cultural and economic investments, the commodification of experiences or the *brandification* of place (Fainstein 2005), may ultimately convert the creative landscape into a place-based consumption object for the city and its visitors.

The following case study sheds more light on the dynamics of student-generated creative landscapes, their impacts and long-term perspectives.

Student activity as an antidote to mass tourism in Venice

Venice, a 'star destination' in peril

The capital of the Veneto Region in the north-east of Italy, Venice is probably the city in the world whose image is most associated with tourism and leisure: a 'romantic', immutable 'city of stones', which most visitors are surprised to discover is a living community and not a theme park. In spite of this image – or rather *because* of it – Venice is struggling to catch up with a booming regional economy. Plagued by lack of accessibility, environmental problems, inflexible physical structure, lack of space and an economy geared to an inelastic and potentially infinite tourist demand, Venice has been losing population, economic activities and institutional capacity at a steady pace in the last 30 years (Musu 2000).

At the same time, the tourist activity of Venice is growing at an almost incontrollable pace. The 'pivot' of a very large tourist region spanning the whole Italian north-east and beyond (Russo 2002), Venice is succumbing to its very popularity. On the one hand, under the pressure of a not very innovative tourism industry, it is becoming impossible to reproduce the cultural dynamism which has characterized it historically: not long ago Venice and its Biennale were the heart and pulse of Italian visual and performing arts, while today creative talents and innovative organizations flee to dynamic cities like Milan or Bologna. On the other hand, increasing tourist demand in such a delicate environment inevitably produces damage. Tourism not only affects the atmospheric and physical elements which make Venice beautiful, but it also leads to declining quality of tourist products. The market is dominated by excursionists (85 per cent of visitors do not spend the night in the city), and only 3 per cent of the tourist expenditure is on culture or leisure (Manente and Andreatta 1998). As a consequence, Caserta and Russo (2002) warn that the very survival of Venice's heritage is at stake, as the city is incapable of generating the resources required for its maintenance and preservation.

Venice has therefore been striving for at least two decades to identify alternative models of development and at the same time fight the equation of tourism with declining urban quality. The mantra is that Venice should valorize its intangible cultural strengths better; preserving not only the physical heritage but also the atmosphere of a city poised between land and sea, and located at the crossroads between different European and Eastern civilizations. Venice is supposed to break the mould of low-value heritage tourism and regain its status as a creative place, where global flows and local resources coalesce to produce new cultural forms. Venice can indeed be considered a global place, and not only for its international tourist dimension, but also for its global reputation and lengthy historic past. In recent decades, however, this global dimension has exclusively focused on tourism; and what is worse, mass tourism has sacrificed global symbolic values on the altar of the vernacular – fast food and fast culture – rather than using its international reputation to generate quality.

Attempts to redress this situation have been modest, with the creation of a 'district for innovation' which has started to stimulate creative firms and advanced services in the old city. However, the pressure from the mass tourist economy keeps hindering a virtuous relationship between the cultural assets of the city and its economic development.

Local universities as urban development agents

Venice has four universities – Ca' Foscari; University Iuav of Venice; The Venice International University and The Academy of Arts – but they have long been among the city's most underrated cultural assets. Most university facilities are located in the old city, and together they host some 25,000 students in a historic core with less than 300,000 inhabitants. Student numbers grew continuously in the period 1950–90, and higher education has effectively been the only growth 'industry' in the old city. However, the expansion of the universities was long resisted by a city hungry for affordable housing and adequate social services. In the absence of any active policy of university development, empty buildings were converted piecemeal to academic use. These were often unsafe and impractical, and spatial dispersion led to inefficient use and student discomfort.

Thus the universities were seen by many Venetians as 'colonizing' the city in the same way as tourism, substituting resident activities with imported, 'ephemeral' ones. The municipality has supported this view, treating higher education as a problem rather than a potential solution. Indeed, the problem was partly misunderstood: Venice was – and still is – losing population and economic activities not because of external 'colonization', but because of the inaccessible, fragile and expensive nature of the city. Venice has missed opportunities to repopulate because it lacks valuable, postindustrial activities which can compete with the high-volume, low-value tourism industry.

Spatial patterns of academic life: students as urban regenerators

Traditionally, few university students lived in the old city. Most students commuted from their home towns in the surrounding region rather than settling in expensive flats in the city. Moreover, Venice did not have a reputation as a lively place, but rather as a city full of elderly people with very little nightlife and no social dynamism. Finally, the Italian housing regulations made it difficult to rent flats on a short-term basis, and in the old city there were no public student residences. All this changed during the 1990s due to the expansion of the universities and the growth in international and extra-regional students looking for housing in the old city. In the same period, house renting rules were relaxed, putting more housing on the market. As a result, Venice experienced for the first time a steady rise in students temporarily residing there, and the first appearance of a 'student life' developing autonomously from the local society, with a hint of cosmopolitanism, something quite new even for Venetians dealing with 12 million visitors every year. A 1994 survey estimated that 7,300 students lived in the centre of Venice, 6,000 of whom rented from the private sector. Ten years later this figure had almost doubled.

In the meantime, the local universities have also embraced the idea of higher education and research as engines of urban renewal. An option to use the old Arsenal to host the whole higher education sector was abandoned; the city favouring complementarity with its inland settlements, reconcentrating most university functions on a triangle: two large poles at the western side of the old city (using refurbished industrial and port premises) and a pole for scientific faculties in the industrial areas on the other side of the bridge over the lagoon, where the city's science park is also located (Figure 10.3).

This plan also rationalizes the presence of students in the city, with the development of additional residence facilities. The pole of St. Margherita–St. Marta, which accommodates the largest share of the university facilities, can be seen as the 'university area' of the city, and reaches sufficient mass in terms of student population to become a sort of laboratory for the 'new Venice'. Student settlements are now more 'sustainable' but still diffuse, so that many parts of the city experience the new vibrancy. Simultaneously, a division is evident between the Venice of tourists, a reconstructed theme-park environment that closes down at night, and the Venice for students; a dynamic social structure which integrates more fully with the cultural and physical texture of the city.

Students recreate the 'libertine' flavour of the historic city and at the same time they become involved in its rebirth because they sustain a cycle of production and consumption of creative activities (music, theatre, video-art), places (cafés, galleries, street-life, bookshops) and events (carnivals, film festivals, revamped traditional fairs, markets) which serve much more than the student community alone. For the first time in a long while, Venice is now 'cool' and highly attractive for a group of young urban consumers.

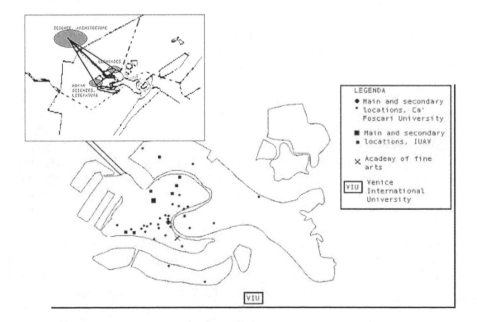

Figure 10.3 The location of university facilities in the historical centre of Venice (larger map) and the reorganization plan for the University in two central poles and one inland pole (smaller map).

The Venetian students as creative cultural landscapers

Student life is a major novelty in the sterile tourism landscape of the last decade, offering an inviting alternative to the visitors who seek to escape the 'gondola cliché'. These are a minority, but still economically very important. In visitor surveys conducted in 1997, Di Maria *et al.* (2004) identified a segment of well-informed tourists, accounting for approximately 10 per cent of the visitor population, predominantly young couples and singles, mainly spending the night in city centre accommodation, in budget hotels, bed and breakfast or with friends and relatives, rather than in expensive hotels. This group attaches great importance to cultural events and activities, restaurants and leisure (and less to monuments and landscapes). They also have the longest average duration of stay and the highest willingness to pay for cultural visits.

While only 15 per cent of all visitors to Venice stay overnight, these 'empathic tourists' see the night-time environment as part of the essential experience of the city. The time of the day when souvenir shops close, tourist groups disappear into their anonymous motorway hotels and the local community devotes itself to leisure, is to them a creative landscape: the city's cultural qualities can only be discovered in these circumstances. Direct observation and interviews with key city managers and operators in the tourist areas confirmed that presently there is a very high participation of foreigners in student bars, restaurants and events. The cultural supply associated with student consumption attracts curious visitors. One such institution is *Teatro Fondamente Nuove*, a small venue in an area of Venice where day visitors barely penetrate, opened by students and mainly catering for the young *fuori sede* or transient citizens, that offers a high-level programme of experimental music and performing arts events.

The student associations themselves organize events during the famous (but sadly over-exploited) Venetian carnival, which are eagerly sought out by the most attentive tourists. Architecture faculties organize world-class exhibitions as side-programmes or follow-ups of the International Biennale of Architecture, one of Venice's few new attractions of the last years. St. Margherita's bars and clubs can also be packed with tourists even during weekends, when some students go home. Moreover, the few tourist attractions which resist the trend towards banalization – first of all the Biennale, including the Arts and Architecture exhibitions, the Film Festival and the Ballet Festival, but also the Peggy Guggenheim Collection and other private galleries, trendy bars and restaurants, and the dance and theatre companies – find in the student community a large pool of seasonal and part-time workers, allowing them to compete with mainstream tourist providers.

What can be seen in Venice is possibly the first step towards the creation of a creative cluster based on a group of activities, from art and culture to creative professions and lifestyles. The key trigger has arguably been the physical centralization of most university facilities in a part of the city, which was a tactical choice (facilitating university user mobility) but paradoxically achieved the opposite strategic effect (rooting students to an area of the city). Furthermore, there are cultural and emotive links between creative producers and consumers, who belong to the same community, strengthening the cluster climate and its resilience.

Of all social agents in the city, the student community is arguably the best equipped to deal with inquisitive tourists – they are not purely commercially motivated, are internationally oriented and cultured enough to establish a mental link with visitors. As one of our interviewees said:

> The Rector of a University, the lead singer of Venice's most famous reggae music band, and the owner of Venice's trendiest restaurant may be sharing drinks with a couple of students happening to sit close to them in a bar, then a couple of foreigners would approach them to ask for information, ending up in a big party somewhere . . . this would be a normal night in Venice.

Today, Venice's student neighbourhoods – St. Margherita/Zattere, Cannaregio and, more recently, Rialto and Giudecca (where a large student residence has been opened) – are among the liveliest and most liveable parts of the city. The student community has been a powerful catalyst for the generation of an alternative image to the ambivalent or predatory face that Venice presents to its tourists; it has sustained and revived commercial activities which were disappearing from Venice under the pressure of tourism, such as bookshops, record shops, cinemas, cafés and traditional taverns. Regeneration extends to important social aspects. Having a boat, the most traditional lifestyle marker of old Venice, has become fashionable again; university students are among the most enthusiastic members of rowing clubs previously endangered by the ageing of their patrons. The dream of many students now is to find a job and a house in the city, while for many original residents, work careers can only be pursued elsewhere.

Challenges for a sustainable integration of visitors in Venice's creative landscape

There seems to be a clear role for the student community as the engine of Venice's 'tourist rebirth'. This is today limited to establishing a virtuous relationship with a small number of visitors, those who fit with the profile of empathic and economically empowered patrons of student-sustained attractions. However, in the future new student-generated attractions may compete with mainstream tourist suppliers, triggering increased product quality and range. An example of this trend may be the decrease of fast food establishments in the old city, as colourful taverns offering traditional local food or pizzas have been sustained in zones of the city where there is student demand.

The question remains whether such 'virtuous' integration of student settlements in the tourism development trajectory of the city is sustainable. Two issues are important here: first, in the short-to-medium term the demand for urban resources from the emerging 'creative tourist' segment could come to compete with student settlement models. Second, the level of rooting of students in the local society, and the possibility of tying the creative economy more solidly into the urban development strategies for the city, are crucially related to the expectations regarding the status of students as future citizens.

Clearly the tourist attractiveness of the student city does not depend directly on the career choices of students. Cultural animation will be present in the city irrespective of the individual decisions of creative producers: some will leave the city in search of work and affordable housing somewhere else, but new members of the creative class will take their place. However, a stable student population is likely affect the long-term development trajectory of the city on various levels, as is argued by Van den Berg and Russo (2004). First of all, the city can only escape from the tourist monoculture if it develops economic alternatives, and this requires human capital, new organizations and social networks that are consistently attracted and integrated into the urban fabric. This also implies being able to offer a convenient 'package' of housing and employment to talented graduates who develop skills as cultural managers, architects, public relations experts, telecom or maritime engineers and environmental scientists. Even when such jobs are available in the old city, housing prices and living costs often mean that they leave anyway. This, in the end, also affects the contribution of students to the economy of the city: they live their Venice experience as a 'temporary cocoon' which is likely to end with their study, and solid social and economic relationships are therefore seldom established. Furthermore, an urban development strategy aiming at transforming a 'lifestyle' or ambience into an economic specialization needs stable firms and social networks.

For this reason, the housing scheme devised by the municipality, providing interest-free loans to young couples willing to buy protected housing, could achieve much more than just a deceleration of the population exodus; indeed, a substantial group of beneficiaries have been university researchers and young creative professionals (architects and art managers), who decided to use this chance to settle in the city which they cherish and in which they have good working opportunities.

However, students – and the creative social capital of the city in general – may face another problem if creative tourism grows. In fact creative tourists use (and to some extent compete for) the same 'urban facilities' as the student population. Student facilities are 'protected' from mass tourism, which uses tourist infrastructure, but creative tourists use student bars and local restaurants, and are now beginning to compete directly for housing. Following the deregulation of the property rental market, anyone who has a spare room can rent it to tourists provided they register at the statistical office. This was a very positive thing for tourism, enabling more visitors to stay overnight and increasing the economic benefits derived from tourism.

Yet it is apparent that students and tourists are now competitors for such housing: not only can tourists pay higher prices, but renting to tourists is also less risky and more flexible than renting to students. If the growth in tourist room rentals continues, students will have to start renting outside the centre, and this could affect the level of animation in the old city. So, while deregulating tourist room rental has been a positive change in a tight accommodation market, some form of regulation should be re-introduced in order to preserve the student city climate. The housing issue and how it will be managed by the city and regional bodies in charge of housing and university policy will be the key playing field for 'creative Venice'.

However, the tourism sector itself also needs to recognize the important benefits gained from student life in terms of the range of products that can be offered to visitors. Creative initiatives, such as private residences being converted into tourist accommodation in the summer months and during the weekends, or sponsoring student cultural events, could be a way to build real synergies between student life and tourism.

Conclusions: students as creative tourism landscapers

In the paradigm shift towards a postfordist economy it is argued that mass tourism models are giving way to more articulated, diversified tourism markets, which benefit from empathic links with local communities. Yet this shift is restricted by the physical and social structure of destinations, which generally continue to cater for the typical 'mass tourist market' by delivering low-quality experiences, largely driven by asymmetric information between hosts and guests (Caserta and Russo 2002).

The Venice case study shows that the student community may be an important player in the restructuring of a traditional (and unsustainable) tourist destination. At the crossroads between student creative lifestyles and tourism, new landscapes of consumption have emerged, opening opportunities to diversify supply and modify the form of visits (longer stays, choice of central accommodation, more dispersed articulation of itineraries) and presumably their economic value (higher quality products, escape from 'tourist traps', resulting in increased visitor satisfaction). This represents a break with the mass tourist supply based on 'gazed heritage' and mainstream cultural facilities, producing an alternative, and equally profitable, model of tourism development. This more creative model would also be more sustainable for the local community and perfectly consistent with Venice's ambition to develop economic alternatives based on its cultural and creative assets.

Figure 10.4 provides a model of the relationship between student life, creativity and tourism. It shows that an active student life is crucial for the

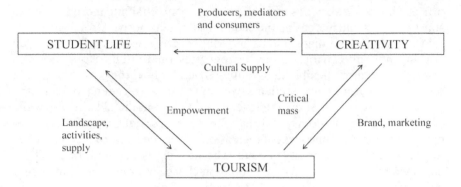

Figure 10.4 Relationship between student life, creativity and tourism.

production, mediation and consumption of creative services and goods, which also benefits the student community in the form of opportunities for cultural consumption. The presence of creative activities also adds value to the tourist destination, enhancing the cosmopolitan and vibrant character of the place. At the same time, tourists become an important group of consumers for the creative economy. Finally, student life diversifies the tourism supply of the city, acting as a proactive group for the emergence of creative landscapes and related activities. This model highlights the importance of the student community; promoting their empowerment and active integration as subjects of urban policy (and not just as objects of higher education policy).

A number of issues will need to be addressed before this model can be implemented. In particular, the increasing influence of tourism in such landscapes is likely to generate a struggle between tourists and the student community, especially as far as the housing market and the use of public space is concerned. The higher capacity of the tourist to pay for goods and services could be a serious problem for the retention of the student community and for the continuity of cultural activities and the night-time economy they produce, mediate and consume. Just as landscapes of creativity can change the previous uses and social structure of areas of the city, so tourism can also change creative landscapes into commodified landscapes, embedded in what Zukin (1990) called 'real cultural capital': the process of capital accumulation driven by real estate, financial and global capital.

Thus, as Van den Berg and Russo (2004) argue, a main challenge for urban policy – as well as higher education planning – is to maintain the 'embedding' of student communities in the social fabric of the city. Rather than through 'hard' physical and economic planning, this should be achieved through the empowerment of the student community as local agents: for instance by enabling them to stay after their studies. This would also give them status and perspective as members of the community. A clear role for local government is also to give support for cultural and tourist communication, which could lead to a more explicit branding of student life as a creative landscape open to aware and empathic visitors.

In conclusion, urban policy should take into consideration the opportunities offered by the emergence of creative landscapes connected with student life without neglecting the risks: gentrification processes, loss of value of the original landscape and excessive specialization of areas; and the conditions for a sustainable process of convergence between creative landscapes and tourism: the empowerment of students and their inclusion in broader community agendas. Creative landscapes should not be consumed as just another good, service or symbol but be shaped by dynamic processes of innovation and interaction between students and the other communities who live in the city.

Note

1 In some cases, as in Bordeaux (quoted in Dubet and Sembel 1994), the creation of suburban 'citadels of studies' has produced unattractive peripheral enclosures rather than achieving the sought-for integral, comprehensive academic community that is found in American campuses.

11 Amsterdam as a gay tourism destination in the twenty-first century

Stephen Hodes, Jacques Vork, Roos Gerritsma and Karin Bras

In the last quarter of the twentieth century Amsterdam was universally known as the 'Gay Capital of Europe' with a thriving gay scene that had a strong magnetic effect on gay travellers from all continents. The celebration of Amsterdam as an attractive destination for gays culminated in 1998 in the Gay Games under the slogan 'friendship through sport and culture' where, according to the Federation of Gay Games (2006), 250,000 tourists visited the city during the event.

At the turn of the century things began to change. Increasingly reports began to appear in the media that Amsterdam was losing its attractiveness as a gay destination and was being bypassed by other cities in Europe such as Barcelona, Berlin and London; cities that were positioning themselves as gay destinations. However, no empirical data were available to substantiate this claim, so in 2005 the Leisure Management Research Group of the Inholland University in Amsterdam decided to launch a study into 'Amsterdam as a gay tourism destination in the twenty-first century' (in this chapter the word 'gay' is used to include gay, lesbian, bisexual and transgender).

At the same time that the reports of Amsterdam's declining attractiveness as a gay destination grew, Florida (2002) published his work on the 'creative class' and the creative industries (the arts, media and entertainment, and creative service providers such as architects, designers, fashion designers, graphic artists and advertising agencies). In the industrial economy, employees migrated to the cities where the industries were located. But in the creative economy, according to Florida, the creative class is 'footloose' and migrates to the cities where they want to live and work. Creative firms will increasingly follow the creative workers, or be founded by them. He argues that the attractiveness of a city for the creative class is crucial for a city with the ambition of functioning as a creative hub.

What are the conditions that cities must fulfill to be considered attractive for the creative class, and to what extent can a city make itself attractive for visitors? Most of the criticism of Florida's theories is focused on his notion of the 'instant makeover' with regard to the ability to manufacture the 'creative city' (Hall 2004), because ultimately the origins of a creative city are dependent on a variety of complex factors. There is enough reason to doubt the possibility of simply creating creative cities, but diverse studies have shown that certain conditions have to be present in order to assure their possible development.

Florida (2002: 249) calls these important conditions 'the three Ts of economic development: Technology, Talent and Tolerance' (Figure 11.1). According to Florida, a city must fulfill all three conditions to be considered attractive to the creative class. Briefly, the three Ts can be described as:

- Tolerance: openness, acceptance of all ethnicities, races and lifestyles;
- Talent: the proportion of the population with a bachelor's degree or higher;
- Technology: a concentration of innovation and technology in the region.

According to Florida (2002: 244) 'The key to economic growth lies not just in the ability to attract the creative class, but to translate that underlying advantage into creative economic outcomes in the form of new ideas, new high-tech businesses and regional growth.' To gauge these capabilities he developed the Creativity Index, a mix of four equally weighted factors: (1) the creative class share of the workforce, (2) the High-Tech Index, (3) innovation measured as patents per capita and (4) diversity. Diversity was measured by the 'Gay Index', which Florida argued was a reasonable proxy for an area's openness to different kinds of people and ideas. The Gay Index measures the over- or under-representation of gay male couples in a metropolitan area relative to the total population. The index is constructed as the fraction of gays who live in a metropolitan area divided by the fraction of the total population who live in that area. The value takes on the

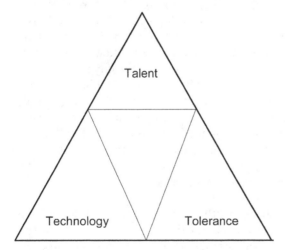

Figure 11.1 The three Ts of economic growth.

properties of an odds ratio whereby a value over one indicates that a gay couple is more likely to locate in the area than the population in general while values below one suggest that gays are under-represented.

The Gay Index generated a great deal of media coverage and was clearly misunderstood in many instances. Florida was not suggesting that gays were more creative than other members of the population and that therefore the number of gays in a region or city explained the degree of creativity of the population. Rather, he was saying that there was a direct relationship between a high density of gays and the openness and tolerance of a place, a necessary condition for the fostering of creativity. In the words of Florida (2002: 255–56):

> Several reasons exist why the Gay Index is a good measure for diversity. As a group, gays have been subject to a particularly high level of discrimination. Attempts by gays to integrate into the mainstream of society have met substantial opposition. To some extent, homosexuality represents the last frontier of diversity in our society, and thus a place that welcomes the gay community welcomes all kinds of people.

The concept of the creative class and the creative industries generated a great deal of interest in both Amsterdam and the Netherlands, resulting in many conferences and symposia, a series of studies (e.g. Gemeente Amsterdam 2004a, 2004b), research projects and a national policy paper from the Ministry of Economic Affairs and the Ministry of Education, Culture and Science (2005) intended to boost the economic utilization of culture and creativity in the Netherlands.

The widely reported but as yet unsubstantiated decline of Amsterdam as a gay destination and the growing interest in the creative class, the creative industries and the related role of the Gay Index were the reasons why we wanted to carry out research linking these developments. The hypothesis on which the study was based is that not only is the number of gays living in a city indicative of the openness and tolerance of the city, but also the number of temporary gay residents – the gay tourists – visiting the city is an indicator of the degree of tolerance and openness. The reasoning behind this is that not only will gays visit cities with an interesting tourism offering, but also they will be inclined to visit those cities that are open and tolerant; where they are made to feel welcome; where their lifestyle is accepted. In fact, one might argue that gay tourism may be an even more sensitive indicator of tolerance than the gay resident population, since tourist arrivals are likely to be affected very rapidly by any signs of intolerance, whereas a change in the residential population is usually slower.

Until relatively recently, however, little research had been done on the development of the gay tourism market, or the popularity of gay destinations over time. Our knowledge of gay tourism is now being improved by a number of general texts on the phenomenon (e.g. Clift *et al.* 2002; Hughes

2006; Waitt and Markwell 2006). However, because of the relatively recent nature of much of the literature, there is relatively little empirical research into the development of gay tourism over time. The Annual Lesbian, Gay, Bisexual and Transgender (LGBT) Community Survey undertaken by Community Marketing Inc. (2005) in the United States is the only significant source of longitudinal data.

The current study therefore attempts to analyze the relationship between the city of Amsterdam and gay tourism in more detail, and tries to trace the link between levels of tolerance and the city's attractiveness as a gay destination. Because of the lack of reliable data it was not possible to calculate the Gay Index for Amsterdam. Therefore, we could not objectively determine if the position of gays in Amsterdam was deteriorating or not. As an alternative, we therefore asked the different target groups their opinions on this issue.

The situation of gay tourism in Amsterdam was analyzed in the current study on the basis of three research questions:

1   What is the relationship between the level of tolerance and the attractiveness of Amsterdam for gay tourists?
2   How important are openness and tolerance for gay tourists when choosing a city travel destination?
3   How does Amsterdam score as a gay destination in the competitive arena of European cities?

The complete research programme focused not only on answering these research questions, but also on increasing the effectiveness of gay tourism marketing efforts (Kenniskring Leisure Management 2006). For this reason particular attention was paid to information collection behaviour by gay travellers, the types of transportation utilized and the specific sites visited during their stay in Amsterdam. These results have not been included in this chapter; instead we concentrate here on the analysis of the views and opinions of Amsterdam as a gay tourism destination.

Research methods

Based on desk research a brief profile of the creative city and the development of gay tourism and gay policy in Amsterdam and the Netherlands was drawn up as a framework for the research. This made it clear that the development of gay tourism to Amsterdam is influenced by a number of internal and external stakeholders, in addition to the tourists themselves. For this reason, the study targeted four main groups, namely:

• foreign gay tourists visiting Amsterdam;
• foreign intermediaries who influence gay travellers such as tour operators, travel agents and gay media;

- gay organizations and enterprises in Amsterdam such as bars, restaurants, discos, hotels and incoming tour operators;
- gay residents living in Amsterdam.

The reason for including the gay inhabitants of Amsterdam in this research was twofold: first, the Gay Index is based on gays living in a city/region and, in order to establish a connection between the Gay Index and foreign gay travellers, it was necessary to have a comparative understanding of how both of these groups saw Amsterdam as a gay destination; and second, to get an understanding of how the gay residents, as 'ambassadors' of Amsterdam as a gay destination, see the city.

Two versions of the questionnaires were used to gauge the opinions of these different groups: a printed and a digital version. All of the questionnaires were pre-tested with the target groups and were adapted where necessary. A specific questionnaire was developed for each target group and, where possible, the questions were standardized to allow for crosstabulation.

The questionnaires were distributed in the following ways:

- *Foreign gay tourists visiting Amsterdam.* Self-completion questionnaires in English, French and Spanish were distributed through the reception desks of gay hotels in Amsterdam and through gay bookshops and information centres in the city. In addition, in the weekend of the Gay Pride/Canal Parade in Amsterdam, self-completion questionnaires were handed out and collected on the streets in Amsterdam.
- *Foreign intermediaries such as tour operators, travel agents and gay media.* An English questionnaire was sent by email to the relevant addresses of the International Gay and Lesbian Travel Association (IGLTA) and to a number of non-IGLTA addresses collected via the Internet.
- *Gay organizations and enterprises in Amsterdam.* Written questionnaires in Dutch were sent to the gay organizations in Amsterdam who were members of the Gay Business Association (GBA) and to additional organizations and enterprises listed in Amsterdam-related gay guides, maps and websites.
- *Gay residents living in Amsterdam.* A digital questionnaire in Dutch was distributed via various gay sports and social clubs and associations in Amsterdam to their members, with a request to forward the questionnaire to other gays outside their clubs/associations who lived in Amsterdam. In addition, self-completion questionnaires in Dutch were distributed via gay bookshops in Amsterdam.

The degree to which the results are representative and reliable per group researched is dependent on the sample size and the response. The minimum number of successful interviews was determined in advance and the goals were met in all of the categories except for that of the foreign intermediaries.

Table 11.1 Sampling frame and achieved sample by target group

Target group	Target sample	Achieved sample
Foreign gay tourists visiting Amsterdam	100	142
Foreign intermediaries	35	27
Gay organizations and enterprises in Amsterdam	35	48
Gay residents living in Amsterdam	100	329

The overall number of gay organizations and enterprises in Amsterdam may be low, but considering that it represents approximately half of all gay organizations and enterprises in Amsterdam, the reliability here is high.

Of the total number of questionnaires filled in by foreign tourists, 25 per cent were from lesbians and 75 per cent from gay men. There were no significant differences in the outcomes for these two groups. There were also no significant differences in the responses of the foreign visitors who visited Amsterdam during the Gay Pride/Canal Parade and those who filled in the questionnaires in the weeks prior to and after the Gay Pride event.

The final results of the surveys were discussed in two focus groups consisting of representatives from the gay community, the city government and promotional organizations. All the major stakeholders in Amsterdam sent representatives. They were able to respond to the results and conclusions and could then interpret the results with recommendations for the sector and the city.

Key survey results

The foreign gay tourists visiting Amsterdam indicated that *tolerance and openness* and a *feeling of being welcome* are the primary reasons for visiting a city as a gay traveller. Asked what made a city in general gay friendly it was not the infrastructure, such as gay bars, discos and restaurants, that scored the highest, but the intangible aspects of the city, such as a feeling of being welcome and tolerance and openness towards gays and lesbians (Figure 11.2). This was also underwritten by all the gay organizations and enterprises in Amsterdam and the vast majority of the gay residents and the foreign intermediaries.

The appreciation of Amsterdam as a gay tourism destination was by and large positive. However, foreign gay tourists tended to be more positive about factors such as the welcome for visitors and the level of tolerance than were local residents (Figure 11.3). This is perhaps not surprising, as local residents are more likely to experience intolerance than tourists are likely to encounter it on a short visit. However, residents and gay organizations in Amsterdam were more positive about the position of equal rights for gays in the city.

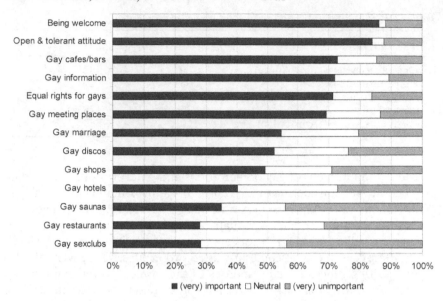

Figure 11.2 Aspects considered important in a 'gay friendly' city by foreign gay tourists.

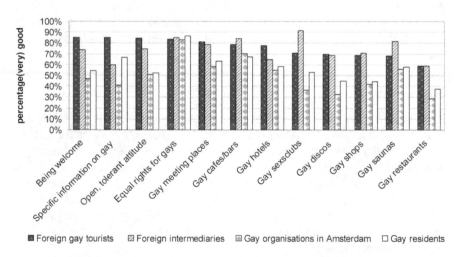

Figure 11.3 Rating of Amsterdam as a gay destination.

The foreign gay tourists and the foreign intermediaries were very positive about Amsterdam as a gay friendly city and rated it on average with a nine on a scale from one to ten. Of the foreign gay tourists, 40 per cent rated Amsterdam with a ten. The gay organizations and enterprises and the gay residents, however, were significantly less positive about Amsterdam as a gay friendly destination; they rated it with an average of seven. Crosstabulation of the results showed that there were almost no significant differences between answers of the gay male and lesbian respondents.

Despite the high rating given by foreign tourists, foreign intermediaries and gay organizations and enterprises in Amsterdam, they all indicated there had been a consistent decline in the number of gay tourists visiting Amsterdam in the period 2000–2005. It is not known what percentage of the foreigners visiting Amsterdam is gay (as is also the case in other cities) and it is therefore not possible to ascertain what the decline is in either percentages or absolute numbers. The number of gay enterprises that saw a decline in gay visitors grew from 11 per cent in 2000 to approximately 50 per cent in 2005, a finding that was also confirmed by the foreign intermediaries. These findings also correspond with those of the tenth Annual LGBT (Lesbian, Gay, Bisexual and Transgender) Community Survey (Community Marketing Inc. 2005). According to the American LGBT respondents (the United States is the most important country of origin for gay travellers visiting Amsterdam), the Netherlands has dropped from third place in foreign countries visited (2003) to fifth place (2004). When asked about their travel intentions for 2006, the Netherlands came in fifth place behind Great Britain, France, Italy and Spain. This apparent decline in gay tourism contrasts with a marked increase in the overall number of foreign arrivals in Amsterdam in 2004 and 2005.

What is the possible explanation for the declining number of gay tourists visiting Amsterdam? Whereas 60 per cent of the gay tourists visiting Amsterdam still considered Amsterdam as the 'gay capital of Europe', the other target groups were far less positive; only 35 per cent of the foreign intermediaries, 20 per cent of the gay organizations and enterprises and 17 per cent of the gay residents saw Amsterdam as the 'gay capital of Europe'. Three-quarters of the intermediaries and 80 per cent of the residents were of the opinion that Amsterdam is being overtaken by the competition and they believed that if Amsterdam does not take action fast it will lose its position as an important gay city destination. Berlin, Barcelona, London, Madrid and Paris were named as the most important competitors by all target groups and this is also underwritten by the findings of the tenth LGBT Community Survey in the USA (Community Marketing Inc. 2005).

In addition to the increased competition from European cities targeting the gay traveller another possible reason for the decline in gay arrivals is the lack of innovation and diversity in what gay Amsterdam has to offer. Even though the gay tourists were positive about the gay scene in Amsterdam, only 40 per cent considered it to be trendsetting. Almost 80 per cent of the foreign intermediaries considered lack of innovation an important reason for the declining interest in Amsterdam as gay destination. Almost all the gay organizations, enterprises and residents were of the opinion that there has been limited innovation in gay Amsterdam over the last 5 years and that there is a strong need for innovation in the gay scene.

The difference between how the gay tourists and the gay residents saw Amsterdam is also evident in how they rated the promotion of Amsterdam as a gay destination. Almost three-quarters of the gay tourists thought that Amsterdam is well promoted as gay destination, compared with only 30 per cent of the residents. Less than 10 per cent of the Amsterdam gay organizations and enterprises were of the opinion that Amsterdam is effectively promoted. Specific gay information was one of the top five aspects that makes a city gay friendly. The appreciation of the gay information in Amsterdam was higher than the gay promotion, but here too the gay tourists were more positive than the residents.

With regard to the safety of gays in Amsterdam the difference between the gay tourists and the gay residents was considerable. The gay tourist saw Amsterdam as a safe city: two-thirds were of the opinion that you can safely walk hand-in-hand in Amsterdam as a gay couple, whereas less than 20 per cent of the residents agreed with this. The foreign intermediaries also saw Amsterdam as a safe city for gays whereas the gay organizations and enterprises were more inclined to agree with the residents. The gay residents and gay organizations and enterprises, and to a lesser degree the foreign intermediaries saw increasing religious fundamentalism as a threat to the safety of gays in Amsterdam. The foreign tourists were divided as to whether the open attitude towards gays in Amsterdam is under threat; one-third agreed, one-third did not agree and one-third was neutral.

Our research shows that the gay visitor to Amsterdam is by and large a loyal client. Of the respondents 45 per cent were first-time visitors, 55 per cent were repeat visitors, while a quarter had visited Amsterdam more than six times. In addition to being regular visitors, the gay tourist stays longer in the city and spends more than the average foreign visitor. Gay tourism is an interesting niche market for Amsterdam given that the economic value of the gay tourist in Amsterdam is 43 per cent higher than the average tourist visiting the city.

Discussion

This study has indicated substantial differences between foreign gay tourists and gay residents in terms of their evaluation of the openness and tolerance of Amsterdam. Across the board, the gay tourists are much more positive than the gay residents are. What is the explanation for the difference? An evident explanation is the information base – the gay resident is far better informed than the gay tourist is and he/she has more up-to-date information on the status quo. In addition the gay resident is more inclined to compare gay Amsterdam today with how it was in the past, whereas the gay tourist is more inclined to compare Amsterdam with his/her city and other cities visited. In terms of the Gay Index we can conclude that the gay resident is a better litmus test concerning the openness and tolerance of a city than the gay tourist. However, the gay tourist may be a better indicator of the competitive position between cities with regard to their *perceived* openness and tolerance, given the prime importance of this aspect in choosing a city destination for gay travellers. It seems that research among gay residents is an effective method to measure the 'openness and tolerance temperature' of a city and to track possible negative or positive changes.

Amsterdam was seen by all the respondent groups as international, relaxed, liberal, pleasant and friendly – all positive associations – yet its position as a unique gay destination is clearly threatened by the increasing number of cities with strong tourism offerings *and* an open and tolerant climate. What was once unique for Amsterdam can now be found in many more European cities. This increasing competition combined with a lack of innovation in the gay scene means that Amsterdam's position as a gay tourist destination is under threat.

City branding research indicates that people are very reluctant to alter their organic perceptions of cities. Research by Anholt-GMI (2005) shows that the image of Amsterdam was not negatively affected by the murder of Theo van Gogh in 2004, nor was the image of Stockholm noticeably negatively affected by the murders of Olaf Palme in 1986 and Anna Lindh in 2003. The gay bashing of Chris Craine, the editor of the gay magazine *Washington Blade*, in Amsterdam in April 2005, also apparently had no negative repercussions on the image of Amsterdam. Amsterdam is a strong city brand, ranking in sixth place in the GMI Anholt research behind cities such as London, Paris, Rome and Barcelona. We can also assume that Amsterdam's gay brand is strong by the positive responses from the gay tourists and the high percentage of repeat visitors among this group (55 per cent).

However, from the focus group sessions held in Amsterdam to discuss the research findings with representatives from the gay community, the city government and promotional organizations, it became clear that today's gay traveller is different from the travellers in Amsterdam's gay heyday. The positioning of Amsterdam as a gay destination, its gay promotion and information, should therefore be readdressed in line with the needs of the modern gay traveller.

What has changed? The focus group sessions indicated that gays to a lesser degree want to visit gay only bars or stay in gay only hotels, but to visit and stay in gay friendly bars and hotels that are inclusive, not exclusive. Gays go out less to gay venues to meet other gays since the advent of (gay) Internet dating. Gays in Amsterdam travel more frequently to cities such as Brussels, Antwerp, London, Barcelona and Berlin, and are therefore less dependent on what Amsterdam has to offer. In addition, there is a growing number of underground gay clubs and parties in Amsterdam that are less visible and accessible to foreign visitors. The way gays travel is changing and Amsterdam needs to respond to the developing needs of this market segment. Research is needed to develop a better understanding of what it means to be an attractive gay city, and how this is likely to change in the future. In addition, further research needs to be done to understand the motivations of the gay non-visitor to Amsterdam.

Conclusions

To conclude, we would like to address the issue of the connection between Amsterdam as a gay destination and Amsterdam as creative city. The research reveals that there is a feeling, especially among gay residents (over 60 per cent), that the openness and tolerance toward gays is declining. The rise of religious fundamentalism is seen as a threat to the safety of gays in Amsterdam. This is an important signal for the city and its residents. Gays are arguably the 'canaries of the Creative Age' (Florida 2002: 256), serving to warn cities of a decline in tolerance and openness.

As we argue above, tolerance and openness are essential preconditions for the development of the creative city. Along with technology and innovation, these two factors ensure that the creative talents in a city are able to exchange ideas freely. It is not just the gay residents, but the interaction between the various cultures, disciplines and concepts that results in the creation of new ideas, insights and concepts. The gay residents are only an *indication* of the degree to which this is possible.

12 Ethnic quarters in the cosmopolitan-creative city

Stephen Shaw

Introduction

A decade ago, Charles Landry and Franco Bianchini (1995: 28) highlighted the contribution of immigrant communities to the cultural and artistic, as well as economic vitality of the creative city: outsiders and insiders at the same time, 'they have different ways of looking at problems and different perspectives'. Where diverse cultures and lifestyles interact productively and without friction, a city can capitalize upon the 'creative buzz' associated with an open-minded, cosmopolitan outlook, as well as upon the energy and drive of ethnic minority entrepreneurs. London, Birmingham, Manchester and other cities in the United Kingdom that are major gateways to immigration have fully embraced this discourse of inclusion and innovation. They highlight these competitive advantages as explicit features of their strategies for promotion. Their counterparts in Canada – especially Toronto, Montréal and Vancouver – have also positioned themselves as globally oriented, creative hubs that connect North America with Europe, Asia, Africa, Latin America and the Caribbean through networks that stimulate cultural as well as commercial exchange.

In both the United Kingdom and Canada, particular places associated with immigrants and transnational communities have been re-imaged and signposted as physical expressions of this new cosmopolitanism. Rejuvenated and re-branded as 'ethnic quarters', they are promoted to the majority culture, and in some cases to international tourists, as 'Chinatowns', 'Little Vietnams', 'Punjabi Villages' and so on, alongside other designations, such as fashion districts, museum quarters and theatrelands. Their core features include exotic cityscapes that are often complemented by flamboyant gateways and other public art. The streets and other public spaces are 'animated' by festivals, performance art and markets, but the main focus is generally upon eating, drinking and shopping. Ethnic minority entrepreneurs have played a leading role in such revitalization. Unlike the examples in Sydney, Australia examined by Collins and Kunz (Chapter 13 of this volume), the cases discussed below were, nevertheless, supported by significant public as well as private investment, through special-purpose agencies and partnerships. Collaboration between city governments and civic society – especially ethnic minority business and cultural organizations – has made these spaces more accessible, safe and attractive for higher-spending visitors to stroll in, by day and after dark (thereby creating the 'controlled edge' referred to by Hannigan in Chapter 3).

Diversity is celebrated by showcasing selected streets within enclaves, a phenomenon that provides an interesting parallel with the valorization of 'gay villages' discussed by Hodes *et al.* (Chapter 11 of this volume) with reference to Amsterdam and other cities that espouse tolerance towards diverse sexual orientations (cf. Florida 2002). Acceptance of this new conventional wisdom resonates with Ulrich Beck's (2002) agenda for 'cosmopolitanism': internalized globalization in which rival ways of life co-exist in individual experience. In Beck's cosmopolis, the spontaneous clash of cultures encourages 'creative reflexivity': the comparison, reflection, criticism, understanding and combination of contradictory certainties. In the cosmopolitan-creative city, an urbane hybridization stimulates innovation. Sandercock (2006: 37) uses Salman Rushdie's (1992: 398) metaphor of a 'mongrel city' to describe a desirable urban condition in which difference, Otherness, multiplicity, heterogeneity, diversity and plurality prevail.

In this egalitarian construct of the 'cosmopolis', respect for difference nurtures creativity and economic success that enables immigrant communities to become established and to prosper in their new homeland. Creative intermediaries and merchants move with relative ease, engage with and negotiate between the different 'worlds'. In this context, an international orientation, together with day-to-day interaction between diverse cultures and lifestyles, stimulates artistic innovation, while traders profit from new commercial opportunities. However, as Binnie *et al.* (2006) argue, a more pessimistic view emphasizes the underlying truth that power and patronage still rest with the dominant culture. Privileged members of a 'cosmopolitan' elite appropriate cultural objects that they select as objects of consumption, features that signify a domesticated, safe and agreeable Otherness (see Hannigan, Chapter 3). Seen from this perspective, the cosmopolite is a voyeur, a parasite without lasting attachments or commitment to places or people (Featherstone 2002). As enclaves of conspicuous consumption, 'Disneyfied Latin Quarters' may say more about the predilections and prejudices of the dominant culture than those of the minority groups that they purport to represent. Segregated from the surrounding urban landscape, they seem 'more likely to contribute to racial, ethnic, and class tension than to an impulse toward local community' (Judd 1999: 53); behind the carnival mask, the less picturesque poverty of low-income groups is never far away.

This chapter considers the place-marketing narratives and urban planning strategies of 'creative cities' in the United Kingdom and Canada, and how they have incorporated the more positive interpretation of cosmopolitanism in a world of flows, characterized by 'floating populations, transnational politics within national borders, and mobile configurations of technology and expertise' (Appadurai 2001: 5). It relates these discourses to the underlying structures of urban governance that accommodate such volatile flows and considers the key drivers of change, as competition between cities intensifies. Through longitudinal case studies, it explores the processes through which two neglected, run-down inner city areas have been transformed into showpiece attractions. The outcomes of these programmes over the past 10 years are then considered with reference to the normative ideal of a creative metropolis that values diversity and trades upon difference.

Trading upon difference: the United Kingdom

In the United Kingdom, the notion of immigrant communities as active agents of regeneration, as opposed to passive recipients of welfare support, has a strong appeal across a wide political and ideological spectrum. Neoliberals have been inspired by Michael Porter's (1995: 57) thesis that public intervention should work in harmony with market forces and build upon the true competitive strengths of firms in the inner city. According to Porter, the prime sources of competitive advantage for many inner city localities include proximity to downtown areas, entertainment and tourist attractions, as well as the entrepreneurial talent and the supply of low-cost labour within immigrant communities. Over the past 10 years, urban authorities and regeneration agencies in the United Kingdom have adopted versions of his model, emphasizing the scope to capitalize on the rising demand for 'ethnic' food, entertainment and other exotic products. The welcome contribution that minority businesses may make to the physical regeneration of a neglected cityscape can be presented as a potent statement of popular capitalism. Wider promotion of cultural spectacles such as Chinese New Year, and the creation of new events such as festivals of 'world' music and dance may provide niche market opportunities. They also offer a visible demonstration that the city as a whole is inclusive and 'cosmopolitan' in the positive sense of the word outlined above: a rhetoric that is more traditionally associated with the Left.

For example, Urry (1990: 144) describes 'cultural re-interpretation of racial difference' in Bradford's guide to the *Flavours of Asia*, a campaign that promoted Asian restaurants and sari centres, and provided visitors with a brief history of Asian religions and immigration to the city. Chan (2004: 184–85) highlights the ethnocentricity of proposals formulated in the 1980s for a 'China Court' in Birmingham. The new development was designed to create a 'comprehensive leisure complex with a genuine Chinese flavour': a unique attraction that would have a particular appeal to conference visitors. More recently, Birmingham's rich ethnic diversity has been promoted with reference to the wide range of music, food and drink offered by transnational communities, which include Pakistanis, Chinese and Afro-Caribbeans (Henry *et al.* 2002). The creation of visitor-oriented enclaves of consumption may, however, contrast sharply with adjacent areas where residents continue to live in a poor and depressingly neglected environment.

Policymakers are acutely aware that the aims of the 'Urban Renaissance' – to make cities more accessible, attractive and safer for everyone – remains unfinished business. 'Massive inequalities persist in our cities. Competition for space pushes up prices for housing, making access for lower income households much harder' (Lord Rogers of Riverside, chair of the Urban Task Force 2005: 3). Even in the most economically buoyant UK cities, disturbing levels of deprivation persist in neighbourhoods less than 15 minutes walk from the centre (Shaw 2006: 7).

In some cases, the somewhat idealistic agenda of the cosmopolitan-creative city may be hard to reconcile with the outcomes of leisure- and tourism-led revitalization and promotion of ethnically mixed neighbourhoods. As Sennett (1994) observed with reference to New York, the absence of a common civic culture may produce a state of mutual withdrawal rather than creative interpenetration of cultures.

The tensions that may arise through the self-conscious creation of an ethnic quarter as a high-profile locale for consumption are illustrated in the case of 'Banglatown' in London's East End. Adjacent to the City of London's Square Mile – one of the world's leading financial centres – this initiative to breathe new life into a 'neglected swathe of inner city neighbourhoods' (Shaw and MacLeod 2000: 166) emphasizes the contribution of immigrant communities (both historical and contemporary) to the cultural vitality of the metropolis. In tune with the Government's strategic guidance, the 'City Fringe Partnership' set out to address the area's problems, while contributing to the strength of London as a whole: 'Inner City action with a World City Focus'. It was acknowledged that this symbiotic relationship had a long history (City Fringe Partnership 1997: 1): 'For hundreds of years, the Fringe has underpinned and complemented the City economy, whilst acting as a point of entry for immigrant communities and refugees.'

'Banglatown' as a cultural quarter in London's East End

Ten years ago, recognition was given to the potential appeal of certain areas on the boundary between the City and Inner London (City Fringe Partnership 1997: 17):

> These cultural areas, unique to the capital and on the doorstep of the City, will be developed to provide a resource for tourists as well as employees and business visitors, helping to enhance the City's reputation as the premier European business location.

Historically, these precincts, known as the 'Liberties', were beyond the jurisdiction of London's mayor, as well as the powerful craft guilds, and thus provided a home for marginalized groups and institutions whose presence was unwelcome inside the city gates. From the fourteenth century, this peripheral space accommodated successive waves of migrants from other areas of the British Isles, along with foreigners that were – to varying degrees and at various times – tolerated and allowed to settle because of the functions they performed (cf. Porter 1996). Protestant Huguenots expelled from France in the sixteenth and seventeenth centuries gave the word *refugee* to the English language, and the area known as Spitalfields became their largest settlement. Their particular creative skills included silk weaving and fine instrument making, and by the early 1700s the area was by far the greatest centre of the textile industry in the capital. However, after two or three generations they ceased to be a distinguishable minority and eventually industrialization made their skills redundant. Most moved away, but others took their place. These included many Irish Catholics, but by the early twentieth century, Brick Lane had become the high street of a large Jewish community that included many poor immigrants from Central and Eastern Europe.

In turn, as the Jewish population moved away in the 1970s, Bangladeshi entrepreneurs acquired their former textile workshops and other businesses. Some prospered and employed local staff, especially recent immigrants. Nevertheless, Spitalfields in the 1970s and 1980s remained one of the most deprived neighbourhoods in the whole of the United Kingdom. Tensions increased as racist groups harassed and assaulted Asians; Brick Lane became the focus of intimidation that was widely reported in the news media. The majority of the new immigrants, escaping famine and poverty in their homeland, found accommodation in low-quality, often high-rise social housing. To address the severe problems of its inner city neighbourhoods, the London Borough (LB) of Tower Hamlets successfully bid for government funding to help revitalize Spitalfields and adjacent Bethnal Green from 1992 to 1997. Then, in parallel with City Fringe, another Partnership called 'Cityside' initiated a programme (1997–2002) to 'strengthen links with the City and encourage diversification of the local economy', especially into leisure and tourism. Cityside's vision was thus 'to achieve a quantum leap in the area's status as a visitor/cultural destination' (LB Tower Hamlets 1996: 13).

Both the City Fringe and Cityside Partnerships recognized that the area would need at least one 'must see' attraction and identified two vacant heritage buildings as suitable sites (Shaw *et al.* 2004). The former Truman's Brewery was to provide the missing flagship attraction: a museum that would memorialize the area's rich multicultural history. Another landmark Victorian building – the 'Moorish' Market – would become 'an ethnic shopping experience' to promote an image of London as an exciting and vibrant multicultural city (LB Tower Hamlets 1996: 14). At the same time, Brick Lane's restaurants would be vigorously promoted to non-Asian customers, especially businesspeople and office staff from the nearby City. Special attention would therefore be paid to the main 'gateways' or access points. Public realm enhancement thus included Eastern-style ornamental entrances at the City end, signage and brighter street lamps, the design of which incorporated 'Asian' motifs. In 1997, Cityside set up a 'town management' group, whose remit included the organization and promotion of events – notably Bengali New Year and the Brick Lane Curry Festival – and it was through this forum that 'Banglatown' came to be used as a brand for the area.

In practice, the regeneration of Brick Lane and the outcomes in terms of land use and functions diverged considerably from Cityside's initial vision. The brewery sold its redundant site to a local entrepreneur, and by 2002 the former industrial building housed over 250 studios for cultural industries, two bars/nightclubs, cafés, galleries, specialist retailers and an exhibition centre. By 2006, the former market building had been converted to studios and loft-style apartments. The rapid rise of 'Banglatown' as a centre for ethnic cuisine greatly exceeded expectations. A survey carried out for Cityside noted that in 1989 there were only eight cafés/restaurants in Brick Lane, with a few additions in the early 1990s. Between 1997 and 2002, however, this rose to 41, of which 16 had opened 2000–2002, making Banglatown 'home to the largest cluster of Bangladeshi/"Indian" restaurants anywhere in the UK' (Carey 2002: 12). All the restaurants reported that their clientele was 'overwhelmingly white', with a clear majority (70 per cent) in the 25–34 age group and predominantly male (ibid.: 4). The boom was facilitated by relaxed planning policies; the central section of Brick Lane was designated a 'Restaurant Zone', where applications for restaurants, cafés, hot food outlets, public houses and bars would be favourably considered (LB Tower Hamlets 1999).

The conversion of the previously run-down, mainly nineteenth-century streetscape of Brick Lane to nightclubs, bars and restaurants has undoubtedly brought wealth to a number of ethnic minority-owned businesses that have created some badly needed employment. Carey (2002) estimated that the Brick Lane restaurants employed around 400 workers, of whom 96 per cent were of Bangladeshi origin and 92 per cent lived in the Borough. Nevertheless, some significant issues and problems identified in recent years have shed doubt on the wisdom of continuing reliance on this sector. The study found that a third of the restaurants interviewed expressed concern over staff turnover, and many felt that low pay and shifts made the work unattractive to younger Bengalis. Thus, it appears that such second generation Bengalis cannot find satisfactory careers in the service economy of Banglatown, and many seek opportunities elsewhere. There is also evidence that Banglatown has displaced from the public realm of Brick Lane some members of the Bengali community who cannot move away from the area. Planning Officer Andrea Ritchie reported (2002) that in a focus group facilitated by the Borough:

> Older Bengali women stressed the point that they had to be escorted by their husbands and that they could not walk along Brick Lane at all because there are just too many men there, with all the visitors and

[restaurant] staff. So, although it is their area, they are socially excluded from it.

Trading upon difference: Canada

Porter's (1995) prescription for 'self-help' revitalization has also been embraced widely and with some enthusiasm in Canada. Since 1993, municipalities have suffered substantial cutbacks in transfer payments from higher levels of government, together with increased responsibilities, especially for transportation, social housing and the environment. Compared to their UK counterparts, the revenue stream of city governments in Canada is very heavily dependent upon the tax base of commercial and residential property that they can attract and retain. For the larger metropolitan authorities in particular, positioning in global markets to lure inward investment, wealthy immigrants and high-yield tourists is therefore an important function of city government. As Mason (2003: 349–50) points out, the neo-liberal agenda has intensified inter-city rivalry and boosterism. In Toronto, Montréal and Vancouver, quality of life – especially for mobile, high-income knowledge workers – has been framed as cultural creativity and social cohesion: a vision that feeds into world-class city aspirations, while emphasizing the 'unique' attributes of the city in question. For example, through the promotion of Toronto as 'Creative City' (City of Toronto 2001) liveability in the inner city has been conjoined with high-value cultural production and consumption that capitalizes on the growth of arts and media-related industries.

At neighbourhood level, encouragement has been given to initiatives in which ethnic minority businesses have been an important driving force, a notable example being Business Improvement Areas (BIAs). If the majority of commercial property owners approve, a compulsory supplementary BIA levy (collected by the municipality, but managed by the BIA board) can be used for place promotion, events, beautification, street furniture and other public realm improvements. In Toronto, the appeal of the city's 'neighbourhoods' – districts associated with particular immigrant groups – has been greatly enhanced through upgrading of the public realm, instigated by the representatives of commercial real estate owners and traders that govern the BIAs. In some cases, however, valorization of cultural identity may be at odds with present-day reality. Hackworth and Rekers (2005) assess the branding of 'Little Italy'; an area that had been Toronto's main *Via Italia* during the first half of the twentieth century. However, between 1971 and 2001, Italian-speaking residents declined from nearly a third to less than 10 per cent. Streetscape improvements (funded 50:50 by the City and the BIA) created an Italian 'café society' ambience that proved commercially successful, especially for restaurant owners. As a reporter for the *Toronto Star* wryly observed (Taylor 2003: B1, quoted in Hackworth and Rekers 2005: 222), '[m]ainly, what's left is Little Italy the

196   *Stephen Shaw*

brand name, the trademark, the logo, the ethnic "swoosh". Very Little Italy.'

As Fowler and Siegel (2002) stress, Canadian cities operate within a hier-archical constitutional framework; legally they are corporations that can be created or abolished by the Provinces, and are sometimes regarded as 'mere administrative agents' of the higher government. Nevertheless, the civic leaders of the major gateway cities of Toronto, Montréal and Vancouver have demonstrated distinctiveness, with significant policy differences – from each other and from their counterparts in the United States – that owe a great deal to urban social movements that overturned the established order in the 1970s and 1980s. Their radical reform of city planning and cultural strategies abruptly halted expansion of urban motorways, and presented a far more participatory and inclusive vision of the new cultural metropolis than their predecessors. In the case of Montréal, the period of reform began in 1986, when the Montréal Citizens Movement (MCM) – a Left-libertarian coalition committed to democratizing urban governance – gained control of Ville de Montréal (Thomas 1995). Cultural and environmental revitaliza-tion was expected simultaneously to attract inward investment and inte-grate diverse groups into the futures envisaged for these cities according to principles of social equity endorsed by Mayor Jean Doré (1986–94). The desire to position Montréal as the 'Geneva of North America' no longer seemed farfetched; the fluency of the local population in two of the United Nations' five official languages, and over a hundred others, was now promoted as a valuable asset in the city's place marketing (Latouche 1994).

Quartier chinois as cultural quarter in Montréal's 'Ethnic Main'

Expressing this new emphasis on the city as a 'great cultural metropolis', Mayor Doré (Ville de Montréal 1992: iii) introduced a new City Plan that emphasized interventionism and pluralism. The Plan was formulated through extensive consultation with representatives of the city's diverse communities and was conceived as 'both an instrument to be used to improve Montréalers' quality of life as well as a social contract uniting them in a common goal: the economic and cultural development of our city'. In this new vision, explicit references were made, not only to Geneva, but also to other European cities that had developed a strong international outlook and an association with the creative arts, especially Lyon and Barcelona. In this context, the regeneration of Montréal's culturally diverse 'Main' and its public realm has been a priority. From the nineteenth century, Boulevard Saint Laurent – the thoroughfare which divided the Anglophone west and Francophone east of the city – became home to many who were neither of British nor French origin. These 'Allophones' included a Chinese community that established the *Quartier chinois* (Chinatown) in the 1860s, together with other groups, notably Italian, Greek, Portuguese and Jewish Montréalers. Although the total population of Allophones remained under

5 per cent of the city's population until 1900, their numbers increased substantially from the early twentieth century (Linteau 2000). From the 1980s – as with Canada's other two major gateway cities – immigrants from Europe became a minority. The population census indicates that Montréal's present-day immigrant population (permanent residents not born in Canada) is 28 per cent, compared with 49 per cent in Toronto and 46 per cent in Vancouver (Statistics Canada 2006), and confirms that many recent immigrants originate in countries where French is spoken as a first or second language, from developing countries in South East Asia, the Caribbean, North and West Africa.

As Germain and Radice (2006: 116) observe with particular reference to the Mile End area to the north of Boulevard Saint Laurent, the city's multi-ethnicity is now inscribed in the fabric of mixed neighbourhoods that have come to replace the old 'ethnic villages' or neighbourhoods. This means that most Montréalers cross paths daily with people from a great variety of cultures, both in the inner city neighbourhoods and in the inner suburbs. Through its policies and programmes, Ville de Montréal has placed a strong emphasis on 'interculturalism' that includes not only the Francophone and Anglophone *solitudes* of the city, but also the Allophone communities, both recent and long-established groups.

In contrast to Mile End and other districts that now have a rich cultural mix, Montréal's *Quartier chinois* has retained a distinctive identity as a 'heritage' area, both for the Chinese community in Québec and for other Montréalers (Shaw 2003). Its survival is, perhaps, more remarkable than it might appear from its visitor-oriented presentation today. As in many other Chinatowns in North America (Guan 2002; Lin 1998; Ng 1999; Yan 2002), it lost customers and merchants as new car-oriented Chinese shopping and entertainment malls were built away from the traditional downtown core. In Montréal, this decline was exacerbated by the isolating effects of an urban motorway in the 1960s, development on adjacent land and the urban blight that followed during the 1970s and 1980s. Within the district, vacant lots and derelict buildings were visible signs of low investment.

Following lobbying from various organizations of Chinese residents, merchants and voluntary societies, a 'Chinatown Development Consultative Committee' was set up by Ville de Montréal in 1990. Inter alia, this established certain principles for development. Incorporated in the City Master-plan for *secteur Ville-Marie* (city centre), the development goals identified by the local Chinese community had included: expansion of the boundaries of Chinatown; consolidation of its commercial core; strengthening housing development, as well as social, cultural and community services within the Chinese community; and improving the quality of life, safety, comfort, image and urban design in the district (Ville de Montréal, 1998: 16–17). By the late 1990s, the beginnings of recovery were apparent. Developments in downtown and the adjacent *Quartier international*, where prestigious head-quarter offices had located, were creating significant opportunities for

Chinatown merchants, especially in business and conference-related tourism. The city planners recognized, nevertheless, that without appropriate management of the area as an urban visitor destination, an upswing would also bring problems of traffic, parking and intrusion by visitors into the everyday life of the community, as well as pressure on residential accommodation through rising land values. In accordance with the guidelines of the Masterplan, the 'Chinatown Development Plan' was adopted to provide a framework for action (Ville de Montréal, 1998: 13), 'so that all efforts can be channelled in the same direction, and it is hoped, stimulate the growth of Chinatown and truly reflect the ever-growing importance of the Chinese community in Montréal and Québec society'. The over-arching aim of the Chinatown Development Plan was to consolidate the commercial core of the *Quartier chinois*. Where necessary, land use (zoning) controls would be used to protect the character of the neighbourhoods to the east and west from incompatible activities, and to prevent displacement, while improvements to existing residential buildings would also be encouraged.

Important outcomes of work carried out under the Action Plan have included the extension and upgrading of the public realm: refurbishment of the pedestrian mall and widening of pavements; murals, street trees and landscaping; conservation of heritage buildings; and the creation of a new Sun Yat Sen Park for the local community. As in other cities, the entrances to the *Quartier chinois* are marked by impressive gateways, and these have been complemented by further 'traditional' and contemporary public art. The Plan stressed that as well as being an attraction for all tourists, the *Quartier chinois* would continue to serve as a venue for a broad range of trading, social and religious activities for the wider Chinese population, and as a potent symbol of their identity, since (Ville de Montréal 1998: 11):

> it provides a major point where Chinese in Québec can meet those from other areas in the north-eastern part of North America (New York, Boston, Toronto), as well as other Quebécers. Chinatown is thus the heart of commercial and cultural exchanges within the community itself, and an eloquent statement of the dense, thriving urban culture that the Administration wishes to promote within downtown areas of Montréal.

As with Banglatown and other UK examples above, minority businesses engaged in servicing the city's expanding visitor economy are thus regarded as prime movers in the regeneration of their locality. Nevertheless, Ville de Montréal rejects the notion that the interests of the area and of the city as a whole are best served by freeing landowners and traders from the 'burdens' of regulation and control by the municipality. Nor does it accept that improvements to streets and other public spaces should be financed, planned and managed by local business associations alone. Indeed, the pressures to redevelop sites associated with the expansion of the nearby city centre are seen as an important rationale for more stringent zoning, traffic and parking controls to ensure that Chinatown continues as a living mixed-use neighbourhood. In 2002 a One Island–One City government was created, merging the 'old' Ville de Montréal with the suburbs into a city of 1.8 million. Since that time (despite the departure of some former Boroughs from the 'megacity' administration in 2005), the ruling party has been a coalition that includes members of the former MCM, and a renewed commitment to interculturalism and to 'improved services, decentralization and enhanced public consultation' (Milner and Joncas 2002: 59). With a markedly greater representation by Allophone Councillors, the symbolic significance of the neighbourhoods along the Boulevard Saint Laurent axis – including the *Quartier chinois* – has further increased in public policy and city promotion as an essential feature of Montréal as a cosmopolitan-creative city.

Conclusion

The more altruistic take on 'cosmopolitanism' is clearly manifest in the policy discourse that has justified the regeneration of Banglatown-Brick Lane in London's East End, and Montréal's historic Chinatown. In both cases, there is an explicit emphasis on celebrating the role of particular minority groups in the creative life of the city. This theme is used to raise the profile of selected streetscapes, and to promote them as new spaces of consumption, designed to attract high-spending visitors. Through a remarkable convergence of policies and practices drawn from very different ideologies, inner city areas are being re-branded as the sights and sites of a mainstream visitor economy in a way that would have seemed inconceivable just 20 years ago. If such transformations are economically successful, inner city neighbourhoods associated with immigrant communities will no longer be regarded as 'revenue sinks' that drain local taxes and discourage inward investors. Instead, they are promoted as cultural attractions that 'celebrate diversity', and complement the bright-lights attractions of the nearby city Central Business District.

A benign interpretation is that cosmopolitanism tuned to leisure and tourism consumption – most notably street festivals, 'ethnic' food, 'world' music and dance – provides a new lexicon of iconic images. In the case of London, such new themes promote the achievements of contemporary urban cultures and the performing arts of minority groups, rather than the more established post-imperial tourism narratives associated with Nelson's Column, Beefeaters, Buckingham Palace and so on. For example, in the preface to the Brick Lane Festival *Official Guide* (2004), Prime Minister Tony Blair wished the festival every success, highlighting its contribution to 'a truly inclusive Britain that takes pride in its diversity'. The attention of 'mass' audiences may provide a welcome boost, broadening their exposure, and bringing a new self-confidence to communities whose low self-esteem has long been reinforced by their social as well as spatial marginalization. Such promotion requires a simple message that can be communicated quickly and effectively to wider audiences, including international tourists, some of whom may be on a relatively short itinerary and know little of the complex social, cultural and historical context of the place concerned.

A more critical reading is that such thematic 'quarterizing' of cities (Bell and Jayne 2004) attempts to stabilize the essential fluidity of urban immigration and to create a false impression of homogeneity within boundaries, freeze-framing a particular identity through signs that can readily be digested by a fast-moving cosmopolitan elite. This carries with it the very real danger that the narrative of a particular place is de-coupled from the bigger and less palatable picture of hardship and under-employment that has been experienced by many recent immigrants from less developed countries in both the United Kingdom and Canada (Shaw 2006). There are justifiable concerns that such formulaic development of ethnic quarters may reinforce rather than challenge historic stereotypes. Thus, they will fail to stimulate the evolving hybridization and 'creative reflexivity' that is the very essence of the mongrel city, a diversity of gaze, rather than a scene of discourse and interaction (Sandercock 2006: 40). Ironically, the signposting of difference may produce an anodyne and relatively homogenous culture of consumption, disconnected from the social life of the local population. As with the proliferation of festival malls and other urban visitor attractions (cf. Bell and Jayne 2004; Hannigan 1998; Neill 2004), in time they may come to resemble 'glocalized' (see Chapter 10 in this volume) versions of a universal brand.

13 Ethnic entrepreneurs, ethnic precincts and tourism

The case of Sydney, Australia

Jock Collins and Patrick Kunz

Introduction

In his seminal work in the field of creativity, diversity and the economic advantage of cities, Richard Florida (2003: 5) argues that creativity – the ability to create meaningful new forms – 'is now the *decisive* source of competitive advantage' (emphasis in original). In his studies of the economic importance of creativity, the social dimensions of the creative class (2003) and creative cities (2005), Florida does not give much weight to the impact of immigration (2005: 40), although he develops and evaluates a Melting Pot index and argues that a cosmopolitan city is more likely to be tolerant of strangers (2005: 6). He concedes (2003: 227) that diversity (including ethnic diversity) means 'excitement' and 'energy': 'Creative-minded people enjoy a mix of influences. They want to hear different kinds of music and try different kinds of food.' (This search for difference is also explored by Hannigan in Chapter 3 and Maitland in Chapter 5 of this volume.) While Florida does not make an explicit link between cities open to immigration and the development of the creative class (2003: 254–55), his general approach implies that there should be some link between ethnic diversity and creativity in major centres of immigation (e.g. New York, Los Angeles, Chicago in the USA, and Toronto, Vancouver, Paris, London etc. in other countries). This issue seems worthy of examination in more detail.

The question whether ethnic diversity impacts on creativity and economic performance is particularly pertinent in Sydney, Australia's largest city with a population of just over four million, nearly two-thirds of whom are first and second generation immigrants from over 180 nations. Sydney, which makes it onto Florida's list of creative cities (2005: 1), is an unmistakably cosmopolitan city whose contemporary smell, taste, feel and look reflects two centuries of immigration (Burnley *et al.* 1997: 16; Burnley 2000; Collins and Castillo 1998). It is not possible to understand the creative edge of contemporary Sydney without accounting for the impact of ethnic diversity on its economy, culture, politics and lifestyle. As Connell (2000: 12–13) put it, '[t]he formal rise of "multicultural Sydney" has paralleled and dwarfed other facets of social change ... Sydney has become a place of wide-ranging social and cultural diversity.'

Part of this cosmopolitan character of Sydney stems from its ethnic economy (Light and Gold 2000: 4) and the immigrant entrepreneurs who have shaped the economic, social and cultural landscape of the city (Collins 2003; Collins *et al.* 1995). The spatial dimensions of this immigrant entrepreneurship are ethnic precincts in downtown and suburban Sydney. Another part, particularly related to Sydney's creative class, stems from the immigrants who call Sydney home, particularly, though not only, those who enter as skilled and professional permanent and temporary immigrants (Burnley 2000). The ethnic diversity of Sydney's population and the linguistic, cultural and religious diversity that accompanies it gives it a comparative economic advantage. In the age of globalization, having workers with a knowledge of the diverse languages, cultural and religious practices of the global market confers on Sydney an economic advantage that Cope and Kalantzis (1997) call 'productive diversity'. As Connell and Thom (2000: 325) argue, '[s]timulating the growth of Sydney as a global service city is the presence of a skilled workforce; an ethnic diversity in which overseas employees can feel relaxed; local amenity, both natural and cultural'. A website of the federal Department of Immigration, Multicultural and Indigenous Affairs (DIMIA 2006) cites many case studies of firms who gained the competitive economic advantages that emerge from a multicultural society. Sydney gained the right to host the 2000 Olympic Games partly because of its cultural diversity: it could claim that for nearly every Olympic team that would march in the opening ceremony there would be a community living in the city (Collins and Lalich 2000). Sydney has become the unquestioned corporate and financial capital of Australia and the site of two-thirds of those multinational corporations establishing Asia-Pacific regional headquarters in Australasia (Daly and Pritchard 2000: 167). It has also become the heart of Australia's cultural industries of the arts, film production, media and entertainment, and is Australia's most popular tourist city (Sant and Waitt 2000: 204–7).

At the same time, the national and international migrants (Burnley 2000) who help create this diversity advantage that Sydney has in the national and global economy are attracted to live and work in Sydney because of the physical and social attributes of the city. As Florida (2005: 68–69) argues, non-market forces are important in the ability of a city to attract and retain the creative class of workers (or *talent*, as he calls them). The sociology of the city thus plays an important role in the economy of the city in this regard:

> talent does not simply show up in a region; rather, certain regional factors appear to play a role in creating an environment or habitat that can produce, attract, and retain talent or human capital. Paramount among these factors . . . is openness to diversity.

> (Florida 2005: 109)

In this way, economic advantages in production interrelate with urban and suburban landscapes of consumption. We argue that it is impossible to understand the creativity and dynamism of the city of Sydney, and its ability to attract talent, without giving due weight to the role of its cosmopolitan lifestyle and ethnic diversity in this regard. The urban landscapes of consumption and recreation in Sydney are attractive in great part because of the way that immigration and the resulting ethnic diversity of the Sydney population has generated 'the world in one city' (Collins and Castillo 1998). This is reflected in Sydney's restaurants and cafés, which offer the diversity of the world's cuisines, and develop new hybrids, cooked and served by entrepreneurs and staff from all parts of the world in neighbourhoods offering a cosmopolitan ambiance as the passing parade of other consumers reflects diversity and tolerance. It is reflected also in the way that the physical landscape of the city demonstrates this ethnic diversity in its religious architecture – cathedrals, mosques, temples – and on the main streets of the downtown and suburban areas where immigrant entrepreneurs create an unmistakably cosmopolitan feel to the public spaces of the city. At the same time, national and international tourists are attracted to, or encouraged to return by, the same lifestyle factors that attract creative workers to the city.

The relationship between immigration, ethnic and cultural diversity, and the creative city thus requires much more investigation than has been the case to date. There are many dimensions of this relationship and, correspondingly, a rich and complex research agenda is required to do justice to this task. This chapter explores one dimension of the way that ethnic diversity intersects with lifestyle, creativity and tourism in Sydney by investigating neighbourhoods that have become ethnic precincts, with four – Chinatown, Little Italy, Auburn ('Little Turkey') and Cabramatta ('Little Vietnam') – chosen as case studies.

While tourism in Sydney has recently acknowledged the existence of the ethnic precincts (cf. TNSW 2002; TNSW 2004c) and the growing importance of cultural urban tourism (cf. TNSW 2004a; TNSW 2004b), the cosmopolitan nature of Sydney's people and the suburban areas in which they live has not attracted the place marketer's eye. Most of the tourist promotion of Sydney to an international audience has concentrated on downtown Sydney's landmarks of Sydney Harbour (the Harbour Bridge, the Sydney Opera House, the historical precinct of the Rocks or the newly renovated Darling Harbour precinct) (Sant and Waitt 2000: 201–9). These are now familiar icons to much of the world's population who tuned into the 2000 Sydney Olympic Games or who watch the New Year's Eve celebrations evolve around the world's time zones on television. Yet, Sydney's suburban ethnic precincts, such as Little Vietnam, Little Turkey, Little Italy, together with downtown Chinatown, give visitors the opportunity to experience the cosmopolitan global city as locals do in the suburbs where most of Sydney's people, including the creative classes, live.

This raises a number of important, related questions. How is the ethnic precinct situated within the cultural urban tourism context of contemporary Sydney? What is the role of the ethnic entrepreneurs in the development of the iconography and ambiance of these ethnic precincts? How do government officials and regulators shape the emergence, character and promotion of these ethnic precincts? And, what do tourists think of the ethnic façade of the ethnic precinct and of the authenticity of their urban tourist experience within Sydney's ethnic precincts?

The chapter attempts to answer these questions. It begins by offering a brief review of the relevant, interdisciplinary literature, before presenting a brief overview of the four ethnic precincts under investigation. Following a brief outline of the research methodology, the remainder of the chapter presents and analyzes the findings of the fieldwork conducted in Chinatown, Little Italy, Auburn and Cabramatta involving ethnic entrepreneurs, government officials, ethnic community organization leaders and customers (tourists and locals). The final section draws together the key findings of this research, with a particular focus on the complexities and contradictions that are inherent in the creative commodification of ethnicity for tourism purposes in global cities such as Sydney.

Literature review

The literature, both specific and broad, associated with the role of ethnic entrepreneurs within ethnic precincts in cultural urban tourism contexts has a number of significant shortcomings. These shortcomings can be summarized into four main points. First, there exists a dearth of information linking ethnic entrepreneurship to tourism phenomena of any sort. This is attributable to the lack of investigation into the field by researchers from both disciplines. Ethnic entrepreneurship research up to, and, including, the 1980s was either entrepreneur-specific (examining an ethnicity, or ethnicities), or industry-specific, based on the clothing industry (cf. Waldinger 1986). Throughout the 1990s, the research spotlight (slowly) shifted and highlighted the service industries. Today, tourism, as a service industry, has finally been acknowledged by ethnic entrepreneurship researchers (cf. Collins 2002; Halter 2003; McEvoy 2003; Rath 2002). Halter (2003) investigates the development of cultural urban tourism to and in the ethnic precincts of Boston, USA, and makes passing reference to ethnic entrepreneurs. McEvoy (2003) presents an excellent insight into the formation of what effectively is a Little India, in the area of Rushholme, Manchester, United Kingdom, and the key role of the ethnic entrepreneurs – not Indians, but Pakistanis – in its emergence as an ethnic precinct. Rath (2002) and Collins (2002) identify the various components of ethnic precincts – producers (ethnic entrepreneurs, specifically), consumers (visitors) and critical infrastructure members (professional place promoters, local councils, chambers of commerce) – through the explicit and implicit use of Zukin's (1995) symbolic economy concept. Ethnic entrepreneurs figure prominently in both studies.

Second, in regard to the urban tourism literature, there is an almost stubborn investigative focus on the 'city' as the 'urban'. Most urban tourism studies assume a single-city or inter-city approach (Pearce 1998a, 1998b, 1999, 2001; Shaw and Williams 1998; Shaw and Williams 2002), though there are some multiple city comparisons (Van den Berg *et al.* 1995). What's more, the 'city' is usually ambiguously defined around the inner-most localities, creating a boundary that excludes much of the wider metropolitan area; areas that are similarly urbanized, areas that often house the greatest proportion of the population, and areas that also have the potential for tourism – in sum, urban spaces that are equally as important and interesting to examine. These issues have precluded the examination of other 'urban' spaces, which similarly have urban tourism-related functions.

Third, research into urban spaces with urban tourism-related functions lacks comprehensive, holistic micro-level investigation. Some research does focus on certain specific issues within ethnic precincts in the city (cf. Chang 2000b; Conforti 1996; Fainstein and Powers 2003; Halter 2003; Schnell 2003). Schnell (2003), for instance, investigates the ambiguities of authenticity – that is, the different meanings of authenticity – within Little Sweden in Lindsborg, Kansas, USA, while Fainstein and Powers (2003) examine the marketing dynamics of New York City's ethnic precincts. Frenkel and Walton's (2000) examination of the Bavarian-influenced town of Leavenworth in Washington, USA, is another good example. The authors (implicitly) aim to illustrate how the urban space functions, but do not present a systematic analysis of how the various parties involved interact with one another and, ultimately, how these social relations between the main actors in the ethnic precinct shape urban space.

Fourth, and linked to the micro-level studies, there exist few such investigations that employ comparative approaches: that is, many ethnic precincts within the same city – intra-city research (cf. Collins 2002; Timothy 2002).

It is clear that there exist many fruitful (and eclectic) paths for further research into the development of ethnic creativity for tourists within the contemporary cosmopolitan city. The aim of this chapter is to examine the complex and contradictory interaction of ethnic entrepreneurs, customers and critical infrastructure members in the development of Sydney's Little Vietnam, Little Turkey, Little Italy and Chinatown in order to examine the commodification of ethnicity in the urban tourism context.

Sydney's ethnic precincts

As the Vietnamese, Turkish, Italian and Chinese immigrants entered Sydney, they demonstrated clear patterns of initial settlement concentration. Chinese and Italian immigration have long histories in Sydney (Collins and Castillo 1998), dating back to the last decades of the nineteenth century when the Chinese settled in Haymarket's Chinatown (cf. Fitzgerald 1997) and Italians settled in Leichhardt in the city's inner-west (cf. Burnley 2001). Sydney has had a Chinatown since the 1860s (Anderson 1990), moving to Campbell and Dixon Streets in the city in the 1940s, where it is still located today (Collins and Castillo 1998; Fitzgerald 1997). Religion and commerce were at the centre of these flourishing communities, with immigrant-owned business enterprises in these neighbourhoods providing the new immigrant arrivals with goods and services in a business environment of cultural under-standing and linguistic familiarity (Burnley 2001; Fitzgerald 1997; Inglis *et al*. 1992; Jupp *et al*. 1990; Solling and Reynolds 1997; cf. Jupp 1995; Turnbull 1999). After a time, immigrant residential mobility meant that later genera-tions of Chinese and Italians moved to different areas of the city, but the immigrant entrepreneurs remained, continuing to give the precinct its ethnic character (cf. Burnley 2001; Fitzgerald 1997; Inglis *et al*. 1992; Viviani *et al*. 1993). Today, 91 per cent of the enterprises in the Little Italy precinct are owned by Italian-born entrepreneurs and 89 per cent of the enterprises in the Chinatown precinct are owned by Chinese-born immigrants.

Following the end of the 'White Australia' policy in the 1960s, new Viet-namese immigrants often settled in Cabramatta in the 1970s (cf. Burnley 2001) and Turkish immigrants settled in Auburn in the 1980s (cf. Jupp *et al*. 1990). The ethnic precinct in Cabramatta (dubbed 'Vietnamatta' by the sensationalist tabloid media) emerged after an inflow of Vietnamese refu-gees from nearby migrant hostels (Collins 1991). Along John Street, which runs along the western side of Cabramatta railway station, a vibrant ethnic precinct has grown up with over 820 ethnic businesses ranging from bakeries, butcheries, cake shops, confectioners, arts and crafts, dress materials and fabrics, bridal wear shops, clothing retailers and manufacturers, electrical goods suppliers, fish markets, general food stores, take-away foods, fruit shops, groceries, hair and beauty salons, herbalists, jewellers, laundries, newspaper proprietors, newspaper publishers, delicatessens and food importers, and manufacturers, to professional services such as doctors, accountants, dentists and lawyers (Burnley 2001: 252). Today, 86 per cent of the entrepreneurs in the Cabramatta precinct are Vietnamese-born immi-grants, while another 10 per cent are Chinese-born immigrants.

Auburn is different from the previous three ethnic precincts in that it is a newly emerging precinct. Up until the 1970s, Auburn was a white working class suburb of predominantly Australian-born individuals or Anglo-Celtic immigrants. In the past three decades – that is, much later than the other precincts – immigrant minorities began to move into the area. It has thus escaped the attention of those interested in ethnic entrepreneur research. Today, Auburn has a large number of residents who are immigrant minori-ties, particularly the Turkish-born. This is reflected in the current composi-tion of the 'entrepreneur-scape' of the suburb: 78 per cent of the enterprises in the Auburn precinct are Turkish-born immigrants, while another 14 per cent are Chinese-born immigrants.

In Sydney then, 'ethnic precincts' are urban or suburban high street agglomerations of ethnic enterprises, clustered together in a space, which formally or informally adopt the symbolism, style and iconography of that ethnic group in their public spaces. Ethnic eating places – restaurants, cafés, take-away food shops – and shopping establishments are the prime attraction to many precinct visitors, although ethnic professionals often occupy the upper floor spaces, providing services to co-ethnics. As Selby (2004: 24) reminds us, 'ethnic tourism enclaves' are presenting many countries with opportunities:

> The potential for 'ethnic tourist attractions' is now widely recognized by local governments and tourist boards, encouraging the development of areas such as Brick Lane in London [see Shaw, Chapter 12 in this volume], Little Italy in Boston and the Balti Quarter in Birmingham.

In terms of developing such enclaves for tourism, important questions arise. How do ethnic restaurants and other immigrant enterprises symbolize their ethnicity in the restaurant/shop décors, menus, signage and divisions of labour? How does a precinct as a whole develop as an 'ethnic' place/space? And, what roles do regulation and regulators play in developing the ethnic precinct, in shaping the symbolic representation of ethnicity within the precinct, and in place marketing the precinct? In order to explore these questions, we conducted extensive fieldwork in each of the four ethnic precincts in Sydney.

Methods

A qualitative methodology informed the project's overall design (Denzin and Lincoln 2000), while multiple exploratory case studies, with embedded cases, served as the research strategy (Yin 2003). Zukin's (1995) three components of ethnic precincts in cultural urban tourism contexts (producer, consumer and critical infrastructure) comprised the conceptual framework for the fieldwork. In order to explore the dynamics of the commodification of ethnicity, we conducted fieldwork in 2004 and 2005 in these four Sydney ethnic precincts, which involved semi-structured interviews (Jennings 2001) with customers (consumers), ethnic entrepreneurs (producers) and regulators (critical infrastructure). These four ethnic precincts comprised the case studies in which all parties involved in (cultural) urban tourism processes could be studied, and their interactions and contradictions explored (see Figure 13.1).

Data collection, conversion and analysis was two-stage (Miles and Huberman 1994). The research sample comprised 240 individuals: five producers, twenty-five consumers and five critical infrastructure members in each ethnic precinct, per stage (the same producers and critical infrastructure members were included in both stages). Respondents were selected through two sampling techniques. Producers and critical infrastructure members were selected by way of purposive sampling (Jennings 2001; Punch 2001), and consumers, by way of purposive-random sampling (Jennings 2001; Patton 1990; Punch 2001). Interviews were variable length, and occurred on-site (ethnic precinct) and off-site, indoors and outdoors (consumers). (A small number of producers and critical infrastructure members, for purposes of appropriateness, were interviewed once only, and were simply contacted in stage two.) Direct observations of the ethnic precincts were used as a secondary mode of data collection (Yin 2003), as were scratch, field, head and analytical notes (Punch 2001; Rehman 2002). Data conversion and analysis occurred sequentially over the two stages of the research. This encompassed all collected data. The data were analyzed through descriptive, interpretive and pattern coding (Miles and Huberman 1994). QSR N6® was used as the data analysis tool.

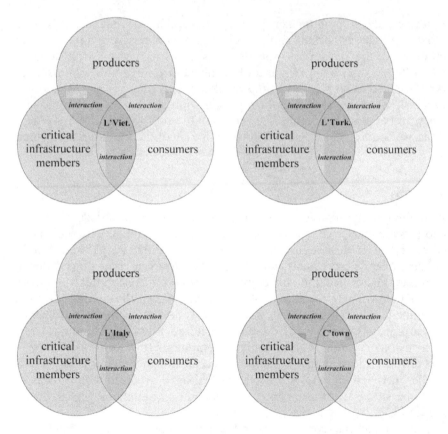

Figure 13.1 Conceptual design application.

Source: adapted from Zukin (1995).

Notes
L'Viet. (Little Vietnam); L'Turk. (Little Turkey); L'Italy (Little Italy); C'town (Chinatown).

Fieldwork in four Sydney ethnic precincts

We divided the customers that we surveyed in the ethnic precincts into three groups: *co-ethnic* (e.g. Chinese in Chinatown), *co-cultural* (e.g. non-Turkish Arabic immigrants in Little Turkey) and *Others* (the rest, including other minority and majority groups). Co-ethnic consumers often accused the Other customers of *gawking*, which angered them a great deal. The Other consumers, in response, steadfastly refuted the claims of gawking. They scoffed at the suggestions and charged that the respective co-ethnic consumers were being overly sensitive. The Other consumers flatly denied being attracted to the relevant ethnic precinct because of the highly visible co-ethnic crowds within. Even so, all the Other consumers later admitted that highly visible co-ethnic crowds did, in fact, contribute to the authenticity of the ethnic precinct's ethnic association.

All groups of consumers identified the ethnic entrepreneurs in the precinct as a key reason for visiting the ethnic precinct; in this sense, for customers in the ethnic precinct, the ethnic entrepreneurs were the ultimate attraction. Co-ethnic consumers were attracted to the ethnic precinct because of the ethnic products and/or services available in the ethnic enterprises located there. They were reluctant to purchase any products and/or services from any other entrepreneurs. But co-ethnic consumers were not attracted to the ethnic precinct because of the highly visible façade of the precinct per se. Rather, they were always strongly opposed to it, viewing it as kitsch, and, on occasion, even inauthentic and offensive. The co-cultural consumers were also attracted to the ethnic precinct because of the ethnic goods or services supplied by the ethnic entrepreneurs therein, but viewed the ethnic façade of the precinct with neutrality: they were neither actively in favour of it, nor actively against it. They came for the goods and services, not the 'ethnic experience'. The Other consumers were mainly attracted to the ethnic precinct to experience ethnic diversity as represented by the highly visible ethnic façade and feel of the precinct itself, which was considered tasteful and authentic – largely available because of the ethnic entrepreneurs and their enterprises. The Other consumers also requested that the signage in the precinct also be in English, that is, dual language signage.

Fieldwork with ethnic entrepreneurs (Zukin's [1995] 'producers') in each precinct revealed that there was a lack of communication between them. This was perceived as regrettable and explained by business time constraints. In Little Turkey and Chinatown, none of the five entrepreneurs interviewed were members of the local organization and/or enterprise association, while in Vietnamatta and Little Italy only one entrepreneur in each precinct was a member of the ethnic enterprise association, but these individuals rarely attended meetings. There was a complete lack of collective action in every one of the ethnic precincts, with price wars between ethnic entrepreneurs introducing competitive, rather than co-operative, social relations between them.

Ethnic entrepreneurs noticed a difference in customer behaviour. In Vietnamatta, Little Turkey and Chinatown, varying degrees of consumer price sensitivity were evident: co-ethnic consumers in each precinct were deemed excessively price sensitive; co-cultural consumers in each precinct were also seen as price sensitive, though to a lesser degree than co-ethnic consumers; Other consumers were consistently regarded the least price sensitive of all consumers and were prepared to pay premium prices for products and/or services because of 'uniqueness'. Some producers in each precinct occasionally found Other consumers to be rude and/or condescending, or arrogant, but ultimately all were prepared to put up with this because of the increased income potential of Other consumer patronage. A small number of producers thought that ethnic entrepreneurs were the ultimate attraction for consumers to the precinct.

Ethnic entrepreneurs in the ethnic precincts were critical of local government and state government in the ethnic precinct (Zukin's [1995] 'critical infrastructure') for their lack of personal support and a lack of communication with the ethnic entrepreneurs. Many ethnic precinct programmes were considered by the entrepreneurs to be poorly promoted, and/or inconveniently timed and/or located, and of low quality. All producers commended the local council for their streetscape revitalization efforts in the relevant ethnic precinct, but felt that it lacked direction. In Vietnamatta and Chinatown, the ethnic entrepreneurs were opposed to the predominantly 'Asian' culturally specific matter of the wider ethnic precinct and wanted a more specific 'Vietnamese' or 'Chinese' focus. In Little Turkey, producers strongly supported the establishment of a more highly visible Turkish feel to clearly differentiate Auburn from other ethnic precincts. In Little Italy, producers strongly supported more specific 'Italian' designed street furniture and the like. Producers were also complimentary of the organizing of ethnic festival(s) in the precincts, though they wanted them with more regularity throughout the year. Vehicle parking availability and vehicular traffic flow problems were the cause of most concern for a number of producers, while in Vietnamatta and Chinatown, crime-related problems were thought to dissuade consumers from frequenting the precincts.

The fieldwork also took into account the viewpoint of those who were part of the regulators and government authorities who oversaw the development, feel and promotion of the ethnic precinct. Critical infrastructure personnel in Vietnamatta, Little Turkey, Little Italy and Chinatown viewed the entrepreneurs as completely unreceptive to their efforts to build the ethnic precinct, since few producers actually participated in their programmes. Critical infrastructure members' remarks about consumers, in the four ethnic precincts, were based around a single topic: satisfaction. The critical infrastructure members were confident that all consumers in the precinct were very satisfied with it, making no distinction about how such views are shaped by the ethnic background of the consumers.

Our research into Sydney's ethnic precincts has also revealed a number of contradictions that underlie the development of the precincts and their marketing to locals and tourists alike. First, there is the problem of credibility and authenticity, which involves who is 'authorized' to claim authenticity, how that authenticity is symbolized, and what employees and employers in ethnic enterprises have to do to generate that authenticity. As Meethan (2001: 27) put it, symbols 'are multivocal, that is they have the capacity to carry a range of different, if not ambiguous and contradictory meanings'. Central to this approach is the view that tourists are not passive dupes, but active agents shaping their tourist experience, though within constraints. Moreover, as Meethan (2001: 94) reminds us, authenticity is a matter of negotiation and ascribed meaning. Because authenticity is subjective, more research is needed in Sydney to investigate the customer or 'demand' side of the commodification of ethnic diversity, including new research on the characteristics of consumers in ethnic precincts that is sufficiently disaggregated to allow distinctions between local, city, national and international tourists, the type of tourism, and the ethnic background and immigrant status of the tourists themselves (if appropriate). Of interest here are the ways and means through which different people gain knowledge of and form expectations about different ethnicities in general, and Sydney's ethnic precincts in particular, including issues related to interpretation.

This leads to the second contradiction, arising from the problem of how legitimate a precinct can be in the eyes of the co-ethnic community, other locals and tourists if it has been developed by deliberate regulation, planning and government intervention. Some of Sydney's ethnic precincts have had a series of 'ethnic facelifts' over time in an attempt to give the precincts an authentic ethnic feel or theme. For Chinatown, for example, the redevelopment of Dixon Street began in 1972, with the introduction of portico and lanterns, and rubbish bins with 'traditional' Chinese symbols. Anderson (1990: 150) argues that Sydney's Chinatown has been revitalized in ways that reflect white Australia's image of Chinese-ness: 'Making the area more "Chinese" . . . [meant] making the area appear more consistent with the architectural motifs and symbols of ancient China.' That is, attempts by city councils in Sydney to 'create' a Chinatown in the image that would attract tourists has led to the introduction of façades, monuments and facelifts that reflect stereotypical images of homogenous, ancient 'Chinese-ness' that exist only in the imagination of the 'white gaze'. Is it possible to develop ethnic precincts, such as Sydney's Chinatown, without necessarily reproducing white stereotypes of the ethnic 'Other'? Attempts to redevelop Sydney's Chinatown have led to internal struggles within the Chinese community over the right to gain representation on the relevant development and planning committees (Anderson 1990). This is not a trivial issue, since there are more than 100 ethnic Chinese community organizations in Sydney alone. Which voice is representative, is authentic? How is this decided? This reality of complex, diverse and changing Chinese communities in Sydney flies in the face of simplistic attempts to construct *the* authentic Sydney Chinatown.

Third, the issue of tourist safety is central to any government tourist strategy. No one wants to go to a place where their or their family's safety is put at risk (as Hannigan points out in Chapter 3, what people usually want is a 'controlled edge'). Control and surveillance are thus embedded in the development of tourism in general (Body-Gendrot 2003), as well as in terms of potential tourist precincts such as ethnic precincts. In relationship to the latter, the historical construction of minority immigrant communities as criminal (Collins *et al.* 2000) gives the places and spaces where they concentrate a criminal reputation. This is reinforced by the way that racism, prejudice and xenophobia construct immigrant minorities as the criminal 'Other' who are a threat to the safety of the host society (Poynting *et al.* 2004). Ethnic precincts of minority immigrant groups are thus constructed as places of crime (gambling, drugs and prostitution) and of criminal gangs at the same moment that they become exotic ethnic places. Cabramatta has developed a reputation as heroin capital of Sydney, while Chinatowns have always been associated with vice, gambling and crime (Kinkead 1993: 36–43).

Conclusion

This chapter has argued that immigration and ethnic diversity play a key role in the development of creativity in Sydney today. This aspect of cosmopolitan diversity requires greater attention in the emerging literature on creative cities. Important here is the way that cosmopolitan lifestyles play an important role in attracting, and retaining, talent or the creative classes who are necessary parts of the economic dynamism that creativity imparts. Critical to this investigation is the way that relations of production (immigrant enterprises), consumption and regulation intersect to generate cosmopolitan neighbourhoods of diversity and tolerance that provide places and spaces for the lifestyles that the creative classes enjoy and increasingly demand. This in turn gives cosmopolitan cities such as Sydney a comparative advantage in attracting more national and international immigrants to swell the ranks of Sydney's creative classes and add to Sydney's economic dynamism in the global economy.

In order to explore some dimensions of the immigration, ethnic diversity and creativity interface, this chapter has investigated the commodification of ethnicity in four ethnic precincts in Sydney and how that is related to the tourist gaze. Ethnic precincts are one important site, though not the only site, of the commodification of ethnicity as part of the creative development and promotion of urban tourism in contemporary cosmopolitan metropolises such as Sydney. Following the schema developed by Zukin (1995), our fieldwork with producers (ethnic entrepreneurs), consumers and critical infrastructure members (regulators and ethnic community representatives) in Chinatown, Little Italy, Vietnamatta and Little Turkey has demonstrated that immigrant entrepreneurs are a necessary, thought not sufficient, component in the emergence of ethnic precincts. In each precinct, the immigrant population was very diverse – they were multicultural suburbs – but the entrepreneur-scape was dominated by the immigrant groups who comprised around 80 per cent or more of all entrepreneurs in the suburb and whose ethnicity was the ethnicity attached to the development of ethnic theming in the precinct.

This study reveals very different consumer responses to, and attitudes about, the commodification of ethnicity depending on the ethnicity of visitors. Those consumers who are of the same ethnicity as the entrepreneurs who give the precinct its ethnic character (co-ethnic consumers) have a slightly different take on the precinct than those consumers who are also of a slightly different immigrant background (co-cultural consumers), but a very different take compared to consumers who are either non-immigrants or who are immigrants from a different region (Other consumers).

The emergence and development of each ethnic precinct is also shaped by planning policies and local development procedures by local and provincial government authorities and ethnic community organizations: the critical infrastructure. The fieldwork revealed that the members of the critical infrastructure often saw the ethnic precinct and its problems in a very different light than the entrepreneurs in the precinct.

Clearly, this study highlights the contradictions that emerge in the social relations between the major groups within the ethnic precinct. Ethnic entrepreneurs in the ethnic precinct were more often in conflict and competition with each other than they were a united force, rarely communicating with one another: the façade of ethnic unity in the precinct was not matched by social relations between the ethnic entrepreneurs who were the precinct's backbone. Customers of the ethnic precinct, locals and tourists, were not homogenous in their response to the development and promotion of ethnic precincts. In this chapter, we have attempted to sketch the different ways co-ethnic, co-cultural and Other consumers view ethnic authenticity in the ethnic precinct. Put simply, it is the Other consumers who are most attracted to the fabrication of the ethnic precinct, though they requested dual language signage to guide them through the precinct. Co-ethnic customers were generally critical of the streetscape revitalization efforts that were thought to be kitsch, even inauthentic and offensive. Co-ethnic consumers often accused their corresponding Other counterparts of gawking, a central contradiction of the ethnic precinct where the chance of experiencing the 'exotic Other' is what is marketed.

Relationships between entrepreneurs and customers in the ethnic precinct are also complex. The ethnic entrepreneurs saw co-ethnic customers as too price sensitive and Other consumers as being sometimes rude, condescending and arrogant. There was a very obvious rift between how the producers saw the efforts of the critical infrastructure in their precinct and how the members of the critical infrastructure saw it. Critical infrastructure members were regularly criticized by the corresponding producers for their lack of personal support. Many programmes were considered to be poorly promoted, inconveniently timed and/or located, and/or of low quality. On the other hand, members of the critical infrastructure praised their own efforts and criticized the producers for a lack of appreciation of, and participation in, their efforts to build and promote the ethnic precinct.

The fieldwork has thus pointed to the contradictory nature of the development of ethnic precincts as part of the creative urban planning process of stimulating urban tourism. Contradictions emerged in each precinct about the authenticity of the ethnic theming that each precinct developed, about the legitimacy of getting advice from one part of the ethnic community and not the others, and about the safety of the ethnic precinct. Moreover, Sydney's place marketers seem fixated on the iconography of downtown Sydney, its harbour, Opera House and beaches. For many international visitors, the cosmopolitan nature of Sydney's people, suburbs and ethnic precincts is a welcome surprise.

Part 4

Creative industries and tourism

14 Economic clustering, tourism and the creative industries in Plymouth

Developing a practical tool for impact assessment

Kevin Meethan and Julian Beer

Urban culture, creative industries and economic growth

Culture, as the other contributions to this volume make clear, has become an increasingly important element in regeneration strategies, and so too has the development of business clusters that benefit from being located in urban areas to create competitive advantage (Landry, 2000; Porter 1995) and also act as attractions which draw in the tourist market. The material presented in this chapter specifically focuses on the relationship and interplay between aspects embodied in one regeneration project: the University of Plymouth's attempts to take the lead in a flagship capital project in the city of Plymouth. This involves the creation of a 'cultural zone' which aims to develop and integrate the creative industries with higher education, to develop entrepreneurial skills, and as a consequence create a vibrant cultural milieu that will also act as tourist destination.

The form and function of cities within the developed economies has changed dramatically over the past three decades (Hall 2000). As we move into a connected and mobile world of global networks and flows of people, capital, commodities and knowledge (Castells 1989, 2000), the boundaries that once existed between the urban spaces of manufacturing and the urban spaces of consumption, the tourist resorts and enclaves of the nineteenth and twentieth centuries have been eroded. Urban centres are no longer places one leaves to go on holiday, but where one can go to for a holiday. The attractions of a city are not of course the same as those of the traditional seaside resort, nor as the more exotic destinations that seek to create an image of utopian paradise on earth. What a city offers is often an eclectic mix of the historic and modern, of culture and consumption which adds up to a varied visitor experience centred on the 'urban buzz' or vibrancy of the particular destination. The urban environment offers a wide range of buildings in the public realm, which as well as accommodating attractions for the visitor often become attractions in their own right. These buildings are not necessarily modern or 'iconic', and more often than not include those that have been adapted in function if not form and are a legacy stemming from the industrial past. Activities and attractions for the visitor in the urban environment include a wide variety of shops, art galleries, museums, parks, promenades and places of entertainment, films, plays, shows and concerts, places to drink and dine: in short, all the trappings of what we can regard as a metropolitan lifestyle that offers a wide range of places and experiences for the visitors as well as the residents. Of course it is easy to imagine some cities as being more successful at attracting visitors than others, perhaps by virtue of their historical legacy (Atkinson 1997; Cohen 1997; Meethan 1996, 1997) or perhaps because they host festivals and other major sporting events (Dimmock and Tyce 2001). For along with natural physical advantages, such as water frontage, rivers or canals, the distinct cultural capital of an urban space (Zukin 1995) is often what distinguishes it from competing destinations. This effectively becomes its unique selling point (see for example Ehrlich and Dreier 1999). A sense of place that can be packaged and sold as a lifestyle opportunity also attracts inward investment in addition to any tourist spend and associated services, and as such becomes a resource to be used and developed for regeneration strategies (Ashworth and Voogd 1990; Bell and Jayne 2004; Bianchini and Parkinson 1993; Judd and Fainstein 1999).

As Bell and Jayne (2004) argue, the most successful cities are those which are most culturally diverse, in that they offer innovative spaces of consumption, which in turn helps to attract inward investment, tourists and consumers. In the United Kingdom, the development of such local regeneration strategies involving elected representatives as well as those from business and other stakeholders, dates from the early 1990s (Coulson 1997). Since then, forms of urban regeneration have increasingly involved more than physical developments and environmental improvement (Zukin 1995), and as tourism becomes more fragmented and niched (Douglas *et al.* 2001) urban centres are well placed to offer a variety of attractions and activities to cater to these needs. It is culture, in the widest sense of the word, which has been viewed as adding the extra 'value added' dimension to urban life. As Hall notes, 'culture is now seen as the magic substitute for all the lost factories and warehouses, and as a device that will create a new urban image, making the city more attractive to mobile capital and mobile professional workers' (2000: 640). Such sentiments are also echoed in official UK government circles as the following quotation from the Office of the Deputy Prime Minister shows:

> There is growing evidence that the most successful modern cities possess distinctive cultural, artistic and sporting assets which have been progressively developed as key strands of the city's overall strategy. Both investors and visitors cite the significance for them of distinctive and iconic assets.

(2003: 16–17)

Culture, the creative industries and economic clustering

The term culture is often the overarching one used to define and describe the physical and varied social activities prevalent in any area. However, an increasingly important component in the fabric of the cultural make-up of any location is the role of those individuals and organizations whose activities contribute to the economic good, as well as contributing to cultural life. In light of this, the UK Department for Culture Media and Sport (DCMS) advocates the use of a standard analytical definition that views culture in both material and non-material terms, but most particularly as a cyclical process of production and consumption that includes visual art, performance, audiovisual, books and the press, sport and health and also heritage and tourism. Based on the logic of the production chain, this definition acknowledges the importance of the totality of interlinked processes involved, which in turn, allows for more efficient and targeted intervention measures (DCMS 2002: 8–9). Such a definition is different to other broader categorizations of culture that encompass 'ways of life' (Meethan 2001). For the purpose of this chapter it needs to be noted that, in terms of policy interventions, the terms 'culture' and 'creative industries' cannot be disaggregated.

The creative industries are characterized by specific forms of organizational and working patterns that rely on extensive networking, self-help and collaboration and thus tend to favour flat hierarchies and partnership, and consequently arc also dominated by micro businesses, the self-employed and sole traders. As such, the sector thrives on innovation, mobility and flexibility. It is at the core of the expanding knowledge-based industries and creates added value from intellectual property rights. In short, the sector exhibits many of the characteristics of the postfordist knowledge economy (Castells 1989). The sector also tends to cluster and thrive in city centres, supporting the economy in a number of ways and contributing also to a diversity of usages and activities, which Castells (2000: 36) has termed the 'milieux of innovation', more usually referred to as economic clustering. This concept, as both an empirical description and as a tool of policy intervention, is predicated on both the proximity of actors within a sector, and their consequent ability to innovate through knowledge exchange in order to compete and move up the value chain (de Berranger and Meldrum 2000; Florida 2002; Gibson 2003; Gibson *et al.* 2002; Hall 2000; Kirat and Lung 1999; Porter 1995). In turn, such interactions rely on the extent to which such relationships are embedded in a shared set of norms and both formal and informal institutional practices (Amin and Thrift 1994; Kirat and Lung 1999). In the United Kingdom the encouragement of clusters across all sectors of the economy was established in central government policy in the 1998 *White Paper on the Knowledge Driven Economy* (Department of Trade and Industry 1998).

The creative industries are also an attractive and popular employment option for younger people, as they allow scope for expression, often at the edge of developing technologies. The sector is also highly skilled, with 32 per cent of the United Kingdom's creative industry's workforce educated to degree level. Another crucial characteristic is that this sector has one of the fastest growth rates in the United Kingdom. The creative industries grew at 6 per cent per annum between 1997 and 2003, twice the growth rate of the overall economy during the same period (DCMS 2006). Given these characteristics it is hardly surprising that the sector now finds itself at the forefront of regeneration, planning and the re-definition of cities in the United Kingdom and across the world. In the United Kingdom and elsewhere, fostering 'creative quarters' or 'zones' as they are now often termed, especially in urban areas to aid regeneration, has been widespread (see Shaw, Chapter 12 of this volume). Examples from the United Kingdom include a wide range of initiatives such as the Creative Quarter in Manchester, the Jewellery Quarter in Birmingham and the Baltic/Sage in Gateshead. In addition, Bell and Jayne (2004: 4) also make the point that the stimulus for such developments is the existence of a political or institutional vision, and the activities of entrepreneurs and other cultural intermediaries that together, create a critical mass (see also Gibson *et al.* 2002). However, regeneration methods are increasingly understood as being more than physical developments and environmental improvement. Building a sustainable city requires that local communities of interest are empowered and supported (Hughes 1998). The production and support of creative people within this environment is now often seen as a vital element of the regeneration process, and in turn, so too is the support of higher education institutions.

Higher education in regeneration and clustering

The role of arts in higher education institutions (HEIs) within cities has not been widely discussed within the creative industry and urban regeneration debate in the United Kingdom. However, in research undertaken by Universities UK (Charles and Benneworth 2001) HEIs are singled out as providing specific, positive added value to a town, city or region. In terms of their direct contribution, the research highlights four main ways in which higher education arts institutions actively and indirectly support local and regional cultural activities: by providing infrastructure, which the community can use; a programme of activities in which the community can participate; expertise and international connections; and through their students, the highly flexible labour force necessary for the development of a strong cultural sector.

Overall, HEIs provide a range of facilities that contribute to local culture, ranging from small, local facilities to internationally acclaimed institutions. Many arts departments and faculties across the country offer galleries, museums and concert venues open to local artists and the general public. A number of higher education institutions have been extremely innovative in their design of new buildings, and have recently won awards for the high quality of design. Higher education institutions are also large and adept project managers for cultural activities and cultural societies within universities are active in developing linkages with local communities, often being instrumental in organizing cultural festivals that serve to strengthen and sustain their cultural profile and involvement by and in the local community. Finally, students often demand a very cosmopolitan market for certain cultural products – meeting that demand results in activities of local benefit, such as the presence of popular music venues, the market for which might not otherwise develop (see also Russo and Arias Sans in Chapter 10 of this volume). Universities in general can contribute to all of these aims, but arts and cultural schools and faculties have a particular and one could argue vital role in providing cultural capital in places. Sheffield and Brighton, which are now recognized internationally for their successful focus on the creative industries sector within regeneration, both provide good, long term examples of the direct, positive impact arts education can have on the local economy and urban re-development and tourism. Before moving on to look at the role of the University of Plymouth in trying to fuel the development of the creative industries in the city of Plymouth, we need to examine the regional and local context.

Plymouth and South West England: economy, tourism and vision

Since the mid 1990s, the creation of regional government offices, and the development of growth partnerships involving businesses and local government agencies to push forward regeneration, has been a significant feature of government policy in the United Kingdom (Coulson 1997). This was consolidated further in 1998 with the establishment of Regional Development Agencies with a remit to develop economic growth and to actively promote the regions through support for marketing.

The economy of the South West region is diverse, and the majority of businesses employ fewer than ten people, with few large scale employers. There are also a number of notable sub-regional disparities that reveal a clear east–west divide. The West of England contributes one-quarter of the region's GDP, with Bristol alone accounting for almost half of this. At the other end, both economically and geographically, Cornwall contributes the least at 7.3 per cent, and is recognized by the European Union as an Objective 1 area, while parts of Devon are recognized as an Objective 2 area. Overall, Wiltshire and Gloucestershire, and the West of England have experienced the highest rates of GDP growth, while growth is slower in the coastal counties of Cornwall, Devon and Dorset. In the South West, regional disparity and the preponderance of Small and Medium Sized Enterprises (SMEs) makes the task of stimulating economic growth particularly difficult for those charged with economic development (SWRDA 2003).

Over the past decade, the South West has experienced significant in-migration of economically active people looking for a greater quality of life, as well as those of retirement age. As Shaw and Williams (1994) have shown, much of this migration consists of lifestyle entrepreneurs engaged in tourism businesses. Conversely, the region is a net exporter of its youth, including graduates, to other parts of the country and this process has been marked in Plymouth (MBM Arquitectes and AZ Urban Design Studio 2003). In general terms, figures from the 2001 Census reveal that Plymouth is the second largest city in the South West region, behind Bristol. However, the 2001 Census also revealed that the population, at 240,000, had decreased by 10,500 or 1.1 per cent since the 1991 Census; while in contrast, population grew in all other local authority areas in the region over the corresponding period.

Historically, Plymouth's economy has been largely dependent on the defence sector, but while defence employment has declined considerably in the last two decades, it remains important to the local economy. In addition to defence, the public sector is prominent elsewhere in Plymouth's economy, with more than 29 per cent working in public administration, education and health related employment. Manufacturing employment is in line with national averages, but employment in distribution, hotels and restaurants and financial services is lower than the national average (GHK Consulting 2004).

Plymouth has had some success in attracting inward investment since the 1970s, and as a result a number of multinationals, especially in the manufacturing sector, are located in the city. However, reliance on foreign or other outside investment has become a particular concern recently, as several high profile businesses in Plymouth have closed their operations or shifted them overseas, leading to heavy job losses. Of those businesses that remain, many are in relatively low value manufacturing or assembly type operations with limited R&D or input of local knowledge. In recent years, Plymouth has proven to be particularly attractive as a location for call centre operations, but again, the benefits are less embedded in the local economy, are more sensitive to low cost completion and raise related concerns (GHK Consulting 2004).

In terms of tourism the South West of England is characterized by a reliance on the more traditional forms of holidaymaking (see Meethan 1998, 2002). With regards to leisure and tourism, as a destination Plymouth attracts mostly short-stay tourists, often attracted to the city as an excursion from longer stay trips in the region, with nearly one-third being in the visiting friends and relatives category. Being a large 'non-resort' city, Plymouth appears to be 'out of place' in the South West, where urban attractions sit uneasily with the dominant 'bucket and spade' tradition or the more rural offerings of the South West as a whole (Meethan 2002).

With a local economy dominated by the defence industries, in particular the naval dockyards, and other large, predominantly public sector, employers, Plymouth avoided the first round of deindustrialization that hit the developed west throughout the 1980s. This relative advantage has tended to hide an underinvestment in its built environment and facilities and recent major studies related to the regeneration of Plymouth have high-lighted this factor as a major disadvantage to its future growth (see for example, GHK Consulting 2004; MBM Arquitectes and AZ Urban Design Studio 2003). In particular, GHK Consulting (2004) note that the quality of the urban environment, despite some exceptions, is a competitive disadvan-tage for the city. Due to a reliance on defence for employment, Plymouth came very late to the game of attracting large inward investment (see for example Ashworth and Voogd 1990; Bianchini and Parkinson 1993), and it is only in very recent years that the city has begun to experience a high level of investment in infrastructure and the built environment. It is therefore true to say that the city has, compared to other urban centres of compara-tive size, lagged far behind in terms of regeneration and reinvestment, not least due to a lack of civic leadership and the absence of one of the key elements needed for such large scale intervention: a strategic plan and vision for the city.

Now that the city is finally engaged in the regeneration game, it has employed master planners to produce *A Vision for Plymouth* or what has been called the 'Mackay Vision' after the lead architect (Mackay *et al.* 2004). The planners MBM Arquitectes and AZ Urban Design Studio (2003: 2), who were also involved in the regeneration of Barcelona, describe the vision not as a fixed blueprint for the exact future of the city, 'but rather a review of the strengths and weaknesses, an assessment of direction, a pointer to opportunity, and an invitation to aspire', effectively providing and proposing a development and public space strategy for the regeneration of Plymouth which had been previously lacking.

In parallel to the document *A Vision for Plymouth*, which deals largely with the physical regeneration of the built environment, in 2002 Plymouth was selected as one of seven City Growth Strategy (CGS) pilot areas in England. CGS is an approach to regeneration developed in the United States which aims to stimulate the regeneration of deprived urban areas by focusing on their economic strengths and competitive advantage, rather than on their social weaknesses. These strategies place a strong emphasis on business growth and wealth creation as the best means of tackling the social and economic problems of city areas. A central tenet of the CGS is to engage the private sector and in particular local businesses in developing and implementing programmes of action. An important component of this process is the identification and development of business clusters.

Creating a cultural zone

It is against this backdrop of strategy and vision that the University of Plymouth sought to take the lead in a flagship capital project to facilitate the creation of a cultural zone, which aims to develop and integrate the creative industries in the city with higher education and developing entrepreneurial skills. Both the CGS and *A Vision for Plymouth* identified that for a city of its size and regional importance, the creative industries sector in Plymouth is underachieving and should be more active and dynamic. Although the creative industries and the cultural sector in the city are reasonably robust, they are fragmented and relatively unsupported and would benefit from clustering to increase vibrancy, diversity and growth. As noted above, places within the United Kingdom as well as elsewhere have already clearly demonstrated that cultural activity and cultural and creative production can help to regenerate a city both economically and socially and can also contribute significantly to the renewal of the overall urban fabric.

The University is strategically placed on a number of counts to support this development. First, its physical location adjacent to, and within easy access of the city centre. Second, the decision to close its outlying campuses and relocate staff and students to the Plymouth site, in particular the faculty of arts. And third, a new-found commitment to contribute more to the social and cultural life of the city, making a substantial investment in a new arts building that would not only provide teaching space, but also public galleries, theatres and cinemas and incubation space for the development of entrepreneurship in the creative industries sector.

In order to achieve this, the University is constructing a landmark building with a creative and cultural focus. This will occupy a brownfield site at the southern end of University of Plymouth's main campus, and will act both as a locus for a North Hill Cultural Zone and as a gateway for the University's interaction with the city.

The intention is not only to house the arts faculty, but also to use the building as a springboard to create a strong synergy between the University's arts programme, academic and research staff working closely together in a single facility, and the incubation and commercial exploitation of cultural and creative sector intellectual property rights. The project's potential impact also fits with other strategic plans and objectives in both the region and city, and has also been part financed by the South West Regional Development Agency and Government Office South West.

Providing a building is one thing, yet the infrastructure needs to be complemented with an overall package that provides more than a physical presence in the city. To facilitate this, the University has brought together a number of long standing arts events and organizations to develop a coherent programme of culturally diverse events, working closely with a number of regional and city-wide musical and arts organizations under the umbrella of the Peninsula Arts Programme. The provision of new, category 'A' galleries (meeting the criteria for conservation of artworks), a cinema, workshops and lecture rooms will allow larger and better presented exhibitions and cinema, providing audiences from Plymouth and beyond with increased quality and range of work. This will substantially raise the city's ability to attract national/international talent/artists and touring shows. This ability to house artists will offer the city an opportunity not merely to host national and international artists in residence, but also to offer artist workspace to its own arts and media graduates, capturing more of the value of funds invested in education and training. The planned new arts facilities and the cultural and creative activities will have a major impact on tourism and visitor spend, such as increased audiences – both from the general and academic public – better artist provision and retention of graduates, improved education, outreach and artist facilities.

The building should not be considered in isolation but in the context of being the catalyst for the development of the North Hill Cultural Zone in the city. This aspect is important at the city and sub-regional levels, and will attract additional inward investment and the migration of people. The city is undergoing significant regeneration fuelled by the impending implementation of the Mackay Vision for the city. The development of the city's North Hill Cultural Zone should be considered as a central component to its regeneration. The Mackay Vision and CGS both recognize its strategic importance in this regard and the University's plans feature prominently in both documents.

The Cultural Zone in Plymouth will maximize the impact of an exciting range of opportunities that are currently present, and it will provide a major focus and catalyst for innovation, creativity and enterprise in the city and the surrounding communities. It also fundamentally aims to increase access to facilities and opportunities and involve the community. The Cultural Zone will be pivotal in enabling Plymouth to succeed in this era of creativity, through developing an environment in which creativity and knowledge can flourish and be exploited. This in turn will increase the range of tourist products in the city and will form a key strand in the re-branding of the place as more than just a short stay or excursion destination. Visitors will also be encouraged to stay longer through the hosting of special events, exhibitions, festivals and conferences. The intention is that the level and quality of cultural, creative and learning opportunities for the people of the city will be greatly enhanced, and this in turn will help retain and nurture its talent, establishing Plymouth as an important regional centre for education, creative production and entrepreneurship. At the moment the target is to create an active and growing creative industries economy and sustainable cluster – in the future the critical mass of creative activities that form that cluster will in turn be a key factor in attracting and retaining undergraduates and postgraduates to the city. The success of the project is inextricably linked to the rest of the city's regeneration strategy and is therefore a prime example of integrated 'cultural regeneration' rather than more narrowly focused 'culture-led regeneration' (Evans 2005a).

In summary, the medium to longer term benefits of the building should also be viewed alongside its potential role as the catalyst for the development of a new Cultural Zone in Plymouth, which in itself could become a significant tourist draw for the city. As well as being assessed for its benefits as an isolated project, in terms of the development of a creative industries cluster the overall project seeks to embed the framework of proximity and innovation to the institutional support structures of the University.

Developing a practical tool for impact assessment

Evans (2005a: 1) notes that 'measuring the social, economic and environmental impacts attributed to the cultural element in area regeneration is problematic' and that 'the evidence of how far flagship and major cultural projects contribute to a range of regeneration objectives is . . . limited'. That is certainly true in the case of the University's new arts building. However, given that the project sought financial support from both the South West Regional Development Agency and the Government Office South West through Objective 2 European Regional Development Funds, outputs, impacts and results had to be modelled and forecast and social, economic and environmental impacts measured. The model that was developed encompassed a range of approaches which combined advocacy and promotion, project assessment, project evaluation, performance indicator setting and impact assessment (see Evans 2005a).

Following concerted advocacy and promotional efforts, political support for the project concept was initially easier to secure than financial support. Given the dearth of evidence of how a major flagship cultural project of this nature would contribute to the local economy, in this case, funding did not naturally follow political support. Therefore, evidence of a quantitative nature that would engage potential public sector and other investors with the project had to be linked to regional and local regeneration strategies, and this was crucial to the success of the project. However, apart from some case study evidence gleaned from the experience of the Dundee Contemporary Arts development in Scotland, linked to the University of Dundee, and graduate retention estimates from Sheffield and Plymouth, there was very little to go on, and the model developed had to be based upon first principles.

Initially, benchmarking was undertaken in the city and subregion of Devon and Cornwall to provide the basis for a series of performance measures. These will also be used to compare the environmental and social performance against the targets for the project and the model included an impact assessment of the proposed activities upon the city for all of the elements of the project including economic, environmental, and social and tourism impacts.

The modelling methodology focused upon three aspects crucially linked to regional socio-economic strategy to engender support. The first related to education and specifically, graduate retention, the second aspect was related to entrepreneurship and business start-up and growth and the third, tourism, with a particular focus upon attracting more cultural visitors to the city. Each aspect within the modelling process had to be thoroughly researched and the evidence provided based upon the little related information and data available from the case studies of Dundee and Sheffield, supplemented by more specific regional and local information and data. In modelling graduate retention, rates based upon the experience of Sheffield were used for comparative purposes and additionally, data from the Higher Education Statistics Agency (HESA) First Destination Survey was utilized for arts graduates from Plymouth to forecast likely impacts outcomes and results.

In modelling the second aspect of entrepreneurship and business start-up and growth, both national and regional statistics looking at the new business start-up rates, turnover and employment levels of creative industry businesses were utilized. Basically, average annual sales per employee across all of the production cycle were calculated and compared against patterns of growth in creative industry businesses including new start-up rates. The incubation and pre-incubation space within the building was then modelled in a business lifecycle approach over 10 years which took into account attrition rates due to business failure, employment characteristics, income and growth patterns to provide the forecasts relating to new sales generated, new jobs created and safeguarded and additional gross value added (GVA).

The third and final aspect of the modelling process concerned visitor numbers and spend correlated to the projected impact of the gallery, cinema, events, festivals and conferences centred in and around the new arts building. Regional and local data were analyzed to provide baselines for visitor numbers and spend for Plymouth, again supplemented by visitor numbers experienced at Dundee Contemporary Arts and historical attendance and visitor numbers from the University's Peninsula Arts programme (a programme of cultural events run by the Faculty of Arts). Projections were then made of new jobs attributed to the increase in tourism, measuring the direct and indirect effect on local GVA. This process provided a number of measurable outcomes, impacts and results relating to new sales generated, new jobs created, and safeguarded and additional GVA.

In effect, the tool developed produced a prospectus, PR and descriptive materials based upon the building design coupled with outputs based on new firm start-up and business support, projected visitor numbers and student retention forecasts. However, project evaluation was built into the process and a series of monitoring procedures were put into place which reflected the projected outputs so that progress against targets can be monitored.

In short, the project objectives are as follows. The arts building will act both as a locus for the North Hill Cultural Quarter in Plymouth and as a gateway for the University's interaction with the city. It will support a wide range of cultural and creative activities, including cinema, performing arts, fine arts and digital media, as well as research, development and knowledge transfer. Pre-incubation/incubation support will be provided for new cultural/creative sector firms. This incubation will be delivered by the University in conjunction with front-line staff of the Business Link Network in Devon and Cornwall. In addition to the general aspects and the building itself, there are also a number of measurable objectives to be achieved: first is the provision of pre-incubation support for up to 600 individuals wishing to start their own businesses. This will include support for the development and assessment of ideas, the development of business plans, in the protection of intellectual property and in the development of teams. In addition, there will also be the provision of high quality pre-incubation and incubation support for 120 firms. Our estimates are that, of those, 55 viable, profitable businesses will be established by 2015. We also forecast that £74 million (€108 million) in gross new sales will be generated during the period from 2007–15, creating 165 new jobs, and that the building, and its associated arts programme, will attract 120,000 (72,000 non-local) visitors to the city, generating £31 million (€46 million) of additional visitor/tourism spend and creating 99 new jobs.

Conclusion

As the importance of creativity and innovation to the success of the knowledge-based economy becomes more widely established, it will become increasingly important that formal education is able to offer a fully rounded creative and cultural education. The new arts building in Plymouth will play a central role in this process for the city, as part of the wider creative industries and other strategies, feeding both the sector and the wider economy of the city. As noted above, the city itself, though well placed geographically to capitalize on the number of tourists to the region, has in the past tended to underperform. The development of a cultural zone should act as an additional 'value added' attraction in addition to the other socio-economic benefits. This is not simply about visitor numbers and, although difficult to forecast and measure now, in the future the aim is not just to measure success in terms of the number of companies or jobs created or the increase in turnover of the creative industries sector, or the increase in tourist spend, but to reflect on the quality of new buildings and diverse new environments, overall graduate retention in the city, the increased range of cultural activities on offer and the vibrancy of the overall economy.

What is different here to many of the examples of regeneration (in the United Kingdom at least) is the central role that the University plays in cultural development. Whereas there are a number of 'soft' outcomes that this development will achieve, such as raising the profile of the city and acting as a spur to further inward investment, the intention is not simply to subsidize the arts, but to create a vibrant and commercially successful economic cluster that will have noticeable benefits to both the city and the region.

15 Creative industries and tourism in the developing world

The example of South Africa

Christian Rogerson

Introduction

The concept of 'creative industries' represents 'a quite recent category in academic, policy and industry discourse' (Cunningham 2003: 1). It has been suggested that the formal origins of the concept can be found in the Blair Labour Government's establishment of a Creative Industries Task Force (CITF) after its election in Britain in 1997 (Flew 2002). The Creative Industries Mapping Document, which was prepared by the newly constituted Department of Culture, Media and Sport (1998), viewed creative industries 'as those activities which have their origin in individual creativity, skill and talent and which have the potential for wealth and job creation through generation and exploitation of intellectual property'. Nevertheless, the boundaries of 'creative industries' remain imprecise; Wood and Taylor (2004: 389) argue that defining creative industries 'is a task fraught with methodological and semantic challenges'.

Recently, research interest concerning the development of creative industries has surged with vibrant debates surrounding creativity, clusters and industrial districts (Brecknock 2004; Caves 2000, 2003; de Berranger and Meldrum 2000; Flew 2003; Florida 2002; Hall 2000, 2004; Landry and Bianchini 1995; Pratt 2004; Santagata 2002, 2004; Scott 1996; Turok 2003). In particular, Scott (2004: 463) draws attention to a growing new creative economy and rising levels of optimism surrounding 'cultural-industrial districts' as 'drivers of local economic development at selected locations, above all in large cosmopolitan cities, but also in many other kinds of geographical contexts'. Over the last few years, creative industries and creative districts 'have become a major new consideration in urban economics and city politics' (Brecknock 2004: 1). Policy initiatives to nurture the category of creative industries have been launched by several cities including Amsterdam, Brisbane, Berlin, Barcelona, Dublin, Helsinki, Manchester, Milan, Montreal, Tilburg and Toronto (Cunningham 2003; Flew 2002; Hall 2000; Leslie 2005; Musterd and Deurloo 2006; Scott 2004). For localities that host significant concentrations of these creative industries, Wu (2005: 3) stresses, the 'beneficial impacts are tremendous', especially in respect of local growth potential. Commonly, creative industries are now added to a distinguished list of 'leading edge' or 'growth sectors' such as financial services, ICT or high-technology, which signal the strength and potential of a local economy (Evans 2005b).

At national level, the creative industry of designer fashion came under the policy spotlight in both the United Kingdom and New Zealand during the 1990s. Unprecedented attention was accorded to designer fashion by the media and national government in campaigns such as 'Cool Britannia'. Moreover, it has been observed that fashion was assigned 'the dual tasks of economic development and re-branding New Zealand as a creative talented nation' (Bill 2005: 7). Several factors are acknowledged as important influences upon the emergence of dynamic creative clusters, including local innovation capacity, availability of venture capital, the role of institutions in mediating collaboration, an appropriate skills and knowledge base, and targeted public policies (Musterd and Deurloo 2006; Wu 2005). In addition, considerable significance is attached to the importance of developing synergies between creative industries and other sectors, including tourism (Evans 2005b: 7).

Against this backcloth of a rising tempo of research and debate concerning creative industries in the developed world, the aim in this chapter is to provide an examination of the nexus between creative industries and tourism in a developing world context by looking at the record of recent developments taking place in South Africa since the 1994 democratic transition. The chapter unfolds through three sections of discussion. First, national level debates around the development of creative industries and tourism are reviewed. Second, the focus shifts to urban scale debates concerning economic development and the role of creative industries and tourism in Johannesburg, South Africa's most important city. Finally, in terms of a conclusion, a synthesis is presented of the existing policy nexus between creative industries and tourism in South Africa.

National level debates in South Africa: tourism and creative industries

Within the developing world, South Africa provides an interesting case study of emerging policy interest around fostering potential synergies between creative industries and tourism. This policy interest concerning both tourism and creative industries is a phenomenon of a changing South Africa in the post-apartheid period. For national government, the creative industries and tourism represent two sectors that are high on the economic policy agenda (Department of Trade and Industry 2005a, 2005b; Jordan 2006; Phumzile-Ngcuka 2006).

The shifting economic role of tourism in South Africa

During the early 1990s, South Africa's tourism industry was in a state of crisis, beset by several problems such as under-investment and the low numbers of international tourism arrivals, and a legacy of sanctions and of apartheid policies (Rogerson and Visser 2004). In addition, it was suggested that another factor behind this crisis was the consequence of mistakes made with past policy frameworks. Certainly, compared to the growth and increasing economic impact of tourism in several other African countries (Egypt, Tunisia, The Gambia or Kenya), it was acknowledged in 1994 by the Department of Environmental Affairs and Tourism (DEAT) that tourism development in South Africa largely had been a 'missed opportunity' (DEAT 2003). Over the first decade of democracy, however, new policy frameworks have been brought into place – most importantly through the appearance of the 1996 *White Paper on the Development and Promotion of Tourism* – in order to maximize the opportunities for tourism to contribute towards economic growth, job creation and enterprise development (Republic of South Africa 1996). Most recently the national Department of Trade and Industry (2005a, 2005b) identified tourism as one of South Africa's strategic priority sectors and has sought to develop a new competitiveness strategy for tourism.

It is clear from the Ten Year Review document published in 2003 by the DEAT that the period since 1994 is regarded by national government as one of considerable achievement, particularly in terms of the renewed growth of international tourism arrivals. Moreover, from being a junior portfolio, DEAT has been re-positioned as a core economic growth department within national government (DEAT 2003). Since 1993 a spectacular growth has occurred in international tourism arrivals in South Africa, building upon the so-called 'democracy dividend', the events of September 11, the Iraq war and subsequent international terrorism activity, which have led to a re-assessment of the country as a relatively safe destination for international travellers (Rogerson and Visser 2004). During 2004 South Africa received the largest number of foreign visitors in its history of recording tourist arrivals. The nearly seven million international tourist arrivals, of which more than two million came from other continents, ranked South Africa thirty-second in terms of international league tables of tourist receipts. Indeed, between 1990 and 2004, South Africa's share of world tourism arrivals quadrupled: a phenomenon that has fundamentally recast the face of the country's tourism industry (Rogerson and Visser 2004, 2007).

In post-apartheid South Africa, tourism is viewed as an essential sector for national reconstruction and development and one that offers 'enormous potential as a catalyst for economic and social development across the whole of the country' (DEAT 2003: 6). The World Travel and Tourism Council (1998) asserted that tourism potentially will be one of South Africa's foremost sectors of national economic growth in the twenty-first century. In 2004, tourism was recognized as a key contributor to national employment creation, GDP and foreign exchange earnings and for the period 1998–2002 the sector recorded both positive growth in employment and contribution to GDP (Monitor 2004). Indeed, as the export earnings calculated from tourism exceed that of gold, in the popular press, tourism has become described as 'the new gold' of South Africa's economy. During 2006 the significant role accorded to tourism in the national economy was re-iterated in the release of the *Accelerated and Shared Growth – South Africa* (ASGI-SA) development strategy document (Phumzile-Ngcuka 2006). Within this new macro-policy framework for South Africa, tourism is profiled, alongside business process outsourcing and bio-fuels development, as an economic sector requiring immediate and unqualified support for expansion from both the public and private sectors.

Looking to the future enhancement of the competitiveness of South Africa's tourism economy, a recent strategic analysis isolates several issues that must be addressed (Department of Trade and Industry 2005a; Monitor 2004). First, is that the existing configuration of the tourism industry needs to be re-shaped. It is recognized that for South Africa to become a globally competitive tourism destination there must be forged and managed a strategic alignment in the industry such that the key players 'cooperate to compete' (Department of Trade and Industry 2005a). Second, the local tourism industry needs to be realigned in terms of re-defining and upgrading its products and services to address the tourism demands in those countries and segments that are most valuable. These international markets include traditional target markets in the developed world such as the United Kingdom, Germany, the Netherlands, emerging markets in Asia such as China and India, and – very importantly – regional African markets (Monitor 2004). It should be understood that the largest number of defined international tourist arrivals in South Africa are from regional or African destinations and travelling to South Africa for purposes not so much of leisure but for shopping or cross-border trading (Rogerson 2004a). Third, the transformation of the tourism industry is identified as a matter of national priority, not merely in terms of redressing racial imbalances of ownership but of providing the basis for the potential innovation of new tourism products through new black-owned enterprises entering the industry (Rogerson 2003). New products in cultural tourism, community-based tourism, township tourism or sports tourism linked to South Africa's hosting of the 2010 FIFA World Cup are seen as some areas of future potential growth (Rogerson and Visser 2004).

Overall, there is policy recognition that the long-term competitiveness of the South African tourism economy is contingent upon addressing these several barriers to innovation in the industry as well as identifying new drivers for growth, particularly as regards the enhancement of existing products and of new product development. Among the potential untapped sources for innovation in South African tourism products is developing stronger linkages between tourism and creative industries.

Policy interest concerning South African creative industries

The terminology of the 'creative city' was first introduced into the lexicon of South African development scholarship by Dirsuweit (1999) in her work on culture and economic development in Johannesburg. Until 2005, however, it could be observed that the term 'creative industries' was utilized little in national level policy debates on South African economic development. One possible reason for neglect is that 'the creative industry in South Africa – and thus to a large extent the symbolic economy – is the least transformed in the country' (Minty 2005: 5).

What has emerged in post-apartheid South Africa is growing recognition of the significance of the parallel (and sometimes overlapping) notion of 'cultural industries' (Dirsuweit 1999; Minty 2005). The importance of promoting cultural industries in South Africa and their potential for economic development was signalled during 1998 by the appearance of a series of reports produced by the Cultural Strategy Group (1998a, 1998b) for the (former) Department of Arts, Culture, Science and Technology. The category of 'cultural industries' was defined broadly to incorporate music, the visual arts, the publishing sector based on creative writing of literature, the audio-visual and media sector, performing arts, the craft sector (including traditional African art, designer goods and souvenirs), cultural tourism and the cultural heritage sector. In addition, the Cultural Strategy Group (1998a: 9) also encompassed within cultural industries the sectors of design and fashion, which were seen as 'sectors where creative input is a secondary but critical means of enhancing the value of other products whose marketability and effectiveness would otherwise be lessened'.

The series of research reports generated by the Cultural Strategy Group are highly relevant for the development of creative industries in South Africa. The core objective of the Cultural Industries Growth Strategy was of 'integrating arts and culture into all aspects of socio-economic development' (Newton 2003: 25) in South Africa. In particular, the Cultural Strategy Group produced detailed reports and recommendations for supporting the craft (Cultural Strategy Group 1998b), film and video (Cultural Strategy Group 1998c), music (Cultural Strategy Group 1998d), and publishing (Cultural Strategy Group 1998e) industries. Taken together, this body of research identified major constraints facing the development of the core segments of South African cultural industries. Key cross-cutting issues for enterprise growth were recognized as a lack of adequate skills, difficulty of market access and a lack of innovative product development. While differentiated for the various sectors under investigation, the major suggested strategic interventions included:

- education and training to improve skill levels
- market development and facilitation of market access
- co-ordination of government initiatives
- the development of partnerships at all levels to implement joint projects
- advocacy for cultural industries (Newton 2003).

Overall, the so-called 'Creative South Africa' initiative sought to introduce and demarcate 'cultural industries as an important sector in its own right' (Cultural Strategy Group 1998a: 7) in terms of national policy debates. The results of this project were directed explicitly at National Government and sought to forge an awareness within Government about the potential of cultural industries for growth, job creation and enterprise development. An outcome of this initiative was that during 2002 cultural industries was one of nine 'priority sectors' for accelerated development, which were selected on the basis of their potential contribution to the national economy in terms of growth, equity and employment creation (Machaka and Roberts 2003). Under the *Accelerated and Shared Growth – South Africa* initiative, which seeks to boost national growth rate for the period 2004–14 to at least 6 per cent, priority sectors were again identified. These are viewed as those sectors in which South Africa 'has a range of comparative economic advantages which, if fully exploited, would lend themselves to higher rates of economic growth' (Department of Trade and Industry 2005b). One of the nine sectors seen as 'medium-term priority' is creative industries (Phumzile-Ngcuka 2006). Within 'creative industries', existing policy thinking by national government currently is directed towards two foci: (1) supporting the film and television production industry and (2) supporting greater linkages or synergies between the craft industry and tourism (Department of Trade and Industry 2005b). In February 2006 the Minister of Arts and Culture announced a ZAR100 million investment (€11 million) to be channelled into support for accelerating the growth of the sector of creative industries (Jordan 2006).

Urban level debates in South Africa: tourism and creative industries

One of the most significant dimensions of development planning in post-apartheid South Africa is the increased responsibility that has been mandated to 'developmental local governments' (Nel and Rogerson 2005; Rogerson 2006). In terms of the local economic development strategies which have been introduced since 1994 by South Africa's urban centres, tourism has been a sector of central importance (Rogerson 2006). In addition, within the major cities of Johannesburg and Cape Town in particular there has been an increase in policy attention devoted to creative industries. South Africa's leading economic city, Johannesburg, is used here to document the growth of tourism and creative industries as part of wider local economic development programming in the country's major urban centres.

Tourism – a new economic focus for Johannesburg

Johannesburg, South Africa's largest city, represents what has been described elsewhere as a 'non-traditional' tourism destination (Law 1993). For many international tourists, it is a city (somewhat unfairly) perceived as unsafe due to high levels of crime, a city in which to pass through the international airport as rapidly as possible and thence to head from this 'gateway' off to the country's national game parks (Cornelissen 2005).

Despite its image problems, Johannesburg remains both one of South Africa's most important tourist centres and one of Africa's leading tourism destinations (Rogerson and Visser 2005, 2007). According to the Johannesburg Tourism Company, total annual visitor numbers to (Greater) Johannesburg are estimated as 6.2 million (Rogerson 2004b). This number is broken down as follows: three million domestic tourists, two million regional tourists or visitors from sub-Saharan Africa and 1.2 million international tourists from long-haul destinations, mainly in Western Europe. In terms of its tourism competitiveness Johannesburg is first and foremost a business tourism destination (Rogerson 2002). The city boasts an excellent network of conference and exhibition centres and has hosted a number of prestigious international conferences, most importantly the World Summit on Sustainable Development in 2002 (Rogerson 2005a). In addition, it is the 'command centre' for the headquarter offices of major South African companies as well as for international companies basing their African operations in South Africa.

A second competitive strength of Johannesburg is for a special form of business tourism, namely shopping, especially for tourists from other countries in sub-Saharan Africa (Rogerson 2002, 2004b). In an apt comparison with other major South African cities it was recently observed that 'Cape Town has Table Mountain; Durban has its beaches; but increasingly Joburg has something that tourists from the rest of Africa want – shops' (ComMark Trust 2006). The focus on cross-border shopping has prompted parallels with Dubai, with Johannesburg emerging as Africa's 'Jobai' (Fraser 2006). In addition to retail shopping, newer tourism products include the promotion of heritage sites associated with the anti-apartheid struggle, the new apartheid museum, Constitution Hill and Soweto, a focal point for international tourists in terms of what has been called variously township tourism, dark tourism or poverty tourism (Ramchander 2007; Rogerson 2004c).

Although Johannesburg is a major focal point for Visiting Friends and Relatives (VFR) domestic tourism (Rogerson and Lisa 2005), what has not so far been viewed as a competitive edge for Johannesburg is the segment of leisure tourism. The city authorities, however, have recognized the potential of tourism as part of job creation initiatives in the city and as part of promoting new economic growth and creating Johannesburg as a world class African city (Rogerson 2005b). In 2001 the important *Joburg 2030* planning document identified tourism as a sectoral priority for the city's development initiatives and in same year the city issued its first ever tourism strategy. The focus in the tourism strategy is to build upon competitive advantages in business tourism and regional tourism, enhance marketing/imaging of the city and address barriers to visitation, most importantly the question of crime and safety (Rogerson 2002). New initiatives to expand tourism include support for product and enterprise development, especially in township areas such as Soweto and Alexandra (Nemasetoni and Rogerson 2005; Rogerson 2004c, 2004d).

In terms of diversification and broadening of the tourism economy, the most important developments surround ongoing initiatives to promote a more 'authentic cultural experience' for tourists in Johannesburg by re-positioning the city not just as a retail and business tourism Mecca but around a cultural tourism theme (Monitor 2005). The emerging 'new Johannesburg' cultural product recognizes that together with its wealth of historical, political and entertainment assets as well as its significance in the anti-apartheid struggle, the city is gaining momentum as a cosmopolitan centre for music, dance, fashion, theatre and the arts (Monitor 2005). A growing synergy and fusion is thus beginning to surface between creative industries in Johannesburg and the expansion of the city's tourism economy.

Creative industries in Johannesburg

The Economic Development Unit of the City of Johannesburg has recognized officially the role and potential of creative industries in contributing towards the goals of *Joburg 2030*, the city's blueprint for economic development over the next three decades (Rogerson 2005b). Indeed, it is significant that the terminology of 'creative industries' is now used widely in planning documents issued by the city and its associated development agencies (see e.g. City of Johannesburg 2005a; Johannesburg Development Agency 2005). The recognition of creative industries is the latest chapter in the implementation of *Joburg 2030*, one important element of which includes support for targeted strategic sectors of the urban economy (Rogerson 2005b). During 2005, alongside new support programmes for business process outsourcing call centres, ICT, freight and logistics, and sport, it was announced that the Economic Development Unit of Johannesburg would also actively support the sector of 'creative industries'.

A Sector Development Programme was prepared for creative industries that focused on sector clustering and support. The Sector Development Programme is a vital component of *Joburg 2030* and is aimed at the removal of constraints and inefficiencies and the harnessing of opportunities in the targeted sectors as well as provision of relevant information to sector participants. The overall goal of the Programme is to enhance the competitiveness of the sectors and attract and retain investment in these sectors, thereby growing the city's economy as a whole (Rogerson 2005b). The programme built upon the foundations laid by the Creative Industries Sector scoping study for Johannesburg, undertaken in 2003, which linked in turn to the national cultural industry study of the Cultural Strategy Group (Newton 2003). Essentially, the definitions used by the Cultural Strategy Group (1998a, 1998b) were applied to re-name the sector in Johannesburg as 'creative industries'. The focus of the scoping investigation was thus primarily upon the segments of television and film – in which (along with Cape Town) Johannesburg is the major national centre – music, performing arts, visual arts, crafts and design. Critically, the research highlighted that as a whole the 'Johannesburg creative industries sector dominates the national profile' (Newton 2003: 42). The sub-sectors that dominate the local economic landscape of Johannesburg are 'craft, performing arts, visual arts, music and film' (Newton, 2003: 47).

The scoping study made a series of recommendations to Council for development of the sector. The most significant recommendations were as follows:

- branding an image for Johannesburg's creative industries so that additional demand is generated;
- addressing the chronic skills shortages in the sector;
- enhancing networks and alliances such that the capacity of the cluster is strengthened to rapidly respond to new demands;
- developing a strong business development infrastructure in terms of providing a business-friendly foundation of physical space, telecommunications, policy support and funding mechanisms;
- dealing with the high levels of crime and urban decay in the inner city which act as deterrents to tourists and the audiences of creative industries.

The City of Johannesburg's creative industries consolidated sector support initiative was announced by the Economic Development Unit in 2005 (City of Johannesburg 2005a). The central goal is described as 'to support both cultural workers with talent but limited institutional support, as well as emerging companies with an entertainment industry focus' (City of Johannesburg 2005a). There are four elements that comprise the support initiative.

First is the establishment of an innovative project styled the Johannesburg Art Bank, which draws upon parallel models already operating in Canada and Australia (City of Johannesburg 2005b). The objective of this project is to furnish support and supplement the income of Johannesburg-based contemporary artists by creating a market for their work over a 5-year period (City of Johannesburg 2005b). The bank functions by purchasing visual art works from local artists and then leases these to companies 'who can refresh their office displays every two years at a fraction of the full cost of buying new art' (City of Johannesburg 2005a).

Second, the city has launched a Creative Industries Seed Fund which will 'provide support to promising creative industries that could benefit both from the provision of business skills and up-front financial assistance to take a viable creative industry business-plan into implementation' (City of Johannesburg 2005a). This project operates as a competition among learners who recently have completed learnerships in craft operations management, music business management or cultural entrepreneurship.

Third, the Economic Development Unit has initiated support for a facility which is targeted 'to incubate start-up filmmakers' (City of Johannesburg 2005a). More specifically, the Film and Video Incubator is aimed at support of new entrepreneurs in the film industry. The project is anchored upon a 'dedicated facility where start-up filmmakers are provided with office space, office infrastructure and specialized equipment at subsidized rentals for approximately 18–24 months' (City of Johannesburg 2005a).

Last, the City is funding a Johannesburg National Arts Festival Fringe Project as a basis for defining appropriate support for a performing arts incubator. This initiative supports selected performing arts companies that have already put on productions at the annual national arts festival (held at Grahamstown) to perform these shows in Johannesburg, thus helping them gain national exposure as well as deepening local demand for quality productions (City of Johannesburg 2005a).

Although the Sector Development Programme for creative industries represents the first co-ordinated explicit support from the city for creative industries, it must be recognized that other significant support interventions have been introduced outside the Programme. Two developments are of special note.

First, is the planning in the inner city of the Newtown Cultural Precinct, a tourism-related project that was a joint initiative between the city and the Gauteng Provincial Government and geared to promote a cluster of creative industries and more especially cultural industries that might enhance the area's tourism potential (Dirsuweit 1999). The planned cultural district represents a cluster of creative activities, entertainment and related industries for the promotion of tourism. Through the promotion of cultural tourism and the making of a cultural district, this historic area of Johannesburg, which contains several museums, theatres and heritage sites, is set to be transformed into the creative capital of South Africa (Rogerson 2004b). In 2005 the provincial government's involvement in the Newtown project was terminated and responsibility passed to the Johannesburg Development Agency (JDA). The JDA views the re-development of Newtown as a major regeneration initiative and seeks to galvanize 'major investment, particularly in the creative industries, culture and tourism' (Johannesburg Development Agency 2005: 17).

Second, assistance has been provided by Council, through the activities of the Johannesburg Development Agency, for the development in the inner city of a 'fashion district'. This support was for establishing a hub for fashion design as part of re-invigorating the city's clothing economy, not on the basis of mass produced goods but of individual fashion items using an African design (Cachalia *et al.* 2004; Rogerson 2004e). Central to the vision has been the notion of promoting the 'Urban Edge of African Fashion', capturing the spirit and vision of a fashion-oriented, trendsetting and outward looking district (Johannesburg Development Agency 2004: 5). In terms of creative industries, this project is highly significant for it goes beyond the group of activities that are the targets of support under the Creative Industries Sector Support Programme. The vision of making Johannesburg Africa's 'fashion capital' is beginning to take shape and it is increasingly recognized that the city 'is becoming a fashion destination in its own right' (Sesikhona Services and ECI Africa 2005: 27).

Conclusion

The South African example confirms an awakening interest in this part of the developing world to the developmental potential of creative industries. More specifically, it discloses policy recognition of the importance of catalyzing synergistic relationships between creative industries and tourism. At the national level of government in South Africa both creative industries and tourism are identified now as priority economic sectors which demand targeted support for maximizing their future potential. At the sub-national tier of government it is significant also that the local economic development potential of creative industries and tourism is high on the policy agenda of the country's major cities, in particular of Johannesburg. In must be concluded that these contemporary developments taking place in the nexus of creative industries and tourism in South Africa potentially require further monitoring as they may provide a source of 'good practice' or learning for other parts of the developing world.

Acknowledgements

Thanks are extended to the National Research Foundation, Pretoria for research funding support under Gun Award 205464.

16 Creative industries and tourism in Singapore

Can-Seng Ooi

As other chapters in this volume have shown, many towns and cities around the world are developing their creative economy, so as to spur economic development, attract investments, rejuvenate their physical environments and spice up their cultural vibrancy (see also Bindloss *et al.* 2003; Crewe and Beaverstock 1998; Dahms 1995; Hutton 2003; Jayne 2004; Tallon and Bromley 2004). Singapore is no exception. With the growth of the creative industries, these towns and cities are also becoming sites of cultural consumption (Crewe and Beaverstock 1998; Hughes 1998). Many 'dying villages' and 'ghost towns' use the creative industries to regenerate and re-market themselves (Dahms 1995). Creative industries development has become aligned with regeneration initiatives in many places (Jayne 2004: 203).

Singapore is not a dying village or a ghost town. Yusuf and Nabeshima (2005: 113) observe that Singapore is the 'most energetic' at pursuing the creative industries in Asia. With more than four million people living on an island of only 680 square kilometres, Singapore is a densely populated and busy place. The island-state is also a thriving financial and trading centre. It has the second highest per capita income in the Asia Pacific region, after Japan. Singapore's wealth is evident from its excellent transportation infrastructure, tightly packed skyscrapers and affluent population. The creative economy is arguably the next 'big thing' in Singapore, as the government pursues various creative industries – arts and culture; design and media – and sees these sectors as necessary for the country's economic survival. Tourism plays a significant role in the new creative economy. This chapter examines the creative industries in Singapore and how tourism fits into the wider scheme of things. With tourism having a stake in the creative economy, clashes of interest with other stakeholders arise; these challenges will be discussed. The efficient strategies behind the creative economy in Singapore must also be understood within the social and political contexts of the country. Unlike places such as Austria, Zurich, New York and Denmark (Center for an Urban Future 2005; Held *et al.* 2005; Ooi 2002; Roodhouse and Mokre 2004), the Singaporean model is top-down, and the authorities are hands-on in wanting to manage creativity. This reflects the Singaporean soft authoritarian regime.

The creative economy in Singapore

The Singaporean government takes an active role in transforming and ensuring the health of the national economy (Low and Johnston 2001). Since Singapore's independence in 1965, its economy has grown and faced many challenges. Today, its economy is moving away from its manufacturing and electronic bases to actively pursue the financial services, telecommunications, life sciences, tourism and the creative industries. This is where the Singaporean government sees Singapore's economic future (MITA 2000: 31):[1]

> In the knowledge age, our success will depend on our ability to absorb, process and synthesize knowledge through constant value innovation. Creativity will move into the centre of our economic life because it is a critical component of a nation's ability to remain competitive. Economic prosperity for advanced, developed nations will depend not so much on the ability to make things, but more on the ability to generate ideas that can then be sold to the world. This means that originality and entrepreneurship will be increasingly prized.

Singapore has no natural resources. It does not even have enough water for its own use. The wealth of this tiny island-state is generated primarily through labour power and by functioning within the global economic system. Not surprisingly then, Singapore is a strong proponent of free trade. The creative economy depends less on natural resources and more on labour, services and brain power. Making money from music, films, concerts, fashion, computer games, architectural services and other creative products is thus attractive for Singapore. In 2001, the Singapore government set up the Economic Review Committee (ERC), consisting of seven sub-committees, with the aim of developing strategies to ensure the continuous economic prosperity of the country. The ERC Sub Committee Workgroup on Creative Industries (ERC-CI) suggests that Singapore should move away from an industrial economy into an innovation-fuelled economy, seeking ways to 'fuse arts, business and technology' (ERC-CI 2002: iii). The city-state must 'harness the multi-dimensional creativity of [its] people' for its 'new competitive advantage' (ERC-CI 2002: iii). The recommendations are not surprising because the Singaporean government has been pushing for the creative turn for some years.

The first creative initiative was taken after the release of the 1989 *Report of the Advisory Council on Culture and the Arts*. Consequently, among other things, the National Arts Council (NAC) was formed in 1991, more support was given to art groups and schools started offering art programmes. Essentially, the government started paying more attention to the arts and culture (Chang and Lee 2003). And in acknowledging the importance of tourism in the arts and culture, and to further the 1989 recommendations, the Singapore Tourism Board (STB)[2] and the Ministry of Information, Communication and the Arts (MICA) took the initiative in 1995 to make Singapore into a 'Global City for the Arts' (Chang 2000a; MITA and STPB 1995; Ooi 2001). In that plan, among other things, Singapore will develop its arts trading sector, get world famous artists to perform, and found the Asian Civilizations Museum, the Singapore Art Museum and the Singapore History Museum. The aim was, and still is, to make Singapore into the cultural centre of Southeast Asia.

In 2000, the MICA pushed the 1995 initiatives further and envisaged Singapore as a 'Renaissance City' (MITA 2000). The plans are more ambitious and one can now see results. The promotion of the arts and culture in Singapore is seen to: 'enrich us as persons'; 'enhance our quality of life'; 'help us in nation-building'; and 'contribute to the tourist and entertainment sectors' (MITA 2000: 30). The 2000 *Renaissance City* report acknowledges that 'the 1989 Report had put in place much "hardware" for culture and the arts and that what is necessary now is to give more focus on the "software" or "heartware"'. It is argued that 'instilling in [the] people a sense of the aesthetics and an interest in [heritage] should be the next step in [the] nation's development' (MITA 2000: 13). The iconic Esplanade – Theatres on the Bay has since opened and the seed of Singapore's parliamentary democracy has been transformed into The Arts House @ the Old Parliament. In moving away from just building infrastructure, the Yong Siew Toh Conservatory of Music was set up at the National University of Singapore. Art schools in Singapore – the Nanyang Academy of Fine Arts and the LASALLE-SIA College of the Arts – have been expanded and their profile increased. Arts festivals and performances have not only become more abundant but have also become more accessible; for instance, the Esplanade offers hundreds of free concerts annually, and besides the Singapore Arts Festival and Singapore Film Festival, there are now also individual festivals for Chinese, Malay and Indian arts and cultures. The government has set aside S$50 million (€25 million) for the various projects and programmes over five years (MITA 2000: 59). Singapore will become more culturally exciting for both residents and tourists.

Building and expanding on the 2000 *Renaissance City* report, the 2002 ERC-CI report (mentioned above) produces the most ambitious and comprehensive blueprint yet on the creative economy, which includes explicit and specific plans to develop the media and design sectors. Borrowing from the United Kingdom, the Singaporean authorities define the creative cluster as 'those industries which have their origin in individual creativity, skill and talent and which have a potential for wealth and job creation through the generation and exploitation of intellectual property' (ERC-CI 2002: iii; Ministry of Trade and Industry (MTI) 2003: 51). Singapore is concentrating on three broadly defined creative sectors (ERC-CI 2002: iii):

Arts and Culture: performing arts, visual arts, literary arts, photography, crafts, libraries, museums, galleries, archives, auctions, impresarios, heritage sites, performing arts sites, festivals and arts supporting enterprises.

Design: advertising, architecture, web and software, graphics industrial product, fashion, communications, interior and environmental.

Media: broadcast (including radio television and cable), digital media (including software and computer services), film and video, recorded music and publishing.

The ERC-CI report provides three aptly named visions for the respective creative sectors: 'Renaissance City 2.0'; 'Design Singapore' and 'Media 21'. These visions of a more creative Singapore are further supported by a Ministry of Trade and Industry paper, 'Economic contributions of Singapore's creative industries' (MTI 2003). The authorities see close linkages between the arts and culture, design and media sectors. The arts and cultural sector is considered the artistic core of the creative economy, and is essential in ensuring the overall economic performance of the various creative industries. The arts and cultural sector is to provide the learning tools and experimentation space for creative individuals, interacting with the media and design sectors (ERC-CI 2002: 10). The arts and culture sector is also considered an investment. Chief Executive Officer of the NAC, Lee Suan Hiang, said (personal communication):

> The government's role is to address market failure. In business and industry, we have R&D [research and development]. R&D is often funded by government because R&D is a cost centre, not a profit centre. In the arts, we also need experimentation. In experimental art, in new art, artists use a new language that the public is not familiar with. Because these are new products and unknown, they are more difficult to market. They need time to gestate and be accepted. So, there is market failure which the government needs to address. This is where NAC comes in. We provide grants to encourage artists to experiment and try new things. Our facilitation is to address market failure. We sometimes need to help expedite certain strategic projects which may take longer if left to the market by removing barriers and providing incentives and seed funding.

The NAC is also working with the MICA to set up a pre-tertiary arts school in 2008, so as to 'identify and nurture the creative talents of young Singaporeans' (National Arts Council (NAC) 2005: 29). These are plans of the comprehensive strategy to:

> build creative capabilities (such as embed arts, design and media into the various levels of education, establish a flagship art, design and

media programme at the National University of Singapore), create 'sophisticated demand' for the arts (promote public art projects, create 'creative towns', where arts, culture, design, business and technology are integrated within community planning and revitalization efforts, introducing a world class Singapore Biennale, and create a new Museum of Modern and Contemporary Art) and develop the creative industries (including cultural tourism, internationalization of recording music, publishing, arts supporting industries, merchandizing Singapore's heritage resources).

(ERC-CI 2002: 15–20)

Tourism will both support and benefit from the creative economy. Tourists will consume many of Singapore's creative products, especially those in the arts and cultural sector. A lively and exciting creative economy will also promote Singapore's image and attract more tourists. In working closely with many other state agencies, the STB has been assigned the tasks of arts marketing and promoting cultural tourism in the creative economy (MITA 2000: 8). These tasks are taken seriously by the STB; it sometimes goes beyond the responsibility of a tourism promotion agency, some may argue.

First, the STB actively seeks out international conferences, exhibitions and events in the various creative industries to be hosted in Singapore. For instance, Singapore will host the International Federation of Interior Architects/Designers Congress in 2009. The design industry is still fledging in Singapore, and such an event will help inch the city-state into the global limelight. While supporting the vision of Design Singapore, the goal for the STB is to bring in high yielding business tourists. Catherine McNabb, STB Director (Strategic Clusters I) said (personal communication):

> We want to secure as many strategic events as possible, events that will reinforce our strategic goals. For instance, we want Singapore to be seen as the design hub of Asia, Singapore as a biomedical hub, we want to strengthen our banking and financial image. We aim to enhance Singapore's brand equity in the key clusters.

The then-Director of Creative Industries Singapore, Baey Yam Keng[3] offered a broader explanation of why the STB should continue to draw grand international events to Singapore (personal communication):

> Recently [2005], we hosted the International Olympic Council meeting. Such events, by themselves, are not profit generating. For example, security is costly. But there are other benefits, not just hotel stays and shopping but also the international branding of Singapore. Such high profile events highlight the Esplanade and the Singapore River; these help to sell Singapore. That is why we have not only to look at dollars and cents but also at the more intangible benefits to Singapore.

So, such big events will not only bring in tourists – which is a primary concern of the STB – it will also promote positive images of Singapore. And the STB actively approaches players in the local industries to stage these international events.

Second, the STB and other state agencies actively seek out opportunities to make Singapore the hub of global and regional organizations, including those in the media, design, telecommunication, pharmaceuticals and financial sectors. As a regional hub for companies and industries, business people will inevitably travel to Singapore. These people will also be high-yielding tourists. In the context of the creative industries, for example, Singapore is fast becoming a regional hub for the global media industry. MTV, Discovery Channel, HBO and the BBC have already made Singapore their regional headquarters. Singapore offers a conducive business environment, which includes political stability, tax breaks, free training of workers and attractive packages for expatriates. But to the STB, it also hopes that Singapore-centred and Singapore-slanted contents will also be promoted in the international media when Singapore is the regional headquarters. For instance, with the MTV Asia 2006 New Year's eve celebrations in Singapore, images of Singapore were telecast throughout the region, creating a happening image for Singapore (see also STB 23 February 2006). As might be expected, the STB sponsors such activities.

Another example that demonstrates how the STB directly supports strategic creative business activities and makes Singapore into an attractive media hub is the Film-in-Singapore scheme for foreign film makers. Wanting to benefit from what films can do for tourism, as proven by what *Braveheart* did for Scotland and *Lord of the Rings* did for New Zealand, the STB wants to use films and television series to create awareness and portray a positive image of Singapore. Under the Film-in-Singapore scheme, the STB will pay for half the costs for film production in the city-state. The STB will also offer advice on where to shoot and will co-ordinate with other authorities (e.g. police, National Park Board and attraction operators) to ensure the smooth shooting of scenes in Singapore.

The STB wants Singapore to be the regional headquarters for international organizations. Such setups will bring in revenues and provide employment; in terms of tourism, there will be more business visitors and the image of Singapore as a business centre will also improve. Thus the STB and other state agencies are constantly hatching schemes to make Singapore even more attractive for these organizations.

Third, and related to the earlier two points, the STB takes the lead in shaping the cultural and physical landscapes of the city. The STB is actively trying to make Singapore into an attractive place to visit, work and live. A vibrant arts and culture scene is considered essential to 'enhance the attractiveness of Singapore to global talent and businesses' (ERC-CI 2002: 10; MITA 2000: 24). The authorities acknowledge that Singapore is inadequate in offering cultural activities to draw highly skilled foreign workers to work in the city-state (Yusuf and Nabeshima 2005: 114). The Economist Intelligence Unit found that Singapore ranks behind Tokyo and Hong Kong as a sought-after place for expatriates because of its dearth of cultural activities (Anon. 2002). Singapore is working hard to move away from its sterile image as a 'cultural desert' (Kawasaki 2004: 22). The STB is at the forefront of many such initiatives, ranging from pushing for longer opening hours for bars to the founding of the three national museums (MITA and STPB 1995; Ooi 2005b). In the latest move, the STB is in charge of seeking the best bids to build two megacomplexes that will house casinos, conference and entertainment facilities. These so-called integrated resorts will drastically alter the cultural scene of the city.

Enlivening the cultural life of the city requires changes to regulations and policies. These changes affect various aspects of social life in Singapore. As a result, during a parliamentary sitting on 13 March 2004, a few Members of Parliament voiced their worries about the loosening of regulations in Singapore to attract tourists and to present a more creative image of Singapore. Member of Parliament Ahmad Khalis bin Abdul Ghani said (*Singapore Parliamentary Hansard* 13 March 2004):

> We have seen discernable moves towards greater easing up of our social scene. The main reason for this easing up is to present Singapore as a more happening place to woo tourists and foreigners. . . . [Some people] are concerned that such moves promote the idea that sexual promiscuity is acceptable, and therefore, this may undermine our family values. . . . [I believe] . . . we do not quite need bar-top dancing or such other types of items to woo more tourists and foreigners.

The then-Minister of State for Trade and Industry, Vivian Balakrishnan, replied that he agrees that Singaporeans 'must not lose our values, and we must not lose our compass' and he continued (*Singapore Parliamentary Hansard* 13 March 2004):

> There was an article that Professor Richard Florida wrote, entitled 'The Rise of the Creative Class'. . . . His research found that cities, which are able to embrace diversity, are able to attract and foster a bigger creative class. These are key drivers in a knowledge-based economy. The larger lesson for us in Singapore is that we need to shift our mindset so that we can be more tolerant of diversity. To achieve this, we have begun to take small but important steps to signal that we need a new respect for diversity and openness to ideas. So these examples that the Members cited, e.g., night spots to open 24 hours, bar-top dancing, and bungee jumping, are just part of that signalling process.

The STB is taking steps to not only promote a trendy image but also to lobby for policy changes to realize a more exciting Singapore. The STB's lobbying has social engineering implications (Leong 1997; Ooi 2005a). In this respect, the STB is not only creating a more exciting environment to enliven the cultural and entertainment sector; it is also challenging the mindsets of many Singaporeans.

Fourth, the arts and cultural sector of the creative economy needs tourism. In 2005, Singapore attracted nearly nine million visitors and generated S\$11 billion (€ 5.5 billion) in tourism receipts (STB 18 January 2006). The STB has a target to triple annual tourism receipts to S\$30 billion (€15 billion), increase annual visitor numbers to 17 million and generate another 100,000 jobs by 2015 (STB 20 January 2005). The Singapore government has allocated S\$2 billion (€1 billion) to the STB to achieve the 2015 goals (STB 11 January 2005). The art and culture sector benefits much from the tourism market. In fact, it is a necessity. The then-Minister for the MICA, George Yeo, was cited in *Asiaweek* (Anon. 1997) as saying: 'As with everything in Singapore, we get more than what we would as a city-state of 3 million people because we serve, maybe, 300 million [from the region]'.

The *Renaissance City* report (MITA 2000) cited a study commissioned by the STB, stating that for S\$1 spent directly on the arts, another S\$1.80 of income would be generated elsewhere in a related industry (MITA 2000: 30). Art festivals, such as the Adelaide Festival of Arts and Edinburgh Arts Festival were used to demonstrate how the arts contribute to the economy (MITA 2000: 31). Tourists coming for cultural performances are thus also very much welcomed. To the authorities, many art and cultural products are only viable because of tourism, in terms of increasing the market and generating revenues. These products, on the other hand, also attract tourists.

Fifth, the STB and tourism offer a framework for Singaporeans, and also the Singaporean creative economy, to imagine themselves. The brand story of Singapore as a creative hub is drawn from 'Uniquely Singapore', the destination branding of Singapore. This brand tells of Singapore as a city that has blended the best of the West and the East, the traditional and the modern. This message has been effectively communicated to not only the world but also Singaporeans. The story fits into the general social engineering agenda set out by the government. The Singaporean government, using the mass media and the education system, promotes the view that Singapore has prospered and developed but Singaporeans are still Asians at heart. Singaporeans are officially constituted by three ethnic groups, namely, Chinese, Malay and Indian (Benjamin 1976; Siddique 1990). The destination brand story accentuates this mix of Asian cultures in a modern context through the simplified and catchy slogan, 'Uniquely Singapore'. Various agencies, the local mass media and even government ministers use this catchy and simple brand story to talk about Singapore. After the launch of 'Uniquely Singapore' in 2004, Singaporeans are even encouraged to seek out 'Uniquely Singapore' products (see Ooi 2004, 2005a). Any destination branding story is, however, only selective in its portrayal – elements are ignored and there are other ways to frame the place. Inadvertently or otherwise, authorities promoting the creative economy in Singapore are also using a similar brand story. The then-Director of Creative Industries Singapore, Baey Yam Keng, said (personal communication):

> The East and West thing is very strong in Singapore. Singapore is based in Asia but because of our colonial days, the way we have connected to the world, the way our education system is structured, we are very close to the West. This is a very nice blend. Creative people like something ethnic, something Chinese, something Japanese and something different. Singapore is where the East and West confer.

The 'Uniquely Singapore' brand story has become a framework to understand and present Singapore's uniqueness, both for tourists and for playing up Singapore in the global creative economy. In a subtle manner and over the past four decades, the STB has helped Singaporeans imagine themselves through what foreign tourists would see as attractive in them (Ooi 2004).

The discussions above show that the creative economy is much broader than tourism, but tourism plays a central part in the scheme of things in Singapore. There are however some serious challenges in using the Singaporean approach of intertwining tourism into the creative economy.

Challenges in the Singaporean approach to the creative economy

A country taking the 'creative turn' requires focused and tremendous efforts in terms of co-ordination between agencies and in seeking resources. Developing the creative infrastructure requires the co-operation of many stakeholders (Roodhouse and Mokre 2004). Furthermore, the management of creativity and culture will always face resistance (Crewe and Beaverstock 1998; Hughes 1998; Jayne 2004; Roodhouse and Mokre 2004). Singapore is no exception even though state agencies are known to work closely together (Schein 1996). For instance, the NAC and the STB approach the arts and culture differently; the former tends to look at arts 'from a non-profit angle' (ERC-CI 2002: 10–11), and the STB approaches art development from a 'business (tourism) angle' (ERC-CI 2002: 10–11). As a result, in the interplay between the different stakeholders and their different interests, many questions arise, such as which programmes are supported? Who decides? How should resources be allocated? What products should be chosen for promotion? As discussed earlier, the holistic and comprehensive approach by the Singaporean authorities of bringing tourism into the creative economy has brought about positive results, but at the same time, there are challenges.

The first challenge is the general concern about quality (see also Santagata, Russo and Segre, Chapter 7 in this volume). How would tourism influence the types of creative and cultural product? For instance, there are criticisms of theatres in London as 'being geared towards the tourist market, resulting in standardization, blandness and the emphasis on spectacle' (Hughes, H. L. 1998: 447). So-called serious plays are being squeezed out by long-running commercially-oriented musicals. The STB involvement in the creative industries is driven by the tourism values it can attain from its involvement. As a result, the STB would like to promote well known products, such as the musicals *West Side Story* and *Mama Mia*. It has successfully lobbied for bringing in the internally renowned Crazy Horse Revue from Paris – with topless women dancing on stage – which caused a stir in Singaporean society. While tourism resources are used to promote the arts and culture in Singapore, it is still debatable whether the types of art and culture product brought in are what locals and other stakeholders desire. Furthermore, there may be longer term consequences for the arts and cultural scene in Singapore when more tourism-oriented productions and cultural activities are lauded in the local media for their commercial successes. This concern is particularly salient in Singapore because the Singaporean authorities always use economic reasoning to convince people of the need for policy changes in the country (Chua 1995; George 2000; Koh and Ooi 2000). Based on the *Renaissance City* vision, the government is quite aware that less commercially-oriented arts must also be nurtured; the NAC supports less commercial forms of art and culture. This does not stop criticism, as artists and segments of the public still think that much more could be done to allay the negative influences of commercialized cultural products.

The second related challenge deals with the functional and economic manner in which the STB raises social causes, so as to further commercialize and commodify creative products in Singapore. As discussed earlier, many Singaporeans are unhappy with the liberalization of social spaces to spice up the cultural life of Singapore. The changes are, on the other hand, welcomed by many other Singaporeans. Among those who welcome the changes, many of them remain ambivalent towards the commercial logic behind the STB-led policy changes. For instance, an unregistered gay activist group in Singapore, People Like Us, is unhappy that gays are given more social spaces only because the authorities want to signal to the world that Singapore has become more open and tolerant. It is still, however, a criminal act to engage in homosexual activities in Singapore, and People Like Us could not be officially registered. The Singaporean authorities do not find it necessary to give formal rights to gays because there are no economic benefits and there may be political costs. To People Like Us, the authorities are treating gays as economic units, not as citizens deserving equal rights.

The third challenge arises from the need by the Singaporean government to manage creative expression in the city-state. It seems that cultural products that are politically and socially sensitive – and lacking in tourism and economic value – are likely to be curtailed. For instance, some plays and films are not allowed; in 2002 the authorities banned the play *Talaq* by

P. Elangovan. The play dealt with rape within an Indian Muslim marriage, and some members of the local Indian community protested. Elangovan (2002) lamented, 'It makes a mockery of Singapore's aim to be a Renaissance City'. In 2005, Martyn See, a young local film maker saw his film, *Singapore Rebel*, banned because it is considered to be 'political'. The 30-minute documentary is about Chee Soon Juan, leader of the Singapore Democratic Party. See was interrogated under the Films Act, which states that it is an offence to import, make, distribute or exhibit a film which contains 'wholly or partly either partisan or biased references or comments on any political matter'. A 'party political film' is an offence punishable by a maximum fine of S$100,000 (€50,000) or a two-year prison sentence. His film equipment and copies of his film were confiscated. Such incidences are difficult to grasp for artists; creative expressions often reflect the embedded social, cultural and political environment. A consequence for some creative workers is self-censorship. For instance, as reported in the *Far Eastern Economic Review*, a local publisher published the book *Crows* by a mainland Chinese author, Jiu Dan, but not before removing reference to the protagonist's affair with a former Singaporean politician (Elangovan 2002). The book might be considered semi-autographical because Jiu Dan lived and studied in Singapore for many years. The soft authoritarian Singaporean government will continue to micro-manage creative expressions, and that may not bode well for cultivating a creative climate. On the other hand, all countries socially engineer their own societies to bring about their own visions of social stability. Singapore is not unique in that manner but it is unique for a society with such a high level of economic development to have such tight political control.

In the promotion of the creative industries, there is a general worry that the creative push is not much more than a means to market the place (Leslie 2005). Such a concern is partly reflected in the three challenges discussed above, for example the opening up of social spaces is only a means to signal a more tolerant Singapore, only tourist-friendly local art events are promoted to the world and some non-commercially oriented creative expressions are banned. Possibly more controversially, the creative push in Singapore is implicitly shaped by the STB. As already seen, the STB marketing messages are taken seriously in Singapore. A destination brand identity is, however, engendered by marketing interests; it is not meant to be an honest reflection of the local society. At the same time in Singapore, the STB has inadvertently provided the brand identity framework for many Singaporeans to uncritically imagine themselves. While Singapore is promoted as unique, the formulation that Singapore offers the best of the East and the West, the traditional and the modern, marginalizes many other realities in Singapore. The brand messages were organized into images that foreigners can understand. They are attempts to assert Singapore's Orientalness for the long-haul tourism markets and are endeavours by the STB to self-Orientalize Singapore (see Ooi 2002, 2005b). As already mentioned, such marketing messages have infiltrated into the general psyche of local residents, politicians and also governmental agencies. These marketing messages are generally taken as accurate. In taking an emergent view of cultural change, one may argue that the adaptation and acceptance of commercialized cultural products and marketing messages can only be expected if they become meaningful to the people (Cohen 1988; Ooi 2004).

Concluding remarks

Many countries and cities are pursuing the creative economy. Each country has its own strategy and model. This chapter has presented the case of Singapore. Tourism in Singapore plays a particularly important part in cultivating the arts and cultural sector. Not only do the tourism authorities market cultural products, the STB also takes the initiatives to create and shape the cultural scene in Singapore. With the co-operation of other state agencies, resources are used to realize tourism goals. As can also be seen in the discussions above, the STB is involved in other creative sectors, for example setting up regional headquarters, hosting industry events and making films in Singapore. Such initiatives can only take place with the generous support of the government. Just as other state agencies see the value of promoting tourism, the STB also sees the value in promoting the Singaporean creative economy. Together, these various agencies and authorities aim to realize the creative dream for Singapore.

Can such a model work in other countries? It makes organizational and economic sense for agencies to cooperate. In Singapore, however, little effort is needed to eliminate the political bickering between agencies and resistance from the public. The Singaporean leaders have ensured cooperation, thus reducing the transaction costs between stakeholders. With the control of the mass media and education system, civil servants, workers, employers and the citizenry in general are often mobilized towards the goals set by the government. Singapore offers a model to integrate diverse commercial and creative interests. But many other countries will not be able to push through their creative dreams in the same manner.

Notes

1 The Ministry for Information and the Arts (MITA) became the Ministry of Information, Communication and the Arts (MICA) in 2003. Except for publication references, the ministry is referred to as MICA throughout this chapter.
2 In 1997, the Singapore Tourist Promotion Board (STPB) became the Singapore Tourism Board (STB). Except in publication references, STB is used in this chapter.
3 Baey Yam Keng resigned from this position in April 2006 to become a Member of Parliament. He is expected to champion the creative economy, especially in the arts and culture, in Parliament.

Conclusions

17 Creativities in tourism development

Greg Richards and Julie Wilson

The diverse approaches analyzed in this volume illustrate that 'creativity' is far from being a uniform process or a panacea for development problems. In fact, it may be far more appropriate to talk about different 'creativities'. There are clearly different models for creative development, different concepts of creativity, different creative experiences, different levels of visibility of creative spaces and spectacles, and creativity may be integrated with or separated from other policy areas. In this concluding chapter we attempt to draw together some of the divergent strands of these different 'creativities', highlighting some of the major issues which emerge from the individual chapters. We pay particular attention to those aspects of creativity that are pertinent to the development of tourism, but in doing so we also cover many themes that are relevant for creative development in general.

Looking at the different approaches to creativity and tourism adopted here, we can identify a spectrum of creativities, ranging from hardware-based to software-based approaches. The hardware-based approaches tend to depend heavily on the development of creative spaces and infrastructure, whereas the software-based approaches depend far more on the development of experiences. Mediating these two extremes are the 'orgware' approaches, which provide the policy, strategy and management frameworks necessary to link the creative software and hardware. It is clear, therefore, that a holistic approach is essential for effective creative development, and that creativity also needs to be place and context specific. In this chapter we first try to outline the context in which creativities are applied to tourism development, and then concentrate on some of the key issues which arise from this process, including the nature of the tourist experience, creative policies, production, marketing and management, and lifestyle and diversity. At the end of the chapter we also consider the impact of creative development strategies, and make suggestions for future research.

The context of creativity in tourism development

The contributions to this volume make it clear that many different destinations around the globe are looking for creative solutions to common challenges in tourism development. As Paul Cloke argues in Chapter 2, it may be possible to identify broad and narrow types of creativity. In the broadest sense, everything that exposes one's identity to some kind of challenge or learning can be creative. In the modern risk society, creativity has become part of our everyday lives – a strategy for shaping multiple identities to cope with the fluidity of modern existence. At the 'narrow' end of the spectrum, development may be restricted to specific creative industry sectors. In between these two extremes, meetings of creative consumers and producers may take place, either formally or informally, pre-arranged or incidental. The location and context of creative objects, subjects and processes is therefore crucial to the whole debate.

As Cloke suggests in Chapter 2, we can combine different elements of tourism experiences in different ways to produce new creative experiences. This can produce changes in the way in which places are lived, conceived or perceived. Such applications of creativity can take tourism into:

1   new realms (use of all the senses)
2   new experiences (transformations)
3   new strategies (placing, directing, staging)
4   the revisiting of old strategies (new ways of seeing old products and processes).

In selecting an appropriate creative strategy, the challenge for the creative destination is finding a context in which to develop synergy between the creative resources available and the creative needs of visitors and residents. The 'easy' model of creative tourism is the recipe prescribed by Florida (2002) – attract the creative class. Even if you cannot (or do not want to) get them to permanently relocate to your city or region, you can convince them to visit and spend their money. This provides a colourful creative backdrop which is attractive to (some) residents and visitors. This is also the model of the student city as creative landscape analyzed by Russo and Arias Sans in Chapter 10.

In contrast, the concept of 'creative tourism' as originally conceived by Richards and Raymond (2000) places the onus on the destination to actively identify those areas of creativity which can be anchored in the destination and developed as USPs (or UUSPs, UESPs or USSPs, as Prentice and Andersen argue in Chapter 6). In the long term, more success is likely to be generated by developing endogenous models of creativity and the indigenous creative class, rather than importing one. However, endogenous creativity requires much more translation and interpretation, because it must be made readable for external audiences. Of course, a creative landscape that is difficult to read can also present a challenge to the skilled creative consumer – hence the small numbers of 'special interest tourists' who visit places well off the beaten tourist trails to make textiles, learn about local gastronomy, or acquire a new language (See Raymond, Chapter 9; Richards 2005).

The analysis of Barcelona presented in Chapter 1 shows that there is a link between skill level, creativity and creative challenge. So-called 'creative' cities such as Barcelona seem to attract a contingent of skilled (or creative) consumers, who have previous experience of the city (or similar cosmopolitan environments) and are seeking creative spaces which enable them to combine the creative resources of the city with their own creative skills. As Maitland shows in Chapter 5, visitors in 'new tourist areas' of London tended to be older, more experienced tourists in search of a 'sense of place' or 'atmosphere' in which the familiar, the mundane and the ordinary were part of the attraction. From this apparently unpromising palette of experiential material, the creative tourist is able to create a richness of experience that may well surpass that of the 'unskilled' tourist. More attention therefore needs to be paid to harnessing the perceptual and imaginative capital of visitors.

When creative strategies are compared to culture-led development strategies, a number of other differences emerge in both structure and content. Although these models may converge in many cases, the ends of the spectrum may be distinguished along a number of dimensions (Table 17.1). These differences between cultural and creative development underline the need to develop new management and marketing approaches, as well as rethinking the nature of the tourist experience. These are all issues which are dealt with later in this chapter.

Creativity and space

What has also become clear from this analysis is the fact that the economic, social and cultural impacts of creativity depend to a large extent on the specific places in which creative strategies are applied. A number of different spatial dimensions can be identified as having an impact on creative development, including national contexts, the urban–rural dimension, and the micro-scale of clusters and enclaves within both urban and rural areas.

The tendency for Florida (2002, 2005) and other writers on creative development to concentrate on urban locations seems to create confusion between general urban processes and the impact of creative processes per se. Florida's analysis also appears to lack important spatial dimensions. By treating whole cities at the macro scale, much of the place-specificity of creative processes is also hidden. A number of chapters in this volume have underlined the importance of enclaves and clusters within cities as generators of creativity, and have also revealed the significant differences in creativity that can exist within these clusters. There seems to be a need to look in more detail at the structure and dynamics of creative spaces, and the way in which these relate to resident populations, tourists and other flows of people and ideas.

Even at a fairly macro scale, there seem to be big differences in the process of creativity and the impact of creative strategies. O'Connor (2005)

Table 17.1 Contexts of creativity in tourism

	Cultural tourism	*Creative spectacles*	*Creative spaces*	*Creative tourism*
Development context	Hardware	Orgware/software		
Spatial context	Backdrop	Activity		
Geographical scale	Global or local	Glocal		
Timescale	Past and present	Present	Present and future	Past, present, future
Cultural context	High culture, popular culture	Arts, performance, festivity	Arts, architecture, design	Creative process
Mode of consumption	Product focus	Performance focus	Atmosphere	Experience, co-makership
Learning orientation	Passive	Passive	Interactive	Active skill development
Reproducibility	Serial	Custom, bespoke, co-production		
Intervention	Economic development	Economic and cultural development	Cultural, social and economic development	Realizing creative potential
Identity	Reinforcing	Pluralizing		
Artistic focus	Aesthetic	IP		
Competitive environment	Competition	Collaboration, co-opertition		
Engagement in consumption process	Abstract	Visual	Multisensory	Flow
Heritage	Historic or contemporary	transcendent		

argues that the development of the creative industries depends on the political, economic and social systems operating in different countries. For example, O'Connor and Xin (2006) point out that China has successfully managed to develop the small and medium sized enterprise (SME) infrastructure needed to support creative industry development, but lacks the free cultural association necessary to stimulate the creative process. In Russia, on the other hand, O'Connor (2005) argues that the weakness of the state has allowed free association to happen, but the constriction of civil society limits networking, access to the media and finance. The current volume illustrates how different places can 'borrow' or copy creative ideas from one another. It is not clear, however, to what extent creative development models are transferable, or indeed translateable, from one context to another. Although Florida (2002) posits creative strategies as a general model of economic development, there are problems with using models derived from North American research to create strategies in other parts of the world (see Evans, Chapter 4). As Shaw points out in Chapter 12, for example, the movement of ethnic enclaves to the suburbs seen relatively frequently in North America is not evident in Europe or Asia. Similarly, Nathan (2005) has argued that many of the UK cities that have followed Florida's recipe in recent years have not performed any better economically (and sometimes have performed worse) than relatively 'uncreative' places. He argues that this has to do with place-specificity. The UK cultural and creative scene is dominated by London, which in common with many European capitals has an inordinate share in the economic and cultural life of the nation. The comparatively interventionist policies in the United Kingdom and other European countries are also very different from the American situation. These factors mean that it is difficult to import models such as Florida's wholesale. This is also the implication of a number of chapters in the current volume, including Shaw (Chapter 12), Collins and Kunz (Chapter 13), Rogerson (Chapter 15) and Ooi (Chapter 16).

In addition to a lack of attention to national differences in the context of creativity, little work has been done on creativity outside major urban environments. Florida (2002) only talks about cities – where he assumes the creatives would want to live. However, as Ray and Anderson (2000) point out, rural areas also have high concentrations of 'cultural creatives'. Drawing distinctions between urban and rural areas in creative terms also tends to ignore the intense flows of people, ideas and capital between the two.

The urban emphasis in the literature on creativity makes it problematic to investigate the relationship between creativities and 'tourisms'. The fact that much tourism takes place outside the ambit of large cities seems to argue for a broader analysis, as suggested by Cloke in Chapter 2. As Collins and Kunz also point out in Chapter 13, the 'city' is also usually taken to mean the inner city, to the exclusion of the wider metropolitan area.

This urban, inner city focus also tends to produce an easy link between the cosmopolitan and creativity. The creative experiences described in many of the chapters in this volume could actually be described as forms of 'cosmopolitan tourism' rather than 'creative tourism', as the main focus lies in the passive consumption of difference in urban settings, with an emphasis on the variety or scale of difference provided by large cities. One therefore needs to ask how far the links made between cosmopolitanism and creativity and innovation are real or assumed.

If the assumption is that large cities are required to stimulate a certain level of creativity, this also problematizes the relationship between the centres of such cities and their suburbs. The argument is usually that most creative activities take place in the city centre (or the surrounding inner city fringe), but the vast majority of the population lives beyond this 'creative core' in the relatively 'uncreative' suburbs. This is particularly problematic in the case of major cities in the developing world, where poor peripheral communities, such as favelas in Brazil and townships in South Africa, are being developed as tourist sites (Ramchander 2007). The main attraction for many tourists visiting these suburbs is the stark reality of everyday life and poverty (Jaguaribe and Hetherington 2004).

But as the Favela Painting Project shows, it is also possible to emphasize the creative capacities of the residents of these areas. Dutch artists Dre Urhahn and Jeroen Koolhaas have helped local people to paint a whole hill-side of a favela on the outskirts of Rio, 'with each house coloured according to an elaborate plan so that the hillside will depict a huge image visible from prominent places in the city center'. Via this creative project, the gaze is shifted from poverty to creativity. 'Normally, outsiders would only come here to buy cocaine. . . . the museum is about giving them another reason to visit the community' (Philips 2007). This particular project may be turning the 'controlled edge' of favela bus tours into a more 'uncontrolled edge', which, as Hannigan hints in Chapter 3, may offer more creative possibilities.

The residents of the urban periphery can also contribute to the 'critical mass' of creativity in terms of the cultural carrying capacity and pulling power of the city, even though they are not actually engaged with the creative economy, or even the cosmopolitan areas of the city. Recent research in Barcelona (Richards 2006a) has indicated that metropolitan residents fall into three groups with respect to their consumption of 'tourist' space: those who live in areas frequented by tourists (and therefore share both the economic advantages and social problems of tourist presence), those who visit 'tourist' areas for their own leisure (and therefore enjoy 'tourist' facilities without the immediate drawbacks) and those who live well away from the tourist areas, and never visit them. In spite of their differing relationships with tourism, all of these groups contribute in some way, however indirect, to the 'cosmopolitan' atmosphere of large cities.

It is clear that much of the discussion on creativity (including much of this volume) tends to focus inordinately on city centres. However, as some of the contributions to this volume have indicated, rural areas are not just locations for traditional crafts, but increasingly for contemporary creative activities as well. The flight of many 'cultural creatives' to rural areas has stimulated the development of creative enterprises and tourism products, and the coverage of rural and natural environments in the media has added new dimensions to the creative performativity of placing. Paul Cloke (Chapter 2) points out how nature itself has become part of the creative dimension of tourism, with the natural spectacles provided by creatures such as whales and dolphins increasingly adding to the tourist attractions of peripheral areas (Garrod and Wilson 2003).

A number of rural development agencies are now turning to creative strategies in the same way as their urban counterparts, and for similar reasons. The challenge, as Crispin Raymond's (Chapter 9) example of a rural creative tourism initiative in New Zealand shows, is the problem of developing 'narrow' creative tourism in rural areas. Rural environments suffer from problems of low population density and therefore offer relatively small markets for creative products. The exception is where very specific creative experiences can be developed which will appeal to people willing to travel relatively long distances and spend large sums of money to participate. An example is the summer schools which have developed around traditional music in Ireland and Scotland (e.g. Kay and Watt 2000) or in a number of festivals elsewhere in the world. Although these events are relatively small scale, they can deliver a relatively large economic, social and cultural impact in local communities. There are also a number of examples where urban-based creative programmes begin to spill out into the surrounding countryside, as Binkhorst shows in Chapter 8. The lack of attention for rural creativity also seems strange in view of the increasing de-differentiation between the 'urban' and the 'rural'. Rural areas are increasingly attracting urban creatives, while many rural residents aspire to the cultural and creative lifestyles of the city.

Creativity and tourist experiences

The idea that the tourist as well as the destination can be creative in their use of the basic building blocks of tourism experience also opens up new perspectives on the nature of tourism itself; in particular on two fundamental concepts in tourism studies – the concept of the tourist 'gaze' and the centrality of authenticity in tourism experiences.

The use of staging in the tourism literature generally has negative connotations, because it suggests 'staged authenticity' (MacCannell 1976), which is seen as devaluing the 'authentic' experience. In creative development, however, placing or staging can become a creative act, which also allows new meanings to be developed. Allowing for multiple meanings in the tourism experience also leaves room for the creativity of the tourist to interact with the placing and staging by producers in the role of performer. As Cloke points out in Chapter 2, the creative tourist can fill the gaps in the narrative left by the producer to develop and complete different storylines. For example, Finnish aficionados of the UK TV series 'Heartbeat' visited the Yorkshire setting of the programme in their vintage Ford Anglia cars to relive the experience portrayed on film. The presence of these creative tourists in turn generated a new placing of 'heritage' objects for local people and other tourists to enjoy.

In this way, 'tourismscapes' (Van der Duim 2005, 2007) can become sites for locating flows of creative texts and narratives. In consuming these texts, the tourist also becomes involved in a creative performance, as exemplified in the development of film tourism: 'Placing imaginative texts is therefore a performative practice which in turn contributes creativity to the place(s) concerned' (Cloke, Chapter 2). This problematizes the concept of authenticity, since the act of placing might be seen as staged authenticity – but if the process, rather than the purity of the product, is important, then creative placing becomes a process of creating emergent authenticity (Cohen 1988).

As Cloke points out, creative performance appears to shift the conceived–lived–perceived register of space into new directions. The conceptualized or conceived space (of a film, for example) becomes an arena of representation, or a lived space for those tourists (re)creating the experience of the film, and this in turn generates a set of spatial practices through which space is perceived – or the film set becomes a site of pilgrimage for the creative consumer. The experiential perception of film space requires creativity on the part of the perceiver – he or she has to make the linkage between the film space and the 'real' space. In fact, the film tourist is often recreating the creative process of film-making – just as the director has to work creatively with sets to create the desired effect, so the film tourist has to imagine the creative process and how this relates to the physical landscape.

Arguably creativities can also expand the idea of the tourist gaze (Urry 1990) into new sensory and creative realms, in which the power relation-ships are not as determined. Recent studies (e.g. Dann and Jacobsen 2003) have pointed out the role of senses other than sight, such as smell and taste. Smell is literally the attraction of the make-your-own perfume workshops run by Galimard perfumeries in rural France (Richards and Wilson 2006). Equally, a creative tour of a wine-producing region can become more than just a series of wine tastings, by offering opportunities for learning about wines and their relationship to a specific region, a people and series of cultural practices (Hjalager and Richards 2002). In some wine regions, tour-ists are even offered the opportunity to make their own wine and bottle it. The fact that the vineyard then offers to store their wine until it is ready to drink can ensure a stream of repeat visits from wine enthusiasts. As noted in the introduction, the involvement of 'signature architects' in the construc-tion of new bodegas is allowing some wine producers to add a further creative dimension through the design value of the prosumption space.

All of this suggests that the shift from 'seeing' to 'being' outlined in Chapter 1 may have important implications for the structure of tourism. By moving beyond the gaze, we also start to ask questions of our own identity and the development of multiple identities presents creative possibilities not available in traditional modes of tourism.

In the development of different aspects of tourist experience, Prentice and Andersen (Chapter 6) go beyond Pine and Gilmore's (1999) concept of the experience economy, because they add a symbolic dimension to the utility and experience values of the 'experience economy'. The investment required to 'do' creative tourism and 'become' a different person means that tourists look for a wider range of benefits – not just having an experi-ence (such as edutainment or immersion in the moment) but also a longer term investment in transformation of self, or the creation of a new identity. Although Pine and Gilmore (1999) see such 'transformations' as a further stage in the development of the experience economy, they fail to deal with the symbolic values of experiences highlighted by Prentice and

Andersen (Chapter 6), who instead prefer to talk about 'engaged creative tourism'.

The ability of the tourist to act creatively also throws new light on the development of tourist performance, or perhaps more accurately the development of tourist consumption skills (Richards 1996b). As Cloke shows in Chapter 2 the required level of creative content depends on the desire of the consumer. Some tourists are content to soak up the creative atmosphere, or to consume the represented space rather than the actual creative practices that constitute that space. Others may require a greater level of practical or conceptual creative challenge. This may be related to Scitovsky's (1976) idea of skilled consumption, which argues that consumers tend to seek an optimum level of stimulation producing a balance between their skill level and the level of challenge. For the skilled consumer, the pursuit of specific creative activities may be very important, whereas for unskilled consumers the experience of a creative landscape as a backdrop to general tourism activity may be sufficient to satisfy them.

As Scitovsky (1976) suggests, finding the right balance between the challenge of the destination and the skills of the audience is crucial to giving them the kind of experience they are seeking. When the challenge (or risk) of an experience is in balance with our skill levels, this is when we experience Csikszentmihályi's (1990) state of 'flow', or achieve an 'optimum experience'. Too little challenge or risk is boring, too much induces discomfort. The goal of the creative destination should therefore be to achieve this kind of balance – or more accurately to allow each tourist the possibility of finding his or her own balance. This is not necessarily a question of constructing complex creative experiences – as Cloke (Chapter 2) and Maitland (Chapter 5) point out, 'everyday life' in the destination can be a sufficient challenge for a suitably skilled creative tourist.

Creativity and policy

Everybody, it seems, wants to jump on the creative bandwagon. As Shaw points out in Chapter 12, the creative metropolis has now become a normative idea. Just as the logic for creative development seems to be fairly similar in different parts of the globe, so the creative policies adopted by the public sector seem to show a high degree of convergence. While creative 'borrowing' of ideas from one's neighbours, or even the other side of the world, may be an easy source of inspiration (Richards and Wilson 2006), it tends to increase the problems of distinguishing oneself from every other place. In addition, as Meethan and Beer note in Chapter 14, in policy terms it is becoming increasingly difficult to distinguish between 'cultural' and 'creative' industries, as the latter in particular has become a trendy label to apply to more traditional cultural policy areas.

There is only limited evidence of resistance, as Evans notes in Chapter 4. Some cities are now trying to distinguish themselves by avoiding the 'creative' label, preferring to stick with the more traditional 'cultural' tag. Cities such Johannesburg and Hong Kong appear to be sticking to the 'culture' label (Rogerson, Chapter 15). But in some cases, the policymakers may be overtaken by events on the ground. Yusuf and Nabeshima (2005), describing the development of the creative industries in East Asia, point out that in Hong Kong annual growth rates of the creative industries have averaged 22 per cent over the past decade. The film industry in Hong Kong now employs 9,000 people, and is being used as one of the spearheads for urban redevelopment, even if this doesn't go under the name of 'creative development'.

Creativity also becomes more difficult to disaggregate from cultural policy in general because as well as entailing specific policies of creative development, it can also be encountered as a part of the general policy of (re)development, or even a residual policy area. For example, if one is trying to generate a 'creative atmosphere', then it is often considered sufficient to create the preconditions to attract the creative class, and they will do the rest.

The development of relatively 'top-down' creative strategies may also compound the problems that Ooi notes in the case of Singapore (Chapter 16) – the challenge of directing the supposedly spontaneous creative process in the direction desired by policymakers. In Barcelona this has become a significant problem, as the few remaining rent-free spaces that can be colonized by young creative producers are being removed from circulation by commercial development. Attempts to organize the 'creative sector' into designated clusters (such as those in Poble Nou) are complicated by resistance from their desired targets to the label 'creative' as well as high commercial rents. Additional resistance may be engendered by attempts to graft the 'knowledge industries' onto areas perceived as creative. Such problems have also been encountered in other cities which prided themselves on the creative stimulus provided by such 'free spaces', notably Amsterdam, Berlin and Toronto (Jones *et al.* 2003; Shaw 2005).

This resistance also links to the 'intrinsic–instrumental' debate noted in Chapter 1. Many in the 'creative sector' not only reject membership of Florida's supposed creative class, but they argue that their activities should be funded through public subsidy, not commercial channels. In the case of creative tourism, the usual logic of public sector intervention is missing, because these activities tend to cater largely for visitors, not 'citizens'. In some places the economic arguments for supporting a relationship between tourism and creativity are too powerful to resist. This is certainly the case in Singapore, where the creativity strategy is actually tourism driven (see Ooi, Chapter 16).

Just as cultural tourism became a 'good' form of tourism that was acceptable to develop as an alternative to conventional 'mass tourism' (Richards 2001), so creative tourism strategies also make appeals to moral correctness. Creative development can be posed as a more sustainable form of tourism, because it is based on the renewable resource of creative energy. The rhetoric of creativity also appeals to certain actors who might otherwise shy away. As Russo and Arias Sans point out in Chapter 10, and Meethan and Beer describe in Chapter 14, universities are beginning to develop creative spaces and creative spectacles, and are beginning to get involved in the incubation of creative businesses.

But because the language of creativity tends to be difficult for many people to penetrate, it may be relatively easy to exclude certain groups from creative policies (whether by omission or design). There seems to be a need to involve residents more actively in creative development and, one might argue, visitors. In any case, as Maitland points out (Chapter 5), there is increasingly little difference between residents and visitors – residents increasingly behave like tourists in their own cities (Franquesa and Morrell 2007). This type of integration between 'locals' and 'tourists' may provide some interesting perspectives on creative development. Prentice and Andersen (Chapter 6) point out the way in which creative development has also involved the locals more in their own culture – providing them with new eyes to see their own environment (Camargo 2007; de Botton 2002).

At the same time, creativity can be viewed as a convenient way for policymakers to deal with a raft of problems caused by the growing fragmentation of society. As culture fragments into diverse 'cultures', creativity seems to be a useful shorthand for dealing with the plethora of demands placed on the hardware and software of the city or region. By shifting the emphasis from cultural products to creative processes, individuals and groups are arguably 'empowered' – everybody is in some way creative, regardless of their membership of a particular social or cultural group. Creative policies may therefore become a politically correct way of avoiding the pitfalls of political correctness. There is no need to have a social policy of positive discrimination for gays or members of ethnic communities, because attracting and retaining a diverse population is now a central plank of macroeconomic policy in the creative economy.

But the issue of diversity may cause problems for creative development strategies in many places. Although Florida-style tolerance may be accepted in many parts of the world, there are still many major tourism destinations which have restrictive legislation which would see them score very low, for example on the 'gay index'. This is a problem which Ooi debates in the case of Singapore (Chapter 16), and which applies to many areas of the world. Visser (2003), for example, has underlined the problems facing Gay Districts in South Africa, and there are considerable barriers to open homosexual practice in many other African and Asian countries. A recent Eurobarometer survey (European Commission 2007) indicated that Southern European countries are also not very gay tolerant: in Cyprus (86 per cent), Greece (85 per cent) and Portugal (83 per cent), the majority of the population feels that homosexuality in their country is still a 'taboo'. There are also signs that established 'tolerant places' such as Amsterdam may become less tolerant over time (Hodes *et al.*, Chapter 11).

Prentice and Andersen (Chapter 6), Binkhorst (Chapter 8) and Raymond (Chapter 9) also suggest that creative tourism is more supply-led than demand-led. This seems to be driven by the competitive climate of regions and cities, who are either trying to develop unique facilities or events to gain a competitive edge, or else are adding these features in a 'me too' fashion. Although this may be counterintuitive for marketing strategists, it is not so unusual in tourism development, where many sectors tend to be supply-led (including beach tourism, cultural tourism, theme parks, skiing and the cruise market).

The contributions to this volume also indicate that there are important distinctions in the way in which space is planned for creative experiences. In particular, there seems to be a clear distinction between the deliberate development of 'creative spaces' and the creative use of space:

- Creative spaces are intentionally designed to facilitate creative use. They tend to be designed to stage creative experiences, based on the deliberate 'placing' of creative resources. Examples include theatres, galleries, exhibition venues, festival sites and creative clusters.
- The creative use of space, on the other hand, tends to involve incidental 'creative resources', such as the diversity of ethnic enclaves, cosmopolitan spaces and lifestyles. Examples include the student landscapes discussed by Russo and Arias Sans in Chapter 10, or the gay community, as outlined in Chapter 11 by Hodes *et al*. As Cloke points out in Chapter 2, such 'incidental' spaces can also be creative if you are stimulated to use your imagination.

The creative experiences of tourists should therefore vary according to the level of deliberate placing involved as well as their own direct involvement in the creative process. As Prentice and Andersen point out in Chapter 6, more deliberate planning of tourist experiences may lead to 'engaged creative tourism'. This also seems to suggest that 'creative spaces' may be capable of delivering experiences which offer more than the 'immersion' offered by Pine and Gilmore (1999).

The spontaneous creative use of space, on the other hand, relies on a more fortuitous coincidence of creative backdrops and tourist consumption. Graeme Evans states in Chapter 4 that creativity can't be planned for. Maitland, by contrast, argues in Chapter 5 that the creative use of space develops through a mix of creativity and everyday life which results from 'subtle, sensitive and limited' intervention. This difference of opinion perhaps hints at the need for 'planned spontaneity' in creative development.

One of the major issues in creative planning is that of scale. The provision of planned creative activities tends to restrict the volume of visitors which can be catered for, because creative activities have to be programmed at discrete times in discrete spaces. Mass creative tourism will therefore probably remain anchored in the development of large scale creative spectacles and spaces which are easier to consume. However, there should be a relationship between the readability of the destination, the level of challenge presented to the tourist, and the perceptual skills of the visitor. Those wanting packaged creativity can go to easily readable destinations, but skilled creative tourists will put together the creative raw materials of the destination in new ways to develop new interpretations of 'local' culture. Tourists with more perceptual capital have a greater ability to fill in the gaps and make sense of the whole.

The idea of allowing tourists to interact more creatively with 'tourismscapes' has already been developed in a playful way by Joël Henry, founder of Latourex (Laboratory of Experimental Travel) and co-author of the Lonely Planet *Guide to Experimental Travel* (Antony and Henry 2005). This concept, based on the 'derive' originated by Guy Debord (1958), sets out a series of playful ways to interact with a place, such as visiting all the ends of the metro lines in a city, visiting a random grid reference or becoming a 'tourist for rent'. Residents may also opt to 'play' the host, for example through the 'dine with the Dutch' scheme (Binkhorst, Chapter 8), or even by adopting a tourist (Grit 2005). In a more serious vein, the idea of 'cultural biographies' has already been used in the Netherlands as a means of opening up more creative relationships between regions and their visitors (Rooijakkers, 1999). A cultural biography comprises the totality of traces that people living and working in a region have left behind over time. These traces can cover a wide variety of forms, including more formal elements of collective memory, such as museums and monuments, and less formal and more intangible elements such as stories, rituals and customs.

Creative production

To develop creative policies and strategies, appropriate products need to be stimulated. If we follow the logic of Florida, it would seem that we need to attract the creative class in order to provide a basis for such production. The presence of a thriving cultural and creative 'scene' should guarantee a solid base for creative production, as well as a local pool of consumers.

However, creative people are constantly in motion. Some creative people will leave their cities or countries in search of work opportunities or affordable housing or production space somewhere else, but new creatives will replace them. This points to the need to anchor 'footloose' creative capital in a particular place (see Santagata *et al.*, Chapter 7; Pratt 2001). This is important not just because the linkage of creativity and place is important for marketing destinations, but also because creative development strategies being implemented by governments mean a growing level of investment in 'coolective', 'fixed' creative capital (or at least 'creative infrastructure'). The problem with these strategies is that the 'creativity' is very mobile.

How can one ensure that the value generated by public sector investment is not lost?

Unlike many other tourism products, the creative tourism product or experience is often directly related to intellectual property (IP). As Santagata *et al.* argue in the case of Venice (Chapter 7), this carries with it dangers of reproduction of tourist products. They therefore recommend that destinations should adopt quality labels in order to protect their IP. However, Binkhorst (Chapter 8) argues that creative experiences are unique, and cannot easily be copied. This seems to suggest that the IP implications of creative tourism development are highly dependent on the type of product being offered (physical, experiential, etc.). In principle, the greater the level of consumer involvement in the experience (or the extent of co-production), the less important the protection of IP may be.

The mode of experience production may also be important in this regard. For example Binkhorst (Chapter 8) also talks about the production of creative experiences through modularization. A greater diversity of experience can be produced through increasing combinations of experience modules. Both the producer and the consumer may begin to exercise creativity in the development and combination of such modules. However, the question also has to be asked whether the production of more 'experience modules' will tend to lead towards standardization.

The development of creative products also raises questions about the sources of inspiration for new products. Is innovation a product of endogenous or exogenous forces, or perhaps a mixture of the two? There is little doubt that exposure to external influences (or openness and tolerance, as Florida would put it) will tend to inject new ideas and products. However, the wholesale importation of the creative class or creative ideas may eventually damage endogenous creativity. This is a potential danger of copying or borrowing creative ideas from other places. If ideas are copied wholesale, then there is little room for local creative input. Who, for example, originates the many festivals and cultural events now being increasingly used as a tool for cultural and economic development in many parts of the world? Are the many Melas in the United Kingdom a product of grass roots creativity, or a local authority-driven means of recognizing diversity? (Smith and Carnegie 2004). This also raises the question of whether creativity is attached to people or places. Is it the community, or creative individuals in the community, or the collective 'spirit' of a place that stimulates creativity?

The development of creative products may stimulate other forms of activity as well. As Rogerson notes in Chapter 15, Johannesburg's attempts to differentiate itself from other South African cities has led it to concentrate on shopping tourism, mimicking the successful example of the Dubai Shopping Festival. Can the development of creativity help to support other forms of tourism, such as shopping tourism, beach tourism or business tourism?

Marketing and managing creativities

The creative products of a destination generally need to be packaged and sold. As a number of authors have pointed out in this volume, creativity can be an important label for marketing destinations. However, there is some debate about the extent to which destinations need to 'brand' themselves as creative. Santagata *et al.* (Chapter 7) argue that some form of branding is necessary in order to support the high quality products required to maintain high quality cultural tourism. Raymond (Chapter 9) takes the opposite view in the case of 'creative tourism', which he argues is not 'a brand to be rolled out in different places like another Guggenheim museum'. Evans (Chapter 4) also points out that many cities are shifting from 'hard branding' of their cultural assets towards the promotion of more flexible and dynamic creative spaces.

This debate hints at one potential problem of branding creative places. Is there a danger that 'creativity' becomes the brand instead of the characteristics of the places it is found in? This has already arguably happened to the European Cultural Capital event, whose brand value has effectively eclipsed the individual cities which host the event (Richards 2001; Palmer Rae Associates 2004).

This problem may be compounded by the fact that creative tourism is recognized as a 'brand' by relatively few of the tourists who consume it. Although many people undertake creative activities or consume the creativity of others when on holiday, they would not necessarily see themselves as 'creative tourists'. This arguably complicates the task of communicating with the potential audience, as Raymond demonstrates in his practical analysis of marketing creative tourism products in Chapter 9. However, Raymond also argues that demand is likely to be greater if creative experiences can be marketed under a brand which has high recognition, as for example 'adventure tourism' is in New Zealand. The question then remains whether 'creative tourism' will become a recognized brand among consumers. As we noted in the introduction to this volume, organizations such as Turisme de Barcelona and VisitBritain seem to think so.

Another issue may be active resistance to a 'creative' brand from the 'creative producers'. The creative industries have enjoyed success as a concept among policymakers precisely because of the commercial connotations. But the perceived vulgarity of commerce may also serve to frighten away some of the very 'creative class' that the policymakers are trying to enlist in their development schemes.

The upshot of all this is that significant challenges arise in marketing, branding and selling creative tourism and other creative products. Raymond (Chapter 9), for example, recognizes that 'creative tourism in its purest form' is likely to remain a minority market. As the Creative Tourism New Zealand experience shows, the number of people willing to invest time and money in taking courses to develop their skills is relatively small. On the other hand, the adding of creative elements to existing touristscapes is likely to reach much bigger audiences, as the Venice example indicates (Russo and Arias Sans, Chapter 10).

In order to reach larger audiences, places often exaggerate the spatial and symbolic significance of their 'creative clusters'. Some places are reduced to a coherent, neatly demarcated zone, in line with the simplification of the thematization process outlined by Gottdiener (1997). Designation and theming processes are attempts to increase the sign value and symbolic capital of specific places. This is now a widespread strategy, encompassing the creation of design districts, creative quarters, cultural avenues (Puczkó and Rátz 2007) and film tourism maps. One could argue that such processes of delimitation, framing, naming and labelling are counter to the concept of creativity. The notion of creativity as a spontaneous process fed by diversity, cultural flows and interaction may not sit easily with packaging. The problem is that promotion of creative clusters requires a simple message that can be communicated quickly and effectively to wider audiences, including international tourists, some of whom may be on a relatively short visit and know little of the complex social, cultural and historical context of the place concerned.

The spread of creative strategies is also accentuated by the way in which cities interact with each other. In many cases, the development of links between cities is supported by a phenomenon which might be termed 'mayors on tour'. The development of networks of city leaders is encouraged by organizations such as City Mayors (www.citymayors.com) and Eurocities (www.eurocities.org). Such networks can help to strengthen the position of individual mayors in their home cities by giving them an international profile (cf. Sweeting 2002). In some cases, however, it seems that these networks can become mutual admiration societies, in which many are seduced by the creative aspects of the cities they visit. It sometimes appears that the tourism consumption preferences of urban policymakers is reproduced in the fabric of their own cities. This may explain why a number of northern European cities seem to be developing Mediterranean-style public spaces, which often become the backdrop for creative activities.

The creative strategies of one's neighbours must also be extremely attractive to those 'wannabee cities' (Short 2004) striving to increase their position in regional and global league tables (e.g. DEMOS 2003; Florida and Tinagli 2004). These league tables make it clear that creative development is still firmly orientated towards global competition. In Chapter 16, Ooi points out that the Singaporean creative industries strategy specifically mentions competition as a driver for creativity, and that the city-state must 'harness the multi-dimensional creativity of [its] people' for its 'new competitive advantage'. Similarly Meethan and Beer (Chapter 14) show that urban regeneration strategies in the United Kingdom are also geared towards competitive advantage.

Markey *et al.* (2006: 24) point out that:

There is an edginess to the idea of competition that fits well within the traditional bravado of economic boosterism and the contemporary rhetoric around neo-liberal economic adjustments. As a result, 'competitiveness' has become a favourite and casual term for economic development consultants and policy-makers.

This trend in urban and regional development seems to have superseded ideas of comparative advantage based on factor endowment, as policymakers have felt themselves compelled to compete in any available development field. This is particularly evident in large cities, which are assumed to have 'a certain scale and capacity to supply the ingredients of competitiveness. In turn, this can lead to a supply-side, "if you build it, they will come" approach to development' (Markey *et al.* 2006: 26). This 'field of dreams' approach which is evident in creative development is very similar to trends seen in cultural tourism and culture-led development in recent years (Richards 2001)

But with so many places jumping on the creativity bandwagon, perhaps the future of creative development lies more in achieving 'collaborative advantage' rather than competitive advantage. As Sweeting (2002: 5) indicates, collaborative advantage concerns the development of synergy between collaborating organizations: 'Collaborative advantage offers participants in collective ventures the prospect of achieving their own aims, the aims of their partners, and, when collaborative advantage is achieved, some creative or higher level output than was originally envisaged.'

There are a number of examples of collaborative creative development initiatives around the globe, but interestingly, most collaboration is international or global in focus. It seems that most places still have problems collaborating with their near neighbours, who they are more likely to view as the 'competition' than places on the other side of the globe. This is in spite of the fact that most creative development strategies also face competition from disembedded creativity, which is global in scope.

It seems that creative development strategies move destinations into a new complex of relationships, organizational structures and resource implications. In this sense, creative strategies may be much more difficult to deal with than traditional cultural strategies, because their implementation requires a raft of new skills, an understanding of the dynamics of co-production and an appreciation of the IP issues of creativity. The increased complexity of creative development can be summarized in a number of basic factors:

- *Increased management challenges.* The need to manage relatively spontaneous and chaotic creative processes throws up additional management problems, and may require more holistic forms of management and more co-ordination between creative sectors.

- *Footloose nature of the creative industries.* Because creative businesses are based on IP, they tend to be more footloose and difficult to tie to specific places. This creates pressure to develop mechanisms to capitalize on increases in IP value.
- *Prosumption.* The development of prosumption and 'co-creation' in the experience economy means managing consumption as well as production. This is clear in creative clusters, where success depends not just on bringing creative producers together, but also on making the creative process transparent and accessible to consumers (Richards and Wilson 2007).
- *Resistance to management.* As noted earlier, not only does the creative sector tend to resist being labelled as such, but creative producers tend to resist being managed. The solution to this problem adopted in many creative firms is to foster a relaxed, informal management style, but this may be difficult to apply in some situations. For the 'creative city' in particular, a relaxed approach to the management of public space may seem the ideal way to create 'atmosphere', but it may also pose significant public order issues.

Lifestyle and diversity

The development of the creative industries has been often related to the growth of specific lifestyles based on creative production and consumption; for example the development of artist's colonies, rural creative tourism businesses or creative clusters in cities.

Lifestyle entrepreneurship (Ateljevic and Doorne 2000) is closely linked to the development of creative activities, because the creative sector is attractive to work in, and many creative occupations blur the boundaries between work and leisure. Lifestyle entrepreneurship is also becoming more closely linked to travel, as creative entrepreneurs discover attractive locations through travel and decide to settle there. Creative activities then provide a means of financial support as well as the lifestyle suited to the new location. This type of lifestyle entrepreneurship is most closely related to active, course-based models of creative tourism.

On the other hand, different forms of mobility help to support the development of creative backdrops for more passive forms of creative tourism. The onset of 'liquid modernity' (Bauman, 2000) underlines the attractiveness of mobility and identity shifts. Short term residential mobility is on the rise, as Russo and Arias Sans show in Chapter 10. But what contribution do these transient populations make to the creativity, vibrancy, atmosphere and attractiveness of the places they visit? Exchange students, mobile professionals, seasonal migrants and working holidaymakers not only make an important economic contribution to the host economy (cf. Richards 2006b), but also add to the intangible culture and atmosphere. Cities often sell themselves on the basis of having a large student population, which arguably supports creative activities which are attractive as 'creative land-scapes' (Russo and Arias Sans, Chapter 10).

Some of these mobile populations may be 'tourists' according to official definitions, but as longer term visitors they may actually act as residents from the point of view of the experience of other 'tourists'. In this sense, they therefore have a role as both producer and consumer of creative experiences. However, these temporary migrants have an interesting relationship to the longer term migrants in a city, particularly ethnic minorities. Although temporary migrants may add to the 'atmosphere' and 'vibrancy' of a destination, they do so in a different way from the long term migrant population. For example, ERASMUS students studying in other European cities will generally have a very similar lifestyle and ethnic background to that of the host population. While the increase in numbers may add to vibrancy, the level of diversity does not increase appreciably.

Such processes are strengthened by many other forms of temporary mobility, such as company placements of managerial and professional staff, which stimulates the development of transnational elites in the global knowledge economy.

If we consider the way in which these creative landscapes are consumed by tourists, it seems that often we may be talking about 'cosmopolitan tourism' rather than 'creative tourism'. What many tourists seem to find attractive in large cities is the hint of creativity, rather than the direct consumption of particular elements of culture or creativity. This links to the idea of the rise of everyday life as spectacle. Tourism producers and consumers are now taking the mundane and spectacularizing it. This happens often when travelling to a new cultural context – but what is often seen as creative could be just the (relatively unchallenging) consumption of difference.

It is also interesting to speculate on the relationship between the lifestyles of creative tourists and their hosts, and the potential for developing 'cultural empathy' between these two groups, as suggested by Santagata *et al.* (Chapter 7). There has been much criticism of the relative lack of diversity present in many cultural and creative developments, and the types of tourist that are attracted to these areas are also fairly uniform in terms of background (Richards 2007). In fact, one could argue that there is more chance of developing 'empathy' between these relatively homogeneous groups of 'hosts' and 'guests' than between groups with very different backgrounds or lifestyles (as in the case of many 'ethnic' enclaves).

Where Florida seems to be right is that the 'creative class' is an important resource for the places its members live in. In terms of tourism, not only do they produce tourist attractions (creative spectacles, spaces, creative tourism) but they can also become tourist attractions in their own right. As Russo and Arias Sans (Chapter 10) and Hodes *et al.* (Chapter 11) show, lifestyles can be important attractions in the creative landscape. Maitland (Chapter 5) also shows specifically that the lifestyles of the creative class are themselves a source of attraction. In 'new' tourist areas of London, creative tourists seeking out the 'everyday' find themselves consuming the lifestyles of the creative class – such as shopping in Tesco Metro or drinking in gastro pubs. In the same way that the 'new cultural intermediaries' are also the most important consumers of cultural tourism (Richards 2001), so creative class members are probably both the most important producers and consumers of creative tourism.

There is also an important link between the creative class and tourism through the lifestyle choices surrounding the phenomenon of 'down-shifting'. Schor (1998) identified a significant group of people in the United States who were opting out of the rat race and electing to work fewer hours or live their lives at a slower pace, usually on lower incomes. This is a pattern also seen in the 'cultural creatives' group studied by Ray and Anderson (2000), who make the link to 'alternative' lifestyles and a search for a more creative way of life. Many of these downshifters have started businesses in rural areas or in low rent inner city areas as a way of taking control of the pace of their lives. Many of these businesses also cater to tourists, and particularly tourists drawn from the 'creative class'. So some members of the creative class are searching for a 'downshifted', creative lifestyle, which they support by selling their creative services to other members of their class, often tourists.

Diversity, distinctiveness and creativity

In much of the literature on creative development, and in many of the chapters in this volume, diversity is seen as a crucial creative resource, essential for developing distinctiveness. Many contributions seem to equate diversity directly with creativity – if one has an open, tolerant and diverse community, it will stimulate creativity (Florida 2002). But how does diversity actually relate to creativity? Santagata *et al.* (Chapter 7) suggest that more diverse experiences are more intense, more creative. Florida (2003: 227) also argues that diversity means excitement and energy, which is good, because 'creative-minded people enjoy a mix of influences'. But where does this cultural mixing take place? Is it necessary for the destination to present a ready-mixed fusion culture, or do the creative tourists themselves do the work of making creative cultural cocktails? This is an important distinction, because the latter suggests that tourists are doing the identity work (Cloke, Chapter 2), whereas the former provides a packaged experience of the destination culture. As Ooi shows in the case of Singapore (Chapter 16), policymakers are packaging the ethnically diverse city as a whole, with the main ethnic groups being mixed into a Singaporean-Asian ethnic soup. In this case, it seems that diversity is being hidden rather than utilized.

Such initiatives also pose difficult questions about the role of the groups who create diversity. For example, do ethnic minorities act as a 'muse' or creative resource for the 'edgy' creatives (usually white middle class) who are attracted by 'diversity'? What level of control do the different groups in the community have over the way their diversity or creativity is presented?

It may be that not only the 'mix' of cultural diversity is important, but the type of cultures involved. As Scitovsky suggests (see above) a certain amount of new information or difference is relatively easy for skilled consumers to deal with. But if the differences are too great, then the experiences are likely to involve greater levels of 'risk' and may become threatening rather than exciting. Perhaps, as Hannigan suggests (Chapter 3) people are actually looking for the 'safe difference' provided by the 'controlled edge'. The controlled edge is marked by theming and signing and the designation of districts, zones and clusters. As Collins and Kunz (Chapter 13) remark, there are perceptual links between ethnic enclaves and criminality. But the sanitization and rebranding of enclaves marks the spread of the controlled edge. The consumer knows that 'Chinatown' or 'Banglatown' (Shaw, Chapter 12) are no problem to visit, because their sign value is easier to read and they are seen as safer for visitors. Evans (Chapter 4) also notes the removal of ethnic celebrations to 'safer' areas outside the ethnic enclave, which raises questions as to whether integration is increased or decreased by such intervention.

The management of the ethnic enclave as a creative district may also invoke resistance to the development of ethnic kitsch on the part of local residents. There is a problem with seeing the 'community' in enclaves as homogeneous and identifying completely with the particular ethnic labelling of the enclave. As Santagata *et al.* (Chapter 7) observe, the zoning of 'creative districts' may impose a constructed identity on local communities, denying them the opportunity to be creative with their own identities.

The relationship between ethnicity and diversity also depends on the scale of analysis. Many cities trumpet their diversity in terms of the numbers of nationalities or languages that they harbour. For example, Melbourne claims to be 'one of the world's most harmonious and culturally diverse communities' with people drawn from 140 nations (City of Melbourne 2007), while Amsterdam lays claim to 'over 150 different nationalities . . . contributing to its diverse and lively culture' (Iamsterdam 2007). Shaw (Chapter 12) argues that ethnic enclaves are often linked to celebrating the role of minority groups in the creativity of the city – the implication being that the city is creative because it has such a diverse population. But at the level of the enclave, there may be a reduction in diversity as people of similar backgrounds crowd together. In this sense, some local authorities promote diversity as 'filling in the gaps' in ethnic enclaves and taking the 'edge' off the threatening/alienating/poverty aspects of relatively homogenous ethnic zones. In Chapter 12 Shaw refers to Sennett's (1994) observation that in diverse communities such as New York, ethnic groups may withdraw rather than engage, leaving a landscape of fragmented enclaves rather than an open, creative city. Florida's concentration on developing creativity at the macro scale also tends to overlook some of these more micro aspects of diversity and space.

There also appears to be a division of opinion as to whether globalization is making places more or less diverse. Some argue that the multiplication of standardized products and services effectively generates 'placelessness', as all destinations become the same (Smith 2007). In this view, local diversity is reduced as small businesses are replaced by global conglomerates. Others argue that the proliferation of global chains may actually increase the range of different products at local level, as well as stimulating new 'glocalized' products (Richards 2007). Take the example of Starbucks. The American coffee giant has introduced a standardized product to countries around the world, where its presence not only spawns local imitators, but it also accentuates the distinctiveness of existing cafés, which then look more exotic to tourists.

Fainstein (2005: 13) poses the question whether cities should want diversity. She argues that diversity is important, but that the argument in favour of diversity can be carried too far. Contrary to the view of Florida, she does not see a clear relationship between diversity and tolerance: 'Sometimes exposure to "the other" evokes greater understanding, but if lifestyles are too incompatible, it only heightens prejudice.' It also seems that diversity is extremely difficult to plan for, because many cities attempting to stimulate diversity actually achieve the opposite – uniform areas of creative class gentrification. This problem, she argues, also extends to the realm of tourism, citing the example of New York, where 'the tourism regime is far less diverse than the city itself, and much of the city remains invisible to it, even if being more inclusionary could improve the city's overall economic health' (ibid.: 10).

Impacts of creative development

Given the far-reaching aims of many creative development strategies, one would expect to see significant economic, social and cultural impacts. However, the relatively short lived nature of most creative strategies means that it is still difficult to judge whether they have delivered on their promises, particularly in terms of the more intangible aspects, such as increased creativity and 'buzz'.

Among the few creative impact studies that have been conducted, the issue of economic impacts tends to attract most attention. This is because economic benefits are usually highest on creative policy agendas, and economic impacts are usually the easiest to gauge.

For example, research by Jones *et al.* (2003) in Toronto indicated that the development of creative clusters in the city had generated a total of 107 per cent increase in retail sales between 1998 and 2000. In addition, the attraction of the creative class and creative businesses to these areas had increased property values substantially, with average house prices rising from 30 per cent of the city average in 1996 to 115 per cent of the city average in 2000. In London, the creative industries are estimated to employ a total of over 500,000 people, and generate a total annual turnover of almost £21 billion (€35 billion) (Creative London 2002).

However, these economic benefits of creativity may be accompanied by other, less desirable impacts. One of the potential problems underlined by many critics is that creative development tends to stimulate gentrification. As property prices rise, local residents may have to move out of the area, to be replaced by the influx of the creative class. As Cameron and Coaffee (2005: 39) have argued, arts-based gentrification is developing rapidly, as the 'first phase' movement of artists into old working-class neighbourhoods is followed by 'second phase' gentrification as capital follows the artistic community and is finally supplanted by a 'third phase' in which 'positive' gentrification is stimulated by public policy as 'an engine of urban renaissance'.

Studies of the 'creative class' in cities such as New York and London also suggest that the creative influx will tend to be more homogeneous than the inner-city populations it tends to replace. In London for example, a recent GLA Economics (2004: 11) report concluded that: 'The proportion of workers of black or ethnic minority (BME) origin in London's creative industries is 11.6 per cent, only around half the proportion in London's workforce as a whole, which is 22.8 per cent.' Importing creativity along the lines suggested by Florida (2002) may therefore leave the city with the problem of creating jobs and homes for the original residents. Those outside the creative class may also be excluded from the cultural fruits of creative development, as 'creativity' is not always designed to be accessible and inclusive.

There are also significant quality of life issues involved in creative development. For example, creative districts may see an improvement in amenity in terms of the development of new restaurants and bars, but at the same time they may suffer from 'loss of character' (Maitland, Chapter 5) as the original residents move out. The fact that many of the 'creative pioneers' who helped to establish a creative cluster may be the first to move out as prices rise is particularly problematic. How is it possible to maintain the 'atmosphere' of diverse creative spaces and at the same time improve the quality of life of residents? It often seems that quality of life for the residents comes last in the order of priorities.

The fact that the creative class is highly mobile also poses challenges for destinations. The development of creative spaces or clusters may involve considerable public sector investment, but how can we be sure that the 'creative capital' attached to the creative individuals in the cluster does not simply move to the next area which is offering economic incentives? It may be that policymakers need to find mechanisms to 'tax' creative revenues derived from public sector investment (perhaps in the same way as a 'graduate tax' taps the return on knowledge capital). It might also be possible to provide incentives for the creative class to stay in the area, perhaps by making sure that the creative sector benefits directly from tourism and other sources of economic development.

This underlines the fact that creative development strategies do not provide a quick fix. There is a need to invest considerable resources over a long period in order to create places which are 'rich with time', as Urry (1994) put it. This may end up being a 25 to 30 year process, as the development trajectories of cities such as Glasgow and Barcelona have indicated. As Prentice and Andersen argue in Chapter 6, the tourist also needs time to change and adjust their expectations to the new reality of the destination. Creative tourism is effectively a process of becoming – becoming a new place or becoming a different person.

However, that very process of becoming 'creative' also poses a general societal risk. The designation of creative regions, cities and districts places a burden of being creative on the people that live there. How do the 'non-creatives' survive in these new creative landscapes? The picture of rising property prices and low access to creative jobs for minorities revealed by creative impact studies seem to suggest that the non-creatives are in danger of becoming an underclass whose job it is to serve the needs of the creative class. This is essentially the argument advanced by Zukin (1995), who saw the rise of the symbolic economy being made possible through an extension of low paid service employment and the global mobility of low cost labour, shadowing the global mobility of the creative elite. As Peck (2005: 756) notes: 'A swelling contingent economy of underlabourers may, in fact, be a necessary side-effect of the creatives' lust for self-validation, 24/7 engagement, and designer coffee.'

Research agenda for creative development

The current volume has highlighted the considerable differences in approach to the role of creativity in tourism development, and diverging opinions on the utility of creative strategies. This is perhaps not surprising in a relatively new area of academic enquiry and public sector intervention, and these differing views on the role of creativity point to an extensive and diverse agenda for future research. In investigating the relationship between tourism, creativity and development, the current volume has probably generated more questions than answers. Hopefully many of these questions can serve as starting points for future research. This section presents a number of areas which we think might provide fruitful avenues of enquiry.

What is creativity?

One of the most basic questions that still needs to be addressed more fully is the nature of creativity itself, and its relationship to tourism. The contributions to this volume underline the problems involved in distinguishing creativity from other aspects of cultural development.

In the first place, the division between cultural and creative industries needs to be defined more precisely. As Rogerson shows in Chapter 15, policy documents in South Africa actually internalize the creative–cultural dichotomy, talking about 'Creative South Africa' as a strategy for the 'Cultural Industries'. On closer inspection, it seems that the cultural industries in this strategy would fall under the creative industries definition of DCMS (1998). Is the use of the word 'creative' in this instance and others simply a marketing tool, or is there a real difference between the cultural and creative industries?

The terms creative and cultural are used interchangeably by many policymakers. There is clearly a lack of precision – the step from cultural to creative industries was supposed to mark an important change in the nature of development, shifting away from the established cultural sectors towards a broader concept based on intellectual property. The mixed use by authors in the current volume suggests that this distinction is not very clear, and that at the moment, creativity may simply be a new synonym for culture.

Another definitional issue which has emerged is the extent to which creativity is related to individual creative producers or to collective forms of creativity. The DCMS (1998) definition of creative industries only talks about individual creativity. However, in the realm of 'creative cities' or 'creative landscapes', there is obviously a collective dimension which is being ignored. Even if the 'creative industries' may reject a collective label for themselves, the production of 'atmosphere' or 'cool' or 'vibrancy' is clearly something which stretches beyond the capabilities of individual members of the creative class. This is a very real issue in the development of 'creative clusters' in ethnic enclaves, for example. Is it the individual ethnic entrepreneurs who make the enclave tick, or are they just the most manageable element of a complex creative system?

Assessing the creative turn in development

The current volume, just as much of the previous literature on the subject, remains divided on the question of whether a radical shift has taken place in development strategies. Vanolo (2006: 13) says it is possible to argue that the:

> urban creativity perspective is not a revolutionary approach towards urban policies: it is just a 'cheap' group of heterogeneous actions (from supporting the local art scene to the organization of public events) that can easily raise public consensus (being basically risk-zero), even in [the face of] of possible feedbacks and outcomes that are certainly difficult to quantify.

In the face of such scepticism, it seems worth investigating the extent to which current creative development strategies represent a real departure from the cultural development or culture-led strategies of the past (Evans, Chapter 4). Such research might involve a consideration of whether the hardware, software and orgware employed in creative development are really radically different from those used in cultural development. One can certainly see a shift in language towards networking, intangible culture and 'ambience', but does this represent a radical departure in the implementation of policy and the role of the different 'creative actors', or simply a repackaging of old ideas? Gibson and Klocker (2004: 431) point to 'an increasingly standardized narrative of "creativity-led" urban economic development', which means that creative solutions are often becoming as standardized as the problems they seek to remedy. It might also be worth investigating the extent to which the rhetoric of creativity 'gurus' is picked up wholesale by policymakers, or whether there is a more critical and thoughtful application of creativity strategies.

Peck (2005) and Gibson and Klocker (2004) also argue that creativity is basically a fad, driven primarily by the creative class themselves, including academics. An analysis of the development and construction of creative policies could be very fruitful in throwing light on this issue. Why do individual cities or regions adopt creative policies? Who are the actors responsible for driving the creative agenda, and what is their relationship to the creative class?

Is creativity more sustainable?

Creative strategies will only be successful if they prove to be more sustainable. Arguably, the fact that creative tourism development utilizes creative rather than created assets implies it should also be more sustainable in environmental terms than traditional forms of tourism. However, much research needs to be done on the sustainability of creativity, particularly in economic, social and cultural terms.

In Chapter 9, Raymond clearly illustrates the problems of economic sustainability that beset many small scale creative developments. Creative enterprises are often attractive for lifestyle entrepreneurs, who may be willing to 'downshift' and put up with lower returns if they can engage in their favourite creative activity. However this may cause problems for long term sustainability and survival, particularly if the enterprise needs to grow. As creativity is currently an unknown quantity for most investors, it may be harder to raise capital than for traditional forms of tourism that are more asset-based. It would therefore be interesting to analyze the inception and survival of creative enterprises, and to gauge if they are any more or less successful than their 'non-creative' counterparts.

In Chapter 4 Evans restates one of the central tenets of the creative city model, which is that the creative city is more sustainable and self-renewing than more traditional cultural capitals. This is an argument repeated implicitly by Santagata *et al.* (Chapter 7) in their analysis of Venice, which Russo (2002) has argued is an example of a city caught in a vicious circle of cultural degradation through tourism. Does creative development allow cities to break out of the downward spiral of increasing tourist numbers and decreasing spend? Does the 'creative tourist class' contribute more to the destination than the classic 'cultural tourist'?

Social sustainability rests on the ability of the host society to benefit from creative development. Yet very few of the studies of creative strategies consider this aspect, because the economic dimension of sustainability tends to be the main focus of enquiry. There is a clear need for social indicators of creative development to be generated. These might include:

• increased creativity and creative activities in the local population;
• growth in tolerance;
• greater social cohesion or social capital.

One of the assumptions of much creative development is that everybody benefits, either because they work in the creative sector, or because there will be 'trickle down' effects. There is a need to investigate the extent to which the benefits of creative development actually trickle down from the creative class to other sectors of society. There are also questions about the extent to which creative clusters and ethnic enclaves exclude people by imposing a specific identity on an area, or through rising property prices. Perhaps in addition to investigating the views of the creative class, there is also a need to talk to 'non-creative' residents to gauge their views on the benefits of creative development. It may be that the attractiveness of the urban is similar for both groups, as Vanolo (2006: 14) suggests:

> The urban milieu, and particularly images of public spaces, crowds, sunny squares and cultural events still remain the very basis of the attractiveness of the urban, both in the eyes of the creative class and in that of the non-creative-one.

But it may be that these two groups differ greatly in their perceptions of what is good for their neighbourhood, thus creating a potentially damaging division of opinion about the benefits of creative strategies.

There is also a possibility that the development of creative strategies may benefit non-residents (tourists, members of the creative class living elsewhere) as well as residents of creative places, which could work against the long term sustainability of such strategies. The question of whether it is actually necessary to be resident in a place to enjoy its creativity is not just a question of visitors or residents, but is also increasingly complex in a highly mobile economic sector. There are numerous examples of creative producers living and working in different cities, or commuting between different locations, or even adopting a seasonal pattern of creative presence, based for example on the weather. It may be that creative strategies end up attracting mainly temporary residents, such as tourists or commuters, rather than the permanent residents needed to sustain the creative infrastructure. This may be particularly important in terms of the long term development of places, because the mobile creatives may be less willing to invest in the social structure and distinctive character of the communities that attract them.

Effectiveness of creative strategies

To date, relatively few studies have tried to evaluate the state of creative development. Some key questions still remained unanswered, such as:

- Are creative destinations more successful?
- Are the creative industries actually growing faster than the cultural industries?
- Is creative tourism growing faster than cultural tourism?
- Are creative regions attracting more tourists than other places?

In the tourism sector, the success of cities in attracting the mobile creative class could be measured indirectly in terms of the growth in tourist numbers, which could be related to various indices of creativity. More detail could be achieved for destinations where visitor profiles are available, particularly if this includes employment data, which might allow the 'creative class' to be identified. Destinations could then be compared in terms of the proportion of 'creative' visitors they receive.

Assessing effectiveness also requires the development of appropriate assessment tools. There is a need to move on from Florida's indices, which have attracted a lot of criticism. Where data permit, more direct measures of creativity, such as involvement in creative professions and creative activities, need to be developed. Where secondary data are lacking, it should be possible to generate primary data by questioning creatives and 'non-creatives' alike. For example, the data available in the United States on the attractiveness of destinations among gay people (see Hodes *et al.*, Chapter 11) could be extended to other areas of the 'creative class'. Regular tourism surveys, such as the ATLAS Cultural Tourism Research Project (www.tram-research.com/atlas) might also provide possible platforms for this. It may well emerge that many 'non-creatives' are totally unaware that they are being rescued by the creative cavalry.

The development of assessment tools also needs to be linked to an evaluation of what 'success' actually means for the creative destination – does it mean more creatives, more creative activity, rising land values, more tourists, more social equity, higher quality of life, or all of these? These 'critical success factors' need in turn to be related to the vision and aims of the city or region concerned.

The creative experience

The assumption of much creative development is that creativity makes places better to live in and more attractive to visit. But as Maitland points out in Chapter 5, there has been relatively little research done on the visitor or resident experience of creativity. In the development of creative tourism, there is an assumption that deeper involvement (Raymond, Chapter 9) or greater empathy (Santagata *et al.*, Chapter 7) produce more meaningful experiences. But is this actually the case? What is the contribution of creative experiences to personal development and levels of engagement?

Assuming that greater involvement *does* produce more satisfying tourist experiences, there is also a need to examine different strategies for developing visitor involvement and creativity. As Cloke suggests in Chapter 2, leaving more creative work for the tourist to do is one way of stimulating him or her to act creatively. The creative landscape should allow tourists to reach their own level, balancing perceptual capital and challenge. How far different types of creative experience achieve this balance could be a fruitful area of enquiry.

The extent to which creativity can be equated with risk is also a question raised by both Cloke (Chapter 2) and Hannigan (Chapter 3). There are many interesting aspects of risk which could be investigated in the context of tourism and creativity. The role of perceived physical risk of some destinations may add a sense of thrill to the journey, but does it also increase one's level of creativity? Does it make for more creative experiences and encounters? The concept of risk is also linked to tourist identities, and many authors have investigated the way in which (post)modern tourists adopt and play with multiple identities. Could this be seen as a creative activity on the part of the tourist? The tensions between risk and the need for security presents a management challenge for the creative destination, particularly where tourists actively seek the paradoxical state of 'safe danger' or the 'controlled edge'.

The relationship between diversity, ethnicity and creativity

Florida placed diversity at the heart of the creative development pantheon. However, evidence about the nature of the relationship between diversity, creativity and development is largely lacking. These days, all cities and regions seem to be able to claim a high level of diversity, but does this necessarily make them more creative?

The question could also be posed whether more diversity actually adds to the distinctiveness of places. As Florida argues, it is the 'mix' produced by diversity that the creative class finds attractive. But if all destinations began to exhibit a similar mix of diverse cultures, would this lead to less differentiation between places? In the case of tourism, it seems that a mixed, 'cosmopolitan' environment is very attractive for tourists, which introduces the suspicion that 'cosmopolitan tourism' might be a more accurate label for many destinations than 'creative tourism'. This might be analyzed by looking at how tourists actually experience the places they visit, and the extent to which they undertake specific creative activities and how they are soaking up the creative atmosphere.

It would also be interesting to look more closely at the role of mobility in the development of creativity, and in particular at the role of mobile populations, such as transnational elites, students and backpackers. Are these people attracted by the 'buzz' of the creative city, or do they also have an important hand in creating that buzz?

Creativity and intellectual property

There seems to be some divergence of opinion about the role of intellectual property (IP) in creative tourism development in the current volume. Santagata *et al.* (Chapter 7) see IP as being central to the process of creative development and an essential mechanism for conserving high quality creative production. Binkhorst (Chapter 8), on the other hand, argues that experiences are unique, and cannot be copied.

What is the IP content of creative tourism? In the case of a physical product, such as a CD or a souvenir model of the Eiffel Tower, the issue of a unique work or design might be relatively clear. But in the case of the creative tourism experiences described by Crispin Raymond in Chapter 9, the IP lies in the 'branding' of creative tourism experiences and the organization of the workshops. These aspects of IP might prove very hard to protect.

What mechanisms can be developed to tie IP to a specific location? Much research remains to be done on the potential for creative funding to be tied to income streams from the future exploitation of creative ideas. As young creatives become more mobile, increasingly sophisticated mechanisms will need to be found to ensure a flow of resources back to the place where ideas were funded or generated.

Cities are also now thinking about protecting the IP linked to public space, generating income from the trade in images of their major public icons. It would be interesting to investigate the effectiveness of such strategies, particularly as tourism promotion has long been based on allowing place images to circulate as freely as possible, stimulating coverage in the media and attracting the attention of potential consumers.

Does clustering favour creativity?

Although it is widely assumed that clustering favours creativity, there is often a lack of empirical evidence to support this assertion. Although it can often be demonstrated that clusters of creative industries perform well economically (Scott 2000), few studies have been able to show how the creative process itself contributes to this performance. The idea that close proximity of creative producers generates creative collaboration is not always supported by the evidence (Hitters and Richards 2002).

The development and designation of creative clusters could also be studied in more detail. What, for example, is the difference between clusters that emerge organically through an economically driven concentration of creative enterprises, and the top-down designation of 'creative districts' by local administrations? The effectiveness of cluster formation and identification could also be examined more closely, for example by comparing different cluster types (creative production centres, creative consumption spaces, ethnic enclaves) or different styles of cluster branding (creative districts, design districts, cultural quarters, ethnoscapes).

In terms of the cluster formation and identification process, it will also be important to consider the symbolic capital attached to the cluster. How is symbolic capital identified, extracted and commodified? Who is authorised to use the creative symbols attached to the cluster? In the case of top-down cluster formation the cluster brand may be developed and protected by the local authority or the cluster management organization, but in the case of more organic forms of creative development, the relationship may be less clear.

Most research on creative clusters to date has tended to consider the production perspective. But the way in which tourists perceive and experience creative clusters or districts in the city is also important. As Collins and Kunz point out in Chapter 13, different visitor groups will have varying relationships with the symbolic capital as well as the tangible and intangible culture of the cluster. Exploring these relationships in more detail may yield important information for the marketing of creative districts. Such research could also be undertaken in rural areas, where the creative image of different regions could be compared and related to their concentrations of creative enterprises (as Raymond describes in the case of the Nelson region of New Zealand in Chapter 9).

The feelings of those who are the subject of clustering and branding will also be relevant in many settings. There are already some signs of resistance to the 'creative' label, even among those who would normally fall within the ambit of the creative industries. Ethnic labelling may be even more problematic, particularly where the ethnic identity promoted for a specific district tends to mask ethnic diversity in the area. It would also be interesting to consider how identification with ethnic branding changes over time, particularly as some ethnic groups disperse and new ones may move into the area, changing the ethnic balance and image.

Planning the creative destination

Can the creative destination be deliberately planned, or should creative processes be left to develop spontaneously? The authors in this volume have differing opinions on the extent to which the creative destination can be planned for. Evans (Chapter 4) says that creativity cannot be planned and Collins and Kunz (Chapter 13) argue that planning may damage the authenticity of creative districts as well as creating problems of legitimacy. On the other hand Santagata *et al.* (Chapter 7) state that planning is essential and should be centralized to provide control over the creative development process. Maitland (Chapter 5) takes an intermediate position, proposing 'subtle' intervention to steer the development process while not stifling creativity.

The effectiveness of different planning regimes could be studied through a transnational comparison of similar clusters in different countries. Similarly, different types of cluster could be compared within one country or region, where the planning context is the same (for example centrally planned or designated versus organic creative clusters). Some work in this direction has already been undertaken by Brooks and Kushner (2001) in their study of different arts districts in the United States. Comparisons of this type are also useful in identifying some of the key success factors in cluster development.

The question of scale also seems to be important here. Are small scale creative developments more effective than mega-projects? Or does the development of signature architecture and tourist icons create more attention for the creative process?

Marketing

There is an interesting debate developing about the role of marketing in the creative destination. Evans suggests in Chapter 4 that the creative city can afford to scrap its tourist board, because it will have outlived its purpose. By contrast, in Chapter 16, Ooi makes the point that in Singapore the tourist board has been given the task of marketing the arts as well as promoting cultural tourism in the creative economy.

The challenge of marketing the creative city is an area that has received relatively little attention to date, and yet it is a potentially fruitful area for enquiry. How effective is the development of 'creative' brands? Is creativity effective in changing the image of a destination?

The sources of brands or images being developed for creative destinations may also vary considerably. Many creative brands seem to be far more geared to external consumption than stemming from internal narratives.

There may be a danger that hegemonic narratives (whether cultural, heritage-based or creative) drown out the local specificity of creative development. It would be interesting to speculate whether thematization and creativity are actually compatible.

Concluding remarks

This volume has underlined not only the opportunities which can be opened up by applying creativities in tourism, but also the considerable challenges of such strategies. The complex relationship between creativities, lifestyles, diversity, tourism experiences and different spaces highlights the need for multidisciplinary approaches to this issue, and for different scales of analysis.

There is also a need to develop more accurate indicators of creative development which can measure the effectiveness of creative development. The information provided in the current volume seems to indicate that the relationship between different indicators of creativity and tourism development is very complex, and difficult to encapsulate in aggregate indices. There seems to be a need for more qualitative research to enable us to move beyond the indices-driven analysis provided by Florida (2002).

It is also clear that a more critical approach needs to be taken to the whole issue of creativities, not just in tourism but in development in general. The uncritical copying of creative strategies from one part of the world to another does not seem to be very effective, particularly in view of the national and local differences in the context of creativity highlighted in the current volume. This is particularly important where the fate of regions and cities and communities is apparently being left to creative strategies.

A major problem with analyzing creative development is that very different approaches are taken to 'creativity' and the 'creative class', depending on the disciplinary perspective adopted. The grounding of Florida's ideas in economics creates a focus on an homogenous 'class' of people, who are in some way attracted by creative (urban) environments. However, this scale of enquiry ignores issues such as social networks, identity and distinction, which a more sociological or geographical enquiry would probably uncover. The development of more multidisciplinary research on the issue of creativity might help to illuminate such areas, as well as providing a firmer basis for more holistic creative policies and management strategies. In particular, more attention should be paid to the growing convergence between the fields of culture, creativity, urbanism and geography. This will be important in identifying the role of transnational elites and other mobile populations in the production and consumption of creativity, and the relationships between the creative economy, the knowledge economy and the symbolic economy.

As Maitland points out (Chapter 5), one problem with Florida's (2002) analysis is an incomplete explanation of what attracts human capital to specific cities. Tourism and mobility studies may have an important role to play here. As we show in Chapter 1, Barcelona is attractive to tourists who can see themselves living there. They may in time become quasi residents (or even permanent residents), adding to the attractive cultural mix of the city. Tourism is therefore a driver of creative development as well as an outcome. Sorting out cause and effect will require careful research, as in the broader case of Florida's theories. However, there is little doubt that tourism, with its growing flows of people between different regions and the consequent de-differentiation between 'locals' and 'tourists', is producing the kind of new juxtapositions between lived, conceived and perceived spaces talked about by Cloke (Chapter 2) and Hannigan (Chapter 3). It seems that tourists are becoming increasingly creative in the consumption of the places they visit, turning the everyday and the mundane of the local into a new experience of the glocal. In contrast, many of the producers of cultural and 'creative' experiences seem to lack this ability.

In our view, what is required to stimulate creative development is not so much the casual transfer of creative strategies from one part of the world to another, but a careful translation of ideas into the local context in which they need to function. This may call for cultural or creative intermediaries, who can undertake the work of translating creative development strategies into the local context, in much the same way that the *animateurs* described by Prentice and Andersen (Chapter 6) operate in France.

While creative strategies to date may be questionable, it may be the case that the application of creativity need not be so rigid or standardized as it seems to have been to date, and it should certainly be applied more sympathetically. Most of the criticism surrounding creative strategies seems indeed to stem from the fact that little account has been taken of place specificity in their design and application. While such criticism is well founded, it also seems that thoughtful application of creative development may produce a more embedded and engaging tourism experience with the potential for a redistribution of power in the tourism system and a new mindset for prosumers.

References

Amendola, G. (1997) *La Città Posmoderna: Magie e Paure della Metropoli Contemporanea*, Bari: Laterza.

Americans for the Arts (2004, 2005) *Congressional Report on the Creative Industries*, Washington D.C.: Americans for the Arts. ww3.artsusa.org/pdf/information_resources/creative_industries/the_congressional_report.pdf (accessed 25 March 2007).

Amin, A. and Thrift, N. (1994) *Globalization, Institutions and Regional Development in Europe*, Oxford: Oxford University Press.

—— (2002) *Cities: Reimagining the Urban*, Cambridge: Polity (2nd reprint).

Anderson, K.J. (1990) 'Chinatown Re-oriented: A critical analysis of recent redevelopment schemes in a Melbourne and Sydney enclave', *Australian Geographical Studies*, 28(1): 137–54.

Anholt-GMI (2005) *The Anholt-GMI City Brands Index*. www.citybrandsindex.com (accessed December 2005).

Anon. (1997) 'Quest for Hardy Blooms: Why art cannot be like hothouse flowers', *Asiaweek*, 5 September 1997.

—— (2002) 'Singapore stages a cultural renaissance', *Financial Times*, 12 October 2002.

—— (2006) *Creative Tourism Portal*, Barcelona Bulletin 245, October 2006, Barcelona: Ajuntament de Barcelona, p. 4.

Antony, R. and Henry, J. (2005) *Lonely Planet Guide to Experimental Travel*, Melbourne: Lonely Planet.

Appadurai, A. (2001, ed.) *Globalization*, Durham, NC: Duke University Press.

Aranda, A. and Colomer, J. (2006) *Nuevas experiencias por Barcelona*, Practicum III, Barcelona: ESADE/St. Ignasi.

Arenas, S. (2006) 'Aquí se come muy bien. Entrevista con Joan Antón Matas Arnalot, Presidente del Gremio de Hostelería de Sitges', *Mon Blau*, 8: 12–15.

Arias Sans, A. (2003) *Framing Queerness in the Post-Industrial City*, unpublished Master Thesis, MEMR Rotterdam.

Ashworth, G. and Hartmann, J.R. (2005, eds) *Horror and Human Tragedy Revisited: The management of sites of atrocities for tourism*, New York: Cognizant.

Ashworth, G.J. and Voogd, H. (1990) *Selling the City: Marketing approaches in public sector urban planning*, London: Belhaven Press.

Ateljevic, I. and Doorne, S. (2000) 'Staying with the Fence: Lifestyle entrepreneurship in tourism', *Journal of Sustainable Tourism*, 8: 378–92.

Atkinson, C.Z. (1997) 'Whose New Orleans? Music's place in the packaging of New Orleans for tourism', in S. Abram, S. Waldren and D. Macleod (eds) *Tourists and Tourism: Identifying with people and places*, Berg: Oxford, pp. 91–106.

Baerenholdt, J., Haldrup, M., Larsen, J. and Urry, J. (2004) *Performing Tourist Places*, Aldershot: Ashgate.

Balibrea, M. (2001) 'Urbanism, culture and the post-industrial city: Challenging the "Barcelona model"', *Journal of Spanish Cultural Studies*, 2(2): 187–210.

Baniotopoulou, E. (2000) 'Art for Whose Sake? Modern art museums and their role in transforming societies: the case of the Guggenheim Bilbao', *Journal of Conservation and Museum Studies*, 7: 1–15.

Barney, J. (1991) 'Firm resources and sustained competitive advantage', *Journal of Management*, 17: 99–120.

Baudrillard, J. (1983) *Simulations*, London: Routledge.

Bauman, Z. (1996) 'From Pilgrim to Tourist – or a short history of identity', in S. Hall and P. Du Gay (eds) *Questions of Cultural Identity*, London: Sage, pp. 18–36.

—— (2000) *Liquid Modernity*, Cambridge: Polity.

Beck, U. (1992) *Risk Society: Towards a new modernity*, London: Sage.

—— (2002) 'The Cosmopolitan Society and its Enemies', *Theory Culture and Society*, 19(1): 17–44.

Beeton, S. (2001) *Film-induced Tourism*. Clevedon, UK: Channel View.

Begg, I. (2002, ed.) *Urban Competitiveness: Policies for dynamic cities*. Bristol: The Policy Press.

Belfiore, E. and Bennett, O. (2006) *Rethinking the Social Impact of the Arts: a critical-historical review*, Centre for Cultural Policy Studies, University of Warwick, Research Paper 9.

Bell, C. (1996) *Inventing New Zealand*, Auckland: Penguin.

Bell, D. (1973) *The Coming of Post-Industrial Society*, New York: Basic Books.

Bell, D. and Jayne, M. (2004a, eds) *City of Quarters: Urban villages in the contemporary city*, Aldershot: Ashgate.

—— (2004b) 'Conceptualising the City of Quarters', in D. Bell and M. Jayne (eds) *City of Quarters*, Aldershot: Ashgate, pp. 1–14.

Benjamin, G. (1976) 'The cutural logic of Singapore's "multiculturalism"', in J.H. Ong, C.K. Tong and E.S. Tan (eds), *Understanding Singapore Society*, Singapore: Times Academic Press, pp. 67–85.

Best, S. (1989) 'The commodification of reality and the reality of commodification: Jean Baudrillard and post-modernism', *Current Perspectives in Social Theory*, 19: 23–51.

Bianchini, F. and Parkinson, M. (1993) *Cultural Policy and Urban Regeneration: The West European Experience*, Manchester: Manchester University Press.

Bill, A. (2005) '"Blood, sweat and tears": disciplining the creative in Aotearoa New Zealand', unpublished paper, College of Design, Fine Arts and Music, Massey University.

Bindloss, J., Parkinson, T. and Fletcher, M. (2003) *Kenya*, Melbourne: Lonely Planet (5th edition).

Binkhorst, E. (2002) *Holland, the American way: Transformations of the Netherlands into US vacation experiences*, PhD Thesis, Tilburg University.

—— (2005) *Creativity in the experience economy, towards the co-creation tourism experience?*, paper presented at the ATLAS annual Conference 'Tourism, creativity and development', Barcelona, November.

Binnie, J. (2004) 'Tales from the city', *New Humanist*, September. www.newhu manist.org.uk/volume119issue5_more.php?id=951_0_32_0_C (accessed 15 February 2007).

Binnie, J., Holloway, J., Millington, S. and Young, C. (2006) *Cosmopolitan Urbanism*, London: Routledge.

Bloomfield, J. and Bianchini, F. (2003) *Planning for the Cosmopolitan City*, COMEDIA and International Cultural Planning and Policy Unit (ICPPU), De Montfort University, Leicester.

Bodineau, P. and Verpeaux, M. (1993) *Histoire de la Décentralisation*, Paris: Presses Universitaires de France.

Body-Gendrot, S. (2003) 'Cities, Security, and Visitors: Managing mega-events in France', in L.M. Hoffman, S.S. Fainstein and D.R. Judd (eds) *Cities and Visitors: Regulating people, markets, and city space*, Oxford: Blackwell, pp. 39–52.

Bonink, C. and Hitters, E. (2001) 'Creative industries as milieu of innovation', in G. Richards (ed.) *Cultural Attractions and European Tourism*. Wallingford: CAB International, pp. 227–40.

Boswijk, A., Thijssen, T. and Peelen, E. (2005) *Een nieuwe kijk op de experience economy, betekenisvolle belevenissen*, Amsterdam: Pearson Education Benelux.

Bourgeon-Renault, D. (2005) 'Du Marketing Expérientiel Appliqué aux Musées', *Cahier Espaces*, 87: 41–47.

Boyle, M. and Hughes, G. (1991) 'The politics of the representation of "the real": discourses from the Left on Glasgow's role as European City of Culture', *Area* 23(3): 217–28.

Braudel, F. (1986) *L'Identité de la France. Espace et Histoire*, Paris: Arthaud-Flammarion.

Brecknock, R.(2004) 'Creative capital: creative industries in the "creative city"', unpublished paper, Brecknock Consulting Australia, Brisbane.

Brick Lane Festival (2004) *Official Guide*, London: Ethnic Minority Enterprise Project.

Brooks, A.C. and Kushner, R.J. (2001) 'Cultural districts and urban development', *International Journal of Arts Management*, 3(2): 4–15.

Bryman, A. (1995) *Disney and his Worlds*, London: Routledge.

Buckley, R. (2004) 'Skilled commercial adventure: the edge of tourism', in T.V. Singh (ed.) *New Horizons in Tourism: Strange Experiences and Stranger Practices*, Wallingford: CABI, pp. 37–48.

Burnley, I.H. (2000) 'Diversity and Difference: Immigration and the multicultural city', in J. Connell (ed.) *Sydney: The emergence of a world city*, South Melbourne: Oxford University Press, pp. 244–72.

—— (2001) *The Impact of Immigration on Australia: A demographic approach*, South Melbourne: Oxford University Press.

Burnley, I.H., Murphy, P. and Fagan, R.H. (1997) *Immigration and Australian Cities*, Sydney: Federation Press.

Butler, R. (2006, ed.) *The Tourism Area Life Cycle, Vol.1: Applications and Modifications*, Clevedon, UK: Channel View.

Butler, R., Hall, M. and Jenkins, J. (1998, eds) *Tourism and Recreation in Rural Areas*, Chichester: Wiley.

Cachalia, F., Jocum, M. and Rogerson, C.M. (2004) '"The urban edge of African fashion": the evolution of Johannesburg's planned fashion district', in D. McCormick and C.M. Rogerson (eds) *Clothing and Footwear in African Industrialisation*, Pretoria: Africa Institute of South Africa, pp. 527–46.

Cachet, E.A., Kroes Willems, M. and Richards, G. (2003) *Culturele identiteit van Nederlandse gemeenten*, Rotterdam: Erasmus Universiteit Rotterdam.

Cahm, E. (1998) 'Mitterrand's *Grand Projets*', in M. Maclean (ed.) *The Mitterrand Years*, Basingstoke: Macmillan, pp. 263–75.

Camargo, P. de (2007) 'Using tourist resources as tools for teaching and creating awareness of heritage in a local community', in Richards, G. (ed.) *Cultural Tourism: Global and local perspectives*, Binghampton, NY: Haworth Press, pp. 239–55.

Cameron, S. and Coaffee, J. (2005) 'Art, Gentrification and Regeneration: From artist as pioneer to public arts', *European Journal of Housing Policy* 5(1): 39–58.

Cantell, T. (2005) 'From Economic Policy to Creative City Ideas – the Helsinki Experience', *MAYA, Estonian Architectural Review*. On-line article www.solness.ee/majaeng/index.php?gid=44&id=553 (accessed 27 February 2007).

Carey, S. (2002) *Brick Lane, Banglatown: A study of the catering sector*, final report, London: Research Works Limited.

Carpenter, H. (1999) *Islington: The Economic Impact of Visitors*, London: Discover Islington.

Caserta, S. and Russo, A.P. (2002) 'More means worse: Asymmetric information, spatial displacement and sustainable heritage tourism', *Journal of Cultural Economics* 26(4): 245–60.

Castells, M. (1989) *The Informational City*, Oxford: Blackwell.

—— (2000) *The Rise of the Network Society*, Oxford: Blackwell (2nd edition).

Cater, C. and Smith, L. (2003) 'New country visions: adventurous bodies in rural tourism', in P. Cloke (ed.) *Country Visions*, Harlow: Pearson, pp. 195–217.

Caves, R. (2000) *Creative Industries: Contracts between art and commerce*, Cambridge, MA: Harvard University Press.

—— (2003) 'Contracts between art and commerce', *Journal of Economic Perspectives*, 17: 73–83.

Center for an Urban Future (2005) *Creative New York*, New York: Center for an Urban Future.

Champsaur, P. (1997) *La France et ses Régions*, Paris: Institut National de la Statistique et des Études Économiques.

Chan, W. (2004) 'Finding Chinatown: Ethnocentricity and Urban Planning', in D. Bell and M. Jayne (eds) *City of Quarters: Urban villages in the contemporary city*, Aldershot: Ashgate, pp. 173–87.

Chang, T.C. (2000a) 'Renaissance Revisited: Singapore as a "Global City for the Arts"', *International Journal of Urban and Regional Research*, 24: 818–31.

—— (2000b) 'Singapore's Little India: A tourist attraction as a contested landscape', *Urban Studies*, 37(2): 343–68.

Chang, T.C. and Lee, W.K. (2003) 'Renaissance City Singapore: a study of arts spaces', *Area*, 35: 128–41.

Charles, D. and Benneworth, P. (2001) *The Regional Mission – The regional contribution of higher education – National Report*, London: Universities UK/HEFCE.

Chartrand, H.H. (1990) 'Creativity and Competitiveness: Art in the Information Economy', *Arts Bulletin*, 15(1): 1–2.

Chatterton, P. (1999) 'University students and city centres – the formation of exclusive geographies. The case of Bristol, UK', *Geoforum* 30: 117–33.

Chrisafis, A. (2004) 'Ibiza on the Liffey: but where are the Irish?', The *Guardian*, 8 November.

Chua, B.H. (1995) *Communitarian Ideology and Democracy in Singapore*. London: Routledge.

CIDA (2007) www.creativeportal.org (accessed 9 January 2007).

City Fringe Partnership (1997) *Revitalising the City Fringe: inner city action with a world city focus*, London: City Fringe Partnership.

City Futures Inc. (2005) *Creative New York*, New York: Centre for an Urban Future.

City of Johannesburg (2005a) 'Five-Year Review', unpublished draft report of the Johannesburg Economic Development Unit.

—— (2005b) 'Joburg 2030 Joburg Art Bank', unpublished report presented to the city Finance Strategy and Economic Development Committee, 5 November.

City of Melbourne (2007) Multicultural communities – Introduction. www.melbourne.vic.gov.au/info.cfm?top=100&pg=901 (accessed 1 February 2007).

City of Toronto (2001) *The Creative City: A workprint*, Toronto: City of Toronto.

Clark, T.N. (2003) *The City as an Entertainment Machine*, San Diego, CA: Elsevier.

Clift, S., Luongo, M. and Callister, C. (2002) *Gay Tourism: Culture, identity and sex*, New York: Thomson.

Cloke, P. (1993) 'The countryside as commodity: new rural spaces for leisure', in S. Glyptis (ed.) *Leisure and the Environment*, London: Belhaven, pp. 53–67.

Cloke, P. and Perkins, H. (1998) 'Cracking the Canyon with the Awesome Four-some: representations of adventure tourism in New Zealand', *Environment and Planning D: Society and Space*, 16: 185–218.

Cloke, P. and Jones, O. (2001) 'Dwelling, place and landscape: an orchard in Somerset', *Environment and Planning A*, 33: 649–66.

Cloke, P. and Perkins, H. (2002) 'Commodification and adventure in New Zealand tourism', *Current Issues in Tourism*, 5: 521–49.

Cloke, P. and Jones, O. (2004) 'Turning in the graveyard: trees and the hybrid geog-raphies of dwelling, monitoring and resistance in a Bristol cemetery', *Cultural Geographies*, 11: 180–208.

Cloke, P. and Perkins, H. (2005) 'Cetacean performance and tourism in Kaikoura, New Zealand', *Environment and Planning D: Society and Space*, 23: 903–24.

Cohen, E. (1972) 'Toward a sociology of international tourism', *Social Research*, 39(1): 164–82.

—— (1979) 'A phenomenology of tourist experiences', *Journal of the British Sociological Association*, 13(2): 179–201.

—— (1988) 'Authenticity and commoditization in tourism', *Annals of Tourism Research*, 15: 371–85.

—— (2004) *Contemporary Tourism*, Amsterdam: Elsevier.

Cohen, S. (1997) 'More than The Beatles: Popular music, tourism and urban regen-eration', in S. Abram, S. Waldren and D. Macleod (eds) *Tourists and Tourism: Identifying with people and places*, Oxford: Berg, pp. 71–90.

Collins, J. (1991) *Migrant Hands in a Distant Land: Australia's post-war immigra-tion*, Sydney: Pluto Press (2nd edition).

—— (2002) 'Ethnic entrepreneurs and the economic, spatial and social development of Sydney and Melbourne', paper presented to 98th Annual Meeting of the Association of American Geographers, Los Angeles, California, 19–23 March.

—— (2003) 'Cosmopolitan capitalists: Immigrant entrepreneurs in Australia', in R. Kloosterman and J. Rath (eds) *Immigrant Entrepreneurs: Venturing abroad in the age of globalization*, Oxford: Berg, pp. 61–78.

Collins, J. and Castillo, A. (1998) *Cosmopolitan Sydney: Explore the world in one city*, Sydney: Pluto Press.

Collins, J. and Lalich, W.F. (2000) 'Cultural diversity and the 2000 Olympic Games', paper presented to UTS Events Management Conference, Sydney, 21–22 April.

Collins, J., Gibson, K., Alcorso, C., Castles, S. and Tait, D. (1995) *A Shop Full of Dreams: Ethnic small business in Australia*, Sydney: Pluto Press.

Collins, J., Noble, G., Poynting, S. and Tabar, P. (2000) *Kebabs, Kids, Cops and Crime: Youth, ethnicity and crime*, Sydney: Pluto Press.

ComMark Trust (2006) 'Struggle history and shopping', *Catalyst*, March 2006. www.commark.org (accessed 19 January 2007).

Commonwealth of Australia (1994) 'Creative Nation: Commonwealth Cultural Policy', October 1994. www.nla.gov.au/creative.nation/contents.html (accessed 9 January 2007).

Community Marketing Inc. (2005) '10th Annual LGBT (Lesbian, Gay, Bisexual and Transgender) Community Survey'. www.communitymarketinginc.com (accessed 13 December 2006).

Conforti, J.M. (1996) 'Ghettos as tourism attractions', *Annals of Tourism Research*, 23(4): 830–42.

Connell, J. (2000) 'And the winner is . . . ' in J. Connell (ed.) *Sydney: The emergence of a world city*, South Melbourne: Oxford University Press, pp. 1–18.

Connell, J. and Thom, B. (2000) 'Beyond 2000: The post-Olympic city', in J. Connell (ed.) *Sydney: The emergence of a world city*, South Melbourne: Oxford University Press, pp. 319–43.

Cope, B. and Kalantzis, M. (1997) *Productive Diversity: Work and management in diverse communities and global markets*, Sydney: Pluto Press.

Cornelissen, S. (2005) 'Producing and imaging "place" and "people": the political economy of South African international tourist representation', *Review of International Political Economy*, 12: 674–99.

Coulson, A. (1997) 'Business partnerships and regional government', *Policy and Politics* 25(1): 31–38.

Creative London (2002) *Creativity: London's Core Business*, London: Creative London.

Crewe, L. and Beaverstock, J. (1998) 'Fashioning the city: Cultures of consumption in contemporary urban spaces', *Geoforum*, 29: 287–308.

Crouch, D. (1999, ed.) *Leisure/Tourism Geographies: Practices and geographical knowledge*, London: Routledge.

Csikszentmihalyi, M. (1990) *Flow: The Psychology of Optimal Experience*, New York: Harper and Row.

Csikszentmihalyi, M. and Hunter, J. (2003) 'Happiness in everyday life: the uses of experience sampling', *Journal of Happiness Studies*, 4: 185–99.

CTNZ (2007) www.creativetourism.co.nz (accessed 19 February 2007).

Cultural Strategy Group (1998a) 'Creative South Africa: A strategy for realising the potential of the cultural industries', unpublished report for the Department of Arts, Culture, Science and Technology, Pretoria.

—— (1998b) 'Cultural Industries Growth Strategy: The South African craft industry report', unpublished report for the Department of Arts, Culture, Science and Technology, Pretoria.

—— (1998c) 'Cultural Industries Growth Strategy: The South African Film sector report', unpublished report for the Department of Arts, Culture, Science and Technology, Pretoria.

—— (1998d) 'Cultural Industries Growth Strategy: The South African Music sector report', unpublished report for the Department of Arts, Culture, Science and Technology, Pretoria.

—— (1998e) 'Cultural Industries Growth Strategy: The South African Publishing sector report', unpublished report for the Department of Arts, Culture, Science and Technology, Pretoria.

Cunnell, D. and Prentice, R.C. (2000) 'Tourists' recollections of quality in museums', *Museum Management and Curatorship*, 18: 369–90.

Cunningham, S. (2003) 'From cultural to creative industries: theory, industry and policy implications', unpublished paper, Creative Industries Research and Applications Centre, Queensland University of Technology, Brisbane.

Cunningham, S. and Hartley, J. (2001) 'Creative Industries – From blue poles to fat pipes', paper presented to Australia Academy of the Humanities, National Summit on the Humanities and Social Sciences, National Museum Australia, Canberra, 26–27 July.

Dahles, H. (1998) 'Redefining Amsterdam as a tourist destination', *Annals of Tourism Research*, 25: 55–69.

—— (2000) 'Tourism, small enterprises and community development', in G. Richards and D. Hall (eds) *Tourism and Sustainable Community Development*, London: Routledge, pp. 154–69.

Dahms, F.A. (1995) '"Dying villages", "counterurbanization" and the urban field – A Canadian perspective', *Journal of Rural Studies*, 11: 21–33.

Daly, M. and Pritchard, B. (2000) 'Sydney: Australia's financial and corporate capital', in J. Connell (ed.) *Sydney: The emergence of a world city*, South Melbourne: Oxford University Press, pp. 167–88.

Dann, G. and Jacobsen, J.K.S. (2003) 'Tourism smellscapes', *Tourism Geographies*, 5: 3–25.

Davies, P. (2004) 'Eco-museums and the Democratisation of Cultural Tourism', *Tourism, Culture and Communication*, 5: 45–58.

DCMS (2002) *Regional Cultural Data Framework: A user's guide for researchers and policymakers*, London: DCMS.

—— (2006) *Creative Industries Economic Estimates Statistical Bulletin*. www.culture. gov.uk/NR/rdonlyres/70156235–38AB8–48F9-B15B-78A326A8BFC4/0/Creativ eIndustriesEconomicEstimates2006.pdf (accessed 25 March 2007).

DEAT (2003) *Tourism 10 Year Review*, Pretoria: Department of Environmental Affairs and Tourism.

Deben, L., Heinemeijer, W. and van der Vaart, D. (2000, eds) *Understanding Amsterdam. Essays on economic vitality, city life and urban form*, Amsterdam: Het Spinhuis.

de Berranger, P. and Meldrum, M. (2000) 'The development of intelligent local clusters to increase global competitiveness and local cohesion: the case of small businesses in the creative industries', *Urban Studies* 37(10): 1827–35.

Debord, G. (1958) *Théorie de la derive*, translated by Ken Knabb and available at www.bopsecrets.org/SI/2.derive.htm (accessed 15 February 2007).

de Botton, A. (2002) *The Art of Travel*, Harmondsworth: Penguin.

Degen, M.A. (2003) 'Fighting for the Global Catwalk: Formalising public life in Castlefield (Manchester) and diluting public life in el Raval (Barcelona)', *International Journal of Urban and Regional Research*, 27(4): 867–80.

DEMOS (2003) 'Manchester is favourite with "new bohemians"'. http://83.22 3.102.49/media/pressreleases/bohobritain (accessed 28 February 2007).

Denzin, N.K. and Lincoln, Y.S. (2000, eds) *The Handbook of Qualitative Research*, Thousand Oaks, CA: Sage (2nd edition).

Department of Culture, Media and Sport (DCMS), United Kingdom (1998) *Mapping the Creative Industries*. www.culture.gov.uk/Reference_library/Publications/archive_1998/Creative_Industries_Mapping_Document_1998.htm (accessed 16 December 2006).

Department of Trade and Industry, South Africa (2005a) *Tourism Sector Development Strategy*, Pretoria: Trade and Investment South Africa.

—— (2005b) *A Growing Economy that Benefits All: Accelerated and Shared Growth Initiative for South Africa (ASGI-SA): Discussion Document*, Pretoria: DTI.

Department of Trade and Industry, United Kingdom (1998) *Building the Knowledge Driven Economy: Government and business in the knowledge driven economy*, Command Paper 4176, London: Department of Trade and Industry.

Dewsbury, J-D. (2000) 'Performativity and the event: enacting a philosophy of difference', *Environment and Planning D: Society and Space*, 18: 473–97.

Di Maria, E., Russo, A.P., Zanon, G. and Zecchin, F. (2004), *Venezia Laboratorio di Cultura: Indagine sulla Dimensione Economica dell'offerta Culturale a Venezia*, Venezia: Marsilio.

DIMIA (2006) *Diversity Australia*, Department of Immigration, Multicultural and Indigenous Affairs, www.diversityaustralia.gov.au (accessed 5 March 2006).

Dimmock, K. and Tyce, M. (2001) 'Festivals and Events: Celebrating special interest tourism', in N. Douglas, N. Douglas and R. Derrett (eds) *Special Interest Tourism*, Milton, Queensland: Wiley, pp. 355–82.

Dirsuweit, T. (1999) 'From fortress city to creative city: developing culture and the information-based sectors in the regeneration and reconstruction of the Greater Johannesburg Area', *Urban Forum*, 10: 183–213.

Donald, B. and Morrow, D. (2003) *Competing for Talent: Implications for Social and Cultural Policy in Canadian City Regions*, Ottawa: Strategic Research and Analysis Directorate, Department of Canadian Heritage.

Doss, E. (1997) 'Making Imagination Safe in the 1950s: Disneyland's Fantasy Art and Architecture', in K.A. Marling (ed.) *Designing Disney's Theme Parks: The Architecture of Reassurance*, Montréal: Centre Canadien d'Architecture, pp. 179–91.

Douglas, N., Douglas, N. and Derrett, R. (2001, eds) *Special Interest Tourism*, Milton, Queensland: Wiley.

DTI (2004) *Creative People, Openness and Productivity: An exploration of regional differences inspired by 'The Rise of the Creative Class'*, London: Department for Trade and Industry.

Dubet, F. and Sembel, N. (1994) 'Les étudiants, le campus et la ville', *Les Annales de la Recherche Urbaine*, 62–63: 224–34.

Edensor, T. (1998) *Tourists at the Taj: Performance and Meaning at a Symbolic Site*, London: Routledge.

Ehrlich, B. and Dreier, P. (1999) 'The New Boston Discovers the Old: Tourism and the struggle for a livable city', in D.R. Judd and S. Fainstein (eds) *The Tourist City*, New Haven, CT: Yale University Press, pp. 155–78.

Eisinger, P. (2000) 'The politics of bread and circuses: Building the city for the visitor class', *Urban Affairs Review* 35(3): 316–33.

Elands, B. and Lengkeek, J. (2000) *Typical Tourists: Research into the theoretical and methodological foundations of a typology of tourism and recreation experiences*, Leiden: Backhuys Publishers.

Elangovan, P. (2002) 'The Renaissance Starts Here?', *Far Eastern Economic Review*, 10 October 2002: 56.

Entriken, J.N. (1991) *The Betweenness of Place: Towards a Geography of Modernity*, Basingstoke: Macmillan.

ERC-CI (2002) *Creative Industries Development Strategy: Propelling Singapore's Creative Economy*, Singapore: Economic Review Committee – Services Subcommittee Workgroup on Creative Industries.

European Commission (2007) *Discrimination in the European Union. Special EUROBAROMETER 263*, Brussels: European Commission.

European Travel Commission/World Tourism Organization (2005) *City Tourism and Culture: The European experience*, Madrid: World Tourism Organization.

Evans, G.L. (1998) 'In search of the cultural tourist and the postmodern Grand Tour', International Sociological Association XIV Congress Relocating Sociology, Montreal, July.

—— (2001) *Cultural Planning: An Urban Renaissance?*, London: Routledge.

—— (2002) 'Living in a World Heritage City: Stakeholders in the Dialectic of the Universal and the Particular', *International Journal of Heritage Studies*, 8(2): 117–35.

—— (2003) 'Hard Branding the Culture City – From Prado to Prada', *International Journal of Urban and Regional Research*, 27(2): 417–40.

—— (2005a) 'Measure for Measure: Evaluating the evidence of culture's contribution to regeneration', *Urban Studies*, 42(5/6): 1–25.

—— (2005b) 'Creative spaces: strategies for creative cities', in J. Swarbroke, M. Smith and L. Onderwater (eds) *Tourism, Creativity and Development: ATLAS Reflections 2005*, Association for Tourism and Leisure Education, Arnhem, 7–10.

—— (2006) *Strategies for Creative Spaces – Case Study London*, London Development Agency.

Evans, G.L. and Foord, J. (2005) 'Rich mix cities: from multicultural experience to cosmopolitan engagement', *Ethnologia Europaea: Journal of European Ethnology*, 34(2): 71–84.

Evans, G.L., Foord, J. and Shaw, P. (2005) *Creative Spaces: Phase I Report*, London: London Development Agency.

Fainstein, S. (2005) 'Cities and diversity. Should we want it? Can we plan for it?', *Urban Affairs Review*, 41(1): 3–19.

Fainstein, S. and Gladstone, D. (1999) 'Evaluating Urban Tourism', in S. Fainstein and D. Judd (eds) *The Tourist City*, New Haven, CT: Yale University Press, pp. 21–34.

Fainstein, S. and Powers, J.C. (2003) 'Tourism and New York's Ethnic and Racial Diversity: An underutilized resource?', paper presented to European Science Foundation Standing Committee for the Social Sciences Exploratory Workshop, 'The Immigrant Tourist Industry: The Commodification of Cultural Resources in Cosmopolitan Cities', Institute for Migration and Ethnic Studies, University of Amsterdam, Netherlands, 6–9 December.

Fainstein, S., Hoffman, L. and Judd, D. (2003) 'Making Theoretical Sense of Tourism', in L. Hoffman, S. Fainstein and D. Judd (eds) *Cities and Visitors. Regulating People, Markets and City Space*, Oxford: Blackwell, pp. 239–53.

Featherstone, M. (1991) *Consumer Culture and Postmodernism*, London: Sage.

—— (2002) 'Cosmopolis: An introduction', *Theory, Culture and Society*, 19(1–2): 1–16.

Federation of Gay Games (2006) www.gaygames.com/en/games/gg5 (accessed 12 December 2006).

Feifer, M.(1985) *Going Places: Tourism in History*, New York: Stein and Day.

Ferrari, S., Adamo, G.E. and Veltri, A.R. (2007) 'Experiential and multisensory holidays as a form of creative tourism', in G. Richards and J. Wilson (eds) *Changing experiences – the development of creative tourism*, Arnhem: ATLAS.

Fitzgerald, S. (1997) *Red Tape, Gold Scissors: The story of Sydney's Chinese*, Sydney: State Library of New South Wales Press.

Flew, T. (2002) 'Beyond *ad hocery:* defining creative industries', unpublished paper presented to the Second International Conference on Cultural Policy Research, Te Papa, Wellington, 23–26 January.

—— (2003) 'Creative industries: from the chicken cheer to the culture of services', *Continuum: Journal of Media and Cultural Studies*, 17: 89–94.

Florida, R.L. (2002) *The rise of the creative class, and how it's transforming work, leisure, community and everyday life*, New York: Basic Books.

—— (2003) *The Rise of the Creative Class*, Melbourne: Pluto Press.

—— (2004) *Cities and the Creative Class*. New York: Routledge.

—— (2005) *Flight of the Creative Class: The New Global Competition Talent*, New York: Harper Collins.

Florida, R. and Tinagli, I. (2004) *Europe in the Creative Age*. London: DEMOS/ Carnegie Mellon University.

Forbes, J. (1995) 'Popular Culture and Cultural Policies', in J. Forbes and M. Kelly (eds), *French Cultural Studies*, Oxford: Oxford University Press, pp. 232–63.

Fowler, E. and Siegel, D. (2002) 'Introduction: Urban Public Policy at the Turn of the Century', in E.P. Fowler and D. Siegel (eds) *Urban Policy Issues*, Oxford: Oxford University Press, pp. 1–16.

Franke, S. and Verhagen, E. (2005, eds) *Creativity and the City: How the creative economy changes the city*, Rotterdam: NAi Publishers.

Franklin, A. (2003) *Tourism. An Introduction*, London: Sage.

Franklin, A. and Crang, M. (2001) 'The Trouble with Tourism and Travel Theory?', *Tourist Studies*, 1(1): 5–22.

Franquesa, J.B. and Morrell, M. (2007) 'Transversal Indicators and Qualitative Observatories of Heritage Tourism', in G. Richards (ed.) *Cultural Tourism: Global and Local Perspectives*, New York: Haworth Press, pp. 169–94.

Fraser, N. (2006) 'Cross-border shopping – moving to "Jobai"', *CitiChat*, 20 March. www.commark.org.

Frenkel, S. and Walton, J. (2000) 'Bavarian Leavenworth and the symbolic economy of a theme town', *The Geographical Review*, 90(4): 559–81.

Friedman, T.L. (2005) *The World Is Flat: A Brief History of the Twenty-first Century*, New York: Farrar, Straus and Giroux.

Garrod, B. and Wilson, J. (2003) *Marine Ecotourism: Issues and Experiences*, Clevedon, UK: Channel View.

—— (2004) 'Nature on the Edge? Marine Ecotourism in Coastal Peripheral Areas', *Journal of Sustainable Tourism*, 12(2): 95–120.

Gemeente Amsterdam (2004a) *Amsterdam als creatieve kennisstad, bewoners en bedrijven over de aantrekkingskracht van Amsterdam*, Amsterdam: Dienst Onderzoek en Statistiek, Gemeente Amsterdam.

—— (2004b) *Amsterdam creatieve kennisstad: een passende ambitie?*, Amsterdam: Dienst Onderzoek en Statistiek, Gemeente Amsterdam.

George, C. (2000) *Singapore the Air-Conditioned Nation*, Singapore: Landmark Books.

Germain, A. and Radice, M. (2006) 'Cosmopolitanism by default', in J. Binnie, J. Holloway, S. Millington and C. Young (eds) *Cosmopolitan Urbanism*, London: Routledge, pp. 112–29.

GHK Consulting (2004) 'Plymouth City Growth Strategy: Draft Consultation', Plymouth: GHK Consulting.

Gibson, C. (2003) 'Cultures at Work: Why "culture" matters in research on the "cultural" industries', *Social and Cultural Geography*, 4(2): 201–15.

Gibson, C. and Klocker, N. (2004) 'Academic publishing as "creative" industry, and recent discourses of "creative economies", some critical reflections', *Area*, 36(4): 423–34.

Gibson, C., Murphy, P. and Freestone, R. (2002) 'Employment and Socio-spatial Relations in Australia's Cultural Economy', *Australian Geographer*, 33(22): 173–89.

Giddens, A. (1990) *The Consequences of Modernity*, Cambridge: Polity.

—— (1991) *Modernity and Self-Identity: Self and Society in the Late Modern Age*, Cambridge/Oxford: Polity in association with Blackwell.

Gilmore, F. (2004) 'Shanghai: Unleashing creative potential', *Journal of Brand Management*, 11(6): 442–48.

Gilmore, J.H. and Pine, B.J. (2002) *The experience IS the marketing*, Louisville, KY: BrownHerron Publishing.

GLA Economics (2004) *London's Creative Sector: 2004 Update*, London: Greater London Authority.

Go, F.M., Lee, R.M. and Russo, A.P. (2004) 'Heritage in the globalizing world: Reconstructing a business model', *Information Technology and Tourism*, 6(1): 55–68.

Goldberger, P. (1996) 'The Rise of the Private City', in J. Vitullo Martin (ed.) *Breaking Away: The Future of Cities*, New York: The Twentieth Century Fund, pp. 135–48.

Gombault, A. and Eberhard-Harribey, L. (2005) 'L'Expérience Louvre-Estuaire: Entre éducation au patrimoine et quête d'identité locale', *Cahiers Espaces*, 87: 68–77.

Gottdiener, M. (1997) *The Theming of America*, Boulder, CO: Westview Press.

Greefe, X. (2006) *La mobilisation des actifs culturels de la France: De l'attractivite culturelle du territoire. . . . a la nation culturellement creative*. Document de travail du Département des études, 1270, May 2006.

Griffiths, R. (1993) 'The politics of cultural policy in urban regeneration strategies', *Policy and Politics*, 21(1): 39–46.

Grit, A. (2005) 'The development of Innovative Cultural Mediators in tourism: Releasers of diversity within the creative city?', paper presented at the Association for Tourism and Leisure Education (ATLAS) Annual Conference, Barcelona, November 2005.

Guan, J. (2002) 'Ethnic Consciousness arises on facing spatial threats to the Philadelphia Chinatown', in A. Erdentug and F. Colombijn (eds) *Urban Ethnic Encounters: The spatial consequences*, London: Routledge, pp. 126–41.

Hackworth, J. and Rekers, J. (2005) 'Ethnic packaging and gentrification: the case of four neighbourhoods in Toronto', *Urban Affairs Review*, 41(2): 211–66.

Halfacree, K. (2005) 'Rural space: constructing a three-fold architecture', in P. Cloke, T. Marsden and P. Mooney (eds) *Handbook of Rural Studies*, London: Sage, pp. 44–62.

Hall, D. (2005, ed.) *Rural Tourism and Sustainable Business*, Clevedon, UK: Channel View.

Hall, P. (1997) 'The University and the City', *GeoJournal*, 41(4): 301–9.

—— (1998) *Cities and Civilization: Culture, Innovation, and Urban Order*, London: Weidenfeld and Nicholson.

—— (2000) 'Creative cities and economic development', *Urban Studies*, 37: 639–49.

—— (2004) 'Creativity, culture, knowledge and the city', *Built Environment*, 30(3): 256–58.

—— (2005) *Acceptance Speech on receiving the 2005 Balzan Prize for the Social and Cultural History of Cities*, Bern, Switzerland.

Halter, M. (2003) 'Destination: Diversity – immigrants and travelers in metropolitan Boston', paper presented to European Science Foundation Standing Committee for the Social Sciences Exploratory Workshop, 'The Immigrant Tourist Industry: The Commodification of Cultural Resources in Cosmopolitan Cities', Institute for Migration and Ethnic Studies, University of Amsterdam, Amsterdam, Netherlands, 6–9 December.

Hankinson, G. (2001) 'Location Branding: A Study of the Branding Practices of 12 English Cities', *Journal of Brand Management*, 9(2): 127–42.

Hannah, J. (2006) 'Tourism Corridors Link Attractions', *Washington Post*, 1 May.

Hannigan, J. (1998) *Fantasy City: Pleasure and profit in the postmodern metropolis*, London: Routledge.

—— (2002a) 'Culture, Globalization, and Social Cohesion: Towards a De-territorialized, Global Fluids Model', *Canadian Journal of Communication*, 27(2): 1–10.

—— (2002b) 'The Global Entertainment Economy', in D.R. Cameron and J.G. Stein (eds) *Street Protests and Fantasy Parks: Globalization, Culture, and the State*, Vancouver: UBC Press, pp. 20–48.

—— (2004) *Boom Towns and Cool Cities: The Perils and Prospects of Developing a Distinctive Brand in the Global Economy*, presented at the Resurgent City Leverhulme Trust Symposium, LSE, London, 19–21 April.

—— (2006) 'From Maple Leaf Gardens to the Air Canada Centre: The Downtown Entertainment Economy in "World Class" Toronto', in D. Whitson and R. Gruneau (eds) *Artificial Ice: Hockey, Culture and Commerce*, Toronto: Broadview Press/Garamond Press, pp. 201–14.

—— (2007) 'A Neo-Bohemian Rhapsody: Cultural Vibrancy and Controlled Edge as Urban Development Tools in the "New Creative Economy"', in T.A. Gibson and M. Lowes (eds) *Urban Communication: Production, Text, Context*, Lanham, MD: Rowman and Littlefield, pp. 61–81.

Hansard (2006) 10 November 2003: Column 32W. www.publications.parliament.uk/pa/cm200203/cmhansrd/vo031110/text/31110w09.htm (accessed 3 July 2007).

Harding, A. (1997) *Curating: the contemporary art museum and beyond*, London: Academy Editions.

Harvey, D. (1989) *The Condition of Postmodernity*, Oxford: Blackwell.

—— (2002) 'The art of rent: globalization, monopoly and the commodification of culture', in L. Panitch and C. Leys (eds) *A World of Contradictions*, Socialist Register. http://socialistregister.com/recent/2002/harvey2002 (accessed 27 February 2007).

Hayllar, B. and Griffin, T. (2005) 'The precinct experience: a phenomenological approach', *Tourism Management*, 26: 517–28.

Hayward, K.J. (2004) *City Limits: Crime, Consumer Culture and the Urban Experience*, London: Glasshouse Press.

Held, T., Kruse, C., Söndermann, M. and Weckerle, C. (2005) *Zurich's Creative Industries – Synthesis Report*, Zurich: Office for Economy and Labour of the Canton of Zurich and City of Zurich.

Henry, N., McEwan, C. and Pollard, J.S. (2002) 'Globalisation from below: Birmingham – postcolonial workshop of the world?', *Area*, 34(2): 117–12.

Hertzberger, H. (1991) *Lessons for Students in Architecture*, Rotterdam: Uitgiverij.

Hitters, E. and Richards, G. (2002) 'Cultural quarters to leisure zones: The role of partnership in developing the cultural industries', *Creativity and Innovation Management*, 11: 234–47.

Hjalager, A-M. and Richards, G. (2002) *Tourism and Gastronomy*, London: Routledge.

Hoffman, L.M., Fainstein, S. and Judd, D. (2003, eds) *Cities and Visitors*, Oxford: Blackwell.

Honigsbaum, M. (2001) 'McGuggenheim?', The *Guardian*, 27 January. www. guardian.co.uk/saturday_review/story/0,3605,429259,00.html (accessed 1 March 2007).

Horkheimer, M. and Adorno, T. (1972) *Dialectic of Enlightenment*, New York: Herder and Herder.

Howell, O. (2005) 'The "Creative Class" and the Gentrifying City. Skateboarding in Philadelphia's Love Park', *Journal of Architectural Education*, 59(2): 32–42.

Howkins, J. (2001) *The Creative Economy*, London: Penguin.

Hughes, H.L. (1998) 'Theatre in London and the inter-relationship with tourism', *Tourism Management*, 19: 445–52.

—— (2006) *Pink Tourism: Holidays of Gay Men and Lesbians*, Wallingford: CABI.

Hughes, R. (1998) Paper presented at the Culture makes Communities Conference. Leeds: Joseph Rowntree Foundation, 13 Febuary.

Huisman, J. (2005) 'Laissez faire met een torenhoge ambitie. Job Cohen over zijn creatieve stad Amsterdam', *Items*, 2: 35–41.

Huntington, S. (1996) *The Clash of Civilizations and the Remaking of World Order*, New York: Touchstone.

Hutton, T.A. (2003) 'Service industries, globalization, and urban restructuring within the Asia-Pacific: new development trajectories and planning responses', *Progress in Planning*, 61: 1–74.

Iamsterdam (2007) 'Diversity in the city'. www.iamsterdam.com/introducing/ people_culture/diversity_in_the (accessed 1 February 2007).

Inglis, C., Elley, J. and Manderson, L. (1992) *Making Something of Myself: Educational attainment and social and economic mobility of Turkish Australian young people*, Canberra: Australian Government Publishing Service.

Ioannides D. (1998) 'Tour operators: The gatekeepers of tourism', in D. Ioannides and K. Debbage (eds) *The Economic Geography of the Tourist Industry: A supply-side analysis*, London: Routledge, pp. 139–58.

ITC/WIPO (2003) *Marketing Crafts and Visual Arts: The role of Intellectual Property: A practical Guide*, Geneva: ITC/WIPO.

Jaguaribe, B. and Hetherington, K. (2004) 'Favela Tours: Indistinct and Mapless Representations of the Real in Rio de Janeiro', in M. Sheller and J. Urry (eds) *Tourism Mobilities: Places to Play, Places in Play*, London: Routledge, pp. 155–66.

Jannson, A. (2003) 'The negotiated city image: Symbolic representation and change through urban consumption', *Urban Studies*, 40: 463–79.

Jansen Verbeke, M. (1986) 'Inner city tourism, resources, tourists and promoters', *Annals of Tourism Research*, 13: 79–100.

Jayne, M. (2004) 'Culture that works? Creative industries development in a working-class city', *Capital and Class*, 84: 199–210.

Jennings, G. (2001) *Tourism Research*, Milton, Queensland: Wiley.

Johannesburg Development Agency (2004) *Development Business Plan JDA 009: Fashion District Development*. www.jda.co.za (accessed 3 July 2007).

—— (2005) *Business Plan 2005–06*, Johannesburg: JDA.

Jones, A. (1998) 'Issues in waterfront regeneration: More sobering thoughts – a UK perspective', *Planning Practice and Research*, 13(4): 433.

Jones, K., Lea, T., Jones, T. and Harvey, S. (2003) *Beyond anecdotal evidence: The spillover effects of investments in cultural facilities*, Toronto: CSCA.

Jordan, P. (2006) Briefing by Minister of Arts and Culture, Pallo Jordan, at the Parliamentary Media Briefing Week, Economic Cluster 11: Sector Investment Strategies. www.info.gov.za/speeches (accessed 19 January 2007).

Judd, D. (1999) 'Constructing the Tourist Bubble', in D. Judd and S. Fainstein (eds) *The Tourist City*, New Haven, CT: Yale University Press, pp. 35–53.

—— (2003a) 'Urban tourism and the geography of the city', *Revista Latinoamericana de Estudios Urbano Regionales EURE*, 29(87): 51–62.

—— (2003b, ed.) *The Infrastructure of Play: Building the tourist city*, London: M.E. Sharpe.

Judd, D.R. and Fainstein, S. (1999, eds) *The Tourist City*, New Haven, CT: Yale University Press.

Jupp, J. (1995) 'Ethnic and cultural diversity in Australia', in Australian Bureau of Statistics (ed.) *Yearbook Australia*, Canberra: Australian Bureau of Statistics, pp. 129–39.

Jupp, J., McRobbie, A. and York, B. (1990) 'Metropolitan ghettoes and ethnic concentrations', University of Wollongong, Centre For Multicultural Studies, Working Paper 90–01.

Kavaratzis, M. (2004) 'From City Marketing to City Branding', *Journal of Place Branding*, 1(1): 58–73.

Kawasaki, K. (2004) 'Cultural hegemony of Singapore among ASEAN countries: Globalization and cultural policy', *International Journal of Japanese Sociology*, 13: 22–35.

Kay, A. and Watt, G. (2000) *The Role of the Arts in Regeneration*, Edinburgh: Scottish Executive Central Research Unit, Research Findings 96.

KEA European Affairs (2006) *The Economy of Culture in Europe*, Brussels: European Commission.

Keane, M.J. (1997) 'Quality and pricing in tourism destinations', *Annals of Tourism Research*, 24(1): 117–30.

Kenniskring Leisure Management (2006) *Onderzoek naar Amsterdam als gaytoeristische bestemming in de 21ste eeuw*, Amsterdam: Hogeschool INHOLLAND.

Kim, W. and Mauborgne, R. (2004) 'Blue Ocean Strategy', *Harvard Business Review*, October: 77–84.

Kinkead, G. (1993) *Chinatown: A portrait of a closed society*, New York: Harper Collins.

Kirat, T. and Lung, Y. (1999) 'Innovation and Proximity: Territories as loci of collective learning processes', *European Urban and Regional Studies*, 6(1): 27–38.

Kirshenblatt-Gimblett, B. (1998) *Destination Culture. Tourism, Museums and Heritage,* Berkeley: University of California Press.

Kirsten, S. (2000) *The Book of Tiki: The Cult of Polynesian Pop in 50's America*, Los Angeles: Taschen America.

Klein, N. (1999) *No Logo*, New York: Picador.

Knox, P. and Taylor, P.J. (eds) (1995) *World Cities in a World-System*, New York: Cambridge University Press.

Koh, G. and Ooi, G.L. (2000) *State-Society Relations in Singapore*, Singapore: Oxford University Press.

Koivunen, Hannele (2005) *Staying Power to Finnish Cultural Exports: The Cultural Exportation Project of the Ministry of Education, the Ministry for Foreign Affairs and the Ministry of Trade and Industry*, Helsinki: Ministry of Education, Publication 2005: 9.

Lamy, Y. (2003) 'Patrimoine et culture: l'Institutionnalisation', in P. Poirrier and L. Vadelorge (eds) *Pour Une Histoire des Politiques du Patrimoine*, Paris: Fondation Maison des Sciences de l'Homme, pp. 45–63.

Landry, C. (2000) *The Creative City. A Toolkit for Urban Innovators*, London: Earthscan.

—— (2005) 'Lineages of the Creative city', in S. Franke and E. Verhagen (eds) *Creativity and the City: How the creative economy changes the city*, Rotterdam: NAi Publishers, pp. 42–54.

Landry, C. and Bianchini, F. (1995) *The Creative City*, London: Demos.

Lange, B. (2005) 'Socio-spatial strategies of culturalpreneurs: The example of Berlin and its new professional scenes', *Zeitschrift fur Wirtschaftsgeographie*, 49(2): 79–96.

Lasansky, D.M. and Greenwood, D.J. (2004, eds) *Architecture and Tourism*, Oxford: Berg.

Lash, S. and Urry, J. (1994) *Economies of Signs and Space*, London: Sage.

Latouche, D. (1994) *Les activites culturelles et artistiques dans la region metropolitaine de Montréal*, Etude realisee pour le Bureau federal de developpement regionale (Québec), Montréal: INRS-Urbanisation.

Lavanga, M. (2004) 'Creative Industries, Cultural Quarters and Urban Development: the Case Studies of Rotterdam and Milan', in G. Mingardo and M. van Hoek (eds) *Urban Management in Europe Vol. II: Towards a Sustainable Development*, Rotterdam: MEMR – The Thesis Series, Erasmus University Rotterdam, pp. 121–57.

Law, C.M. (1993) *Urban Tourism: Attracting visitors to large cities*, London: Mansell.

LB Tower Hamlets (1996) *Eastside Challenge Fund Submission*, London: LB Tower Hamlets.

—— (1999) *Brick Lane Retail and Restaurant Policy Review* (1 March), London: LB Tower Hamlets.

Lee, M. (1997) 'Relocating location: cultural geography, the specificity of place and the City of Habitus', in J. McGuigan (ed.) *Cultural Methodologies*, London: Sage, pp. 126–41.

Lefebvre, H. (1974) *The Production of Space*, Oxford: Blackwell.

—— (1991) *The Production of Space*, Oxford: Blackwell (2nd edition).

Lengkeek, J. (1994) *Een meervoudige werkelijkheid: een sociologisch-filosofisch essay over het collectieve belang van recreatie en toerisme*, Proefschrift, Mededelingen van de Werkgroep Recreatie 20, Wageningen: Landbouwuniversiteit Wageningen.

—— (1996) *Vakantie van het leven: Over het belang van recreatie en toerisme*, Amsterdam: Boom.

—— (2000) 'Imagination and Differences in Tourist Experience', *World Leisure Journal*, 42(3): 11–17.

Leong, W.T. (1997) 'Commodifying ethnicity: State and ethnic tourism in Singapore', in M. Picard and R.E. Wood (eds) *Tourism, Ethnicity and the State in Asian and Pacific Societies*, Honolulu: University of Hawaii Press, pp. 71–98.

le Reun, G. (1999) 'Boutiques en réseau', *Espaces*, 156: 33–35.

Leslie, D. (2005) 'Creative cities?', *Geoforum*, 36: 403–5.

Light, I. and Gold, S.J. (2000) *Ethnic Economies*, San Diego: Academic Press.

Lin, J. (1998) 'Globalization and the revalorizing of ethnic places in immigration gateway cities', *Urban Affairs Review*, 34 (2): 313–39.

Linteau, P-A. (2000) *Histoire de Montréal depuis la confédération*, Montréal: Boréal (2nd edition).

Lloyd, R. (2002) 'Neo-bohemia: Art and neighborhood redevelopment in Chicago', *Journal of Urban Affairs*, 24(5): 517.

—— (2006) *Neo-Bohemia: Art and Commerce in the Postindustrial City*, New York: Routledge.

Logan, J.R. and Molotch, H.L. (1987) *Urban Fortunes: the Political Economy of Place*, Berkeley: University of California Press.

Loosely, D. (1998) 'Cultural policy', in A. Hughes and K. Reader (eds), *Encyclopedia of Contemporary French Culture*, London: Routledge, pp. 128–31.

Lop, I. and Valls, C. (2005) *Informe de la Primera Fase del Projecte Joves i Nit: Una Primera Aproximació*, Barcelona: Fundació Jaume Bofill.

Lorimer, H. (2005) 'Cultural geography: the busyness of being "more-than-representational"', *Progress in Human Geography*, 29: 83–94.

Low, L. and Johnston, D.M. (2001) *Singapore Inc. Public Policy Options in the Third Millenium*, Singapore: Asia Pacific Press.

Lugbull, J-J. (1999) 'L'Image de nos objets est celle de nos musées', *Espaces*, 156: 16–17.

McCabe, S. (2005) 'Who is a tourist? A critical review', *Tourist Studies*, 5(1): 85–106.

MacCannell, D. (1976) *The Tourist. A new theory of the leisure class*, New York: Schocken Books.

McEvoy, D. (2003) 'The evolution of Manchester's "curry mile": From suburban shopping street to ethnic destination', paper presented to Metropolis Conference, Vienna, September.

Machaka, J. and Roberts, S. (2003) 'The DTI's new "Integrated Manufacturing Strategy"?: comparative industrial performance, linkages and technology', *South African Journal of Economics*, 71: 1–26.

Mackay, D., Zogolovitch, R. and Harradine, M. (2004) *A Vision for Plymouth*, Plymouth: University of Plymouth.

McKercher, B. and Du Cros, H. (2002) *Cultural Tourism: The Partnership between Tourism and Cultural Heritage Management*, Binghampton, NY: Haworth Press.

Maitland, R. (2000) *The Development of New Tourism Areas in Cities: Why is ordinary interesting?*, keynote paper given at Finnish University Network for Tourism Studies Opening Seminar 2000/01: Managing Local and Regional Tourism in the Global Market, Savonlinna, Finland, 18 September.

—— (2006) 'Culture, city users and the creation of new tourism areas in cities', in M. Smith (ed.) *Tourism, Culture and Regeneration*, London: CABI, pp. 25–34.

—— (2007) 'Cultural tourism and the development of new tourism areas in London', in G. Richards (ed.), *Cultural Tourism: Global and Local Perspectives*, Binghampton, NY: Haworth Press, pp. 113–30.

Maitland, R. and Newman, P. (2004) 'Developing Metropolitan Tourism on the Fringe of Central London', *International Journal of Tourism Research*, 6: 339–48.

Manente, M. and Andreatta, L. (1998) *Il fatturato del Turismo nel Centro Storico di Venezia*, Quaderno CISET 19/98, Venice: University of Venice.

Markey, S., Halseth, G. and Manson, D. (2006) 'The Struggle to Compete: From comparative to competitive advantage in Northern British Columbia', *International Planning Studies*, 11: 19–39.

Marrero Guillamón, I. (2005) 'Del Manchester Catalán al Soho Barcelonés? La renovación del barrio del Poblenou en Barcelona y la cuestión de la vivienda', *Scripta Nova*, Vol. VII, 146(137), 1 August 2005. www.ub.es/geocrit/sn/sn-146 (137).htm (accessed 25 March 2007).

Marshall, A. (1920) *Principles of Economics*, London: Macmillan.

Martin, G.P. (2005) 'Narratives great and small: neighbourhood change, place and identity in Notting Hill', *International Journal of Urban and Regional Research*, 29(1): 67–88.

Martinotti, G. (1993) *Metropoli: la Nuova Morfologia Sociale della Città*, Bologna: Il Mulino.

—— (1999) 'A city for whom? Transients and public life in the second-generation metropolis', in R. Beauregard and S. Body-Gendrot (eds) *The Urban Moment*, London: Sage, pp. 155–84.

Mason, M. (2003) 'Urban regeneration rationalities and quality of life: comparative notes from Toronto, Montréal and Vancouver', *British Journal of Canadian Studies*, 16(2): 348–62.

MBM Arquitectes and AZ Urban Design Studio (2003) *Draft Interim Planning Statement: A Vision for Plymouth*, Plymouth: City of Plymouth.

Meethan, K. (1996) 'Consuming (in) the "Civilised City"', *Annals of Tourism Research*, 13(2): 322–40.

—— (1997) 'York: Managing the tourist city', *Cities*, 14(6): 333–42.

—— (1998) 'New Tourism for Old? Policy developments in Cornwall and Devon', *Tourism Management*, 19(6): 583–94.

—— (2001) *Tourism in Global Society: Place, culture, consumption*, Houndmills: Palgrave.

—— (2002) 'Selling the Difference: Tourism Marketing in Devon and Cornwall', in R. Voase (ed.) *Tourism in Western Europe: A collection of case histories*, Wallingford: CABI.

Merrifield, A. (1993) 'Place and space: a Lefebvrian reconciliation', *Transactions IBG*, 18: 516–31.

—— (2000) 'Henri Lefebvre: a socialist in space', in M. Crang and N. Thrift (eds) *Thinking Space*, London: Routledge, pp. 167–82.

Miettinen, S. (2004) *Helsinki – Opuwo – Helsinki: Encounters with crafts, tourists and women in the local communities of Namibia*, Helsinki: Like.

Miles, M. and Huberman, M.A. (1994) *Qualitative Analysis: An Expanded Sourcebook*, Thousand Oaks, CA: Sage (2nd edition).

Miles, S. and Paddison, R. (2005) 'Introduction: the rise and rise of culture led urban regeneration', *Urban Studies*, 42 (5/6): 833–39.

Milner, H. and Joncas, P. (2002) 'Montréal: getting through the megamerger', *Inroads*, 11: 49–63.

Ministry of Economic Affairs and Ministry of Education, Culture and Science (2005) *Our Creative Potential*, Den Haag: Ministry of Economic Affairs and Ministry of Education, Culture and Science.

Ministry of Trade and Industry (MTI) (2003) 'Economic Contributions of Singapore's Creative Industries', in MTI (ed.) *Economic Survey of Singapore First Quarter 2003*, Singapore: MTI, pp. 51–75.

Minty, Z. (2005) 'Breaching borders: using culture in Cape Town's Central City for social change', *Isandla Development Communication*, 2 (8/9), Isandla Institute, Cape Town.

MITA (2000) *Renaissance City Report: Culture and the Arts in Renaissance Singapore*, Singapore: Ministry of Information and the Arts.

MITA and STPB (1995) *Singapore, Global City for the Arts*, Singapore: Singapore Ministry of Information and the Arts; Singapore Tourist Promotion Board.

Molotch, H., Freudenburg, W. and Paulsen, K.E. (2000) 'History repeats itself, but how? City character, urban tradition, and the accomplishment of place', *American Sociological Review*, 65: 791–823.

Mommaas, H. (2004) 'Cultural clusters and the post-industrial city: Towards the remapping of urban cultural policy', *Urban Studies*, 41(3): 507–32.

Monitor (2004) *Global Competitiveness Project: Summary of Key Findings of Phase 1*, Johannesburg: South African Tourism.

—— (2005) *Final Report to ComMark Trust on the Feasibility Assessment of Cultural Tourism in Johannesburg*, Johannesburg: Monitor Group.

Montgomery, J. (2003) 'Cultural quarters as mechanisms for urban regeneration. Part 1 – Conceptualising cultural quarters', *Planning Practice and Research*, 18(4): 293–306.

—— (2004) 'Cultural quarters as mechanisms for urban regeneration. Part 2 – A review of four cultural quarters in the UK, Ireland and Australia', *Planning Practice and Research*, 19(1): 3–31.

Moreno, S. (2002) *Jean Leon, el rey de Beverly Hills, la aventura de un español en el Corazón del Hollywood dorado*, Barcelona: Ediciones B, S.A.

Moreno, Y.J., Santagata, W. and Tabassum, A. (2005) 'Material Cultural Heritage, Cultural Diversity, and Sustainable Development', *EBLA Centre Working Paper* 07/2005, University of Turin.

Mormont, M. (1990) 'Who is rural? Or, how to be rural: a sociology of the rural', in T. Marsden, P. Lowe and S. Whatmore (eds) *Rural Restructuring*, London: David Fulton, pp. 21–44.

Municipality of Sitges (2006) *Any Rusiñol, 2006–2007*, Sitges.

Muñoz, F. (2006) *Urbanización. Paisajes Comunes, Lugares Globales*, Barcelona: Gustavo Gili.

Musterd, S. and Deurloo, R. (2006) 'Amsterdam and the preconditions for a creative knowledge city', *Tijdschrift voor Economische en Sociale Geografie*, 97: 80–94.

Musu, I. (ed.) (2000) *Sustainable Venice: Suggestions for the Future*, Dordrecht: Kluwer.

Nathan, M. (2005) *The Wrong Stuff: Creative Class theory, diversity and city performance*, London: IPPR.

National Art Council (NAC) (2005) *National Arts Council Annual Report FY2004/05*, Singapore: NAC.

Neill, W. (2004) *Urban Planning and Cultural Identity*, London: Routledge.

Nel, E. and Rogerson, C.M. (2005, eds) *Local Economic Development in the Developing World: The Experience of Southern Africa*, New Brunswick, NJ: Transaction Press.

Nelson, P. (1970) 'Information and consumer behaviour', *Journal of Political Economy*, 2: 311–29.

Nemasetoni, I. and Rogerson, C.M. (2005) 'Developing small firms in township tourism: emerging tour operators in Gauteng, South Africa', *Urban Forum*, 16: 196–213.

Newman, P. and Smith, I. (1999) 'Cultural production, place and politics on the south bank of the Thames', *International Journal of Urban and Regional Research*, 24(1): 9–24.

Newton, M. (2003) Joburg Creative Industries Sector Scoping Report, unpublished report for the Economic Development Department, City of Johannesburg.

Ng, W.C. (1999) *The Chinese in Vancouver, 1945–80: The pursuit of identity and power*, Vancouver: UBC Press.

Nichols Clark, T. (2004, ed.) *The City as Entertainment Machine*, Oxford: Elsevier.

Nijs, D. (2003) 'Imagineering: Engineering for imagination in the Emotion Economy', in P. Peeters, F. Schouten and D. Nijs (eds) *Creating a fascinating world*, Breda: Breda University of Professional Education, pp. 15–32.

Nijs, D. and Peters, F. (2002) *Imagineering. Het creëren van belevingswerelden*, Amsterdam: Boom.

Noy, C. (2006) *Narrative Community: Voices of Israeli Backpackers*, Detroit, MI: Wayne State University Press.

Nye, R.B. (1981) 'Eight Ways of Looking at an Amusement Park', *Journal of Popular Culture* 15, 63–75.

O'Connor, J. (1999) *The Definition of Cultural Industries*, Manchester: Manchester Institute for Popular Culture.

—— (2005) '"Creative Exports": Taking "Cultural Industries" to St. Petersburg', *International Journal of Cultural Policy*, 11(1): 45–59.

O'Connor, J. and Wynne, D. (eds) (1996) *From Margins to the Centre: Cultural Production and Consumption in the Post-Industrial City*, Aldershot: Arena.

O'Connor, J. and Xin, G. (2006) 'A new modernity? The arrival of "creative industries" in China', *International Journal of Cultural Studies*, 9(3): 271–83.

O'Donnell, A. (2004) 'Just who works for the new creative class?', *The Age*, 2 August: 11.

OECD (2005) *Education at a glance*, Paris: OECD.

Office of the Deputy Prime Minister (2003) *Cities, Regions and Competitiveness*. www.corecities.com/coreDEV/Publications/CR&%20C%20-%202nd%20Report%20from%20the%20Working%20Group.htm (accessed 25 March 2007).

Ooi, C.S. (2001) 'Dialogic heritage: Time, space and visions of National Museum of Singapore', in P. Teo, T.C. Chang and K.C. Ho (eds) *Interconnected Worlds: Tourism in Southeast Asia*, Amsterdam: Elsevier, pp. 171–87.

—— (2002) *Cultural Tourism and Tourism Cultures: The Business of Mediating Experiences in Copenhagen and Singapore*, Copenhagen: Copenhagen Business School Press.

—— (2004) 'Brand Singapore: The hub of New Asia', in N. Morgan, A. Pritchard and R. Pride (eds) *Destination Branding: Creating the Unique Destination Proposition*, London: Elsevier, pp. 242–62.

—— (2005a) 'State–civil society relations and tourism: Singaporeanizing tourists, touristifying Singapore', *Sojourn – Journal of Social Issues in Southeast Asia*, 20: 249–72.

—— (2005b) 'The Orient responds: Tourism, Orientalism and the national museums of Singapore', *Tourism*, 53: 285–99.

Orbasli, A. (2001) *Tourists in Historic Towns: Urban Conservation and Heritage Management*, London: E&FN Spon.

Ory, P. (2003) 'Pour une histoire des politiques du patrimoine', in P. Porrier and L. Vadelorge (eds) *Pour Une Histoire des Politiques du Patrimoine*, Paris: Fondation Maison des Sciences de l'Homme, pp. 27–32.

Page, S.J. (1995) *Urban Tourism*, London: Routledge.

—— (2002) 'Urban tourism: evaluating tourists' experience of urban places', in C. Ryan (ed.) *The Tourist Experience*, London: Continuum, pp. 112–35.

Palmer Rae Associates (2004) *European Capitals/Cities of Culture. Study on the European Cities and Capitals of Culture and the European Cultural Months (1995–2004), Part I and II*, Brussels: Palmer Rae Associates/EC.

Pappalepore, I. (2007) 'Marketing a post-modern city: a shift from tangible to intangible advantages', in G. Richards and J. Wilson (eds) *Changing places – the spatial challenge of creativity*, Arnhem, Netherlands: ATLAS (in press).

Patronato Municipal de Turismo de Sitges (2005a) *Sitges, Joya del Mediterráneo* (promotional material).

—— (2005b) *Sitgestiucultura '05* (promotional material).

—— (2006) *Sitges, the art of living* (promotional material).

Patton, M.Q. (1990) *Qualitative Evaluation and Research Methods*, 2nd edn, Newbury Park: Sage.

Pearce, D.G. (1998a) 'Tourism development in Paris: Public intervention', *Annals of Tourism Research*, 25(2): 457–76.

—— (1998b) 'Tourist districts in Paris: Structure and functions', *Tourism Management*, 19(1): 49–65.

—— (1999) 'Tourism in Paris: Studies at the microscale', *Annals of Tourism Research*, 26(1): 77–97.

—— (2001) 'An integrative framework for urban tourism research', *Annals of Tourism Research*, 28(4): 926–46.

Peck, J. (2005) 'Struggling with the Creative Class', *International Journal of Urban and Regional Research*, 29(4): 740–70.

Peyroutet, C. (1998) *Le Tourisme en France*, Saint Amand Montrond: Nathan.

Philips, T. (2007) 'Notorious slum becomes open-air gallery', the *Guardian*. www.guardian.co.uk/brazil/story/0,2009348,00.html (accessed 21 March 2007).

Phillips, M. (2002) 'The production, symbolization and socialization of gentrification: impressions from two Berkshire villages', *Transactions IBG*, 27: 282–308.

Phillips, M., Fish, R. and Agg, J. (2001) 'Putting together ruralities: towards a symbolic analysis of rurality in the British Mass Media', *Journal of Rural Studies*, 17: 1–27.

Phumzile-Ngcuka, P. (2006) Media Briefing on Background Document: A Catalyst for Accelerated and Shared Growth–South Africa, 6 February. www.info.gov.za/speeches (accessed 19 January 2007).

Pine, B.J. and Gilmore, J.H. (1999) *The Experience Economy: Work is theater and every business a stage*, Boston, MA: Harvard Business School Press.

Poirrier, P. (1998) *Société et Culture en France Depuis 1945*, Paris: Seuil.

Pontier, J-M. (1996) 'A decade of decentralization reviewed by the Council of State', in Institut International d'Administration Publique (ed.) *An Introduction to French Administration*, Paris: La Documentation Française, pp. 163–76.

Poon, A. (1993) *Tourism, Technology and Competitive Strategies*, Wallingford: CABI.

Porter, M.E. (1995) 'The competitive advantage in the inner city', *Harvard Business Review*, May/June: 55–71.

—— (1998) 'Clusters and the New Economics of Competition', *Harvard Business Review*, November/December, pp. 77–90.

Porter, R. (1996) *London: A Social History*, Harmondsworth: Penguin.

Posner, R.A. (2003), *Economic Analysis of Law*, 6th edn, New York: Aspen Law & Business.

Poynting, S., Noble, G., Tabar, P. and Collins, J. (2004) *Bin Laden In The Suburbs: Criminalising the Arab Other*, Sydney: Sydney Institute of Criminology.

Poyraz, M. (2003) 'Animateur de quartier: un métier ethnique?', in G. Hennebelle (ed.) *Existe-t-il des Métiers Ethniques?*, Condé-sur-Noireau: Corlet, pp. 175–84.

Prahalad, C.K. and Ramaswamy, V. (2003) 'The new frontier of experience innovation', *MIT Sloan Management Review*, 44(4): 12–18.

—— (2004) *The Future of Competition: Co-creating unique value with customers*, Boston, MA: Harvard Business School Press.

Pratt, A. (2004) 'The cultural economy', *International Journal of Cultural Studies*, 7(1): 117–28.

Pratt, A.C. (2001) 'The relationship between the city, cultural tourism and the cultural industries', in E. Belda (ed.) *Turisme i cultura*, Barcelona: Interarts, pp. 35–44.

—— (2004) 'Creative clusters: towards the governance of the creative industries production system?', *Media International Australia*, 112: 50–66.

Prentice, R.C. (1996) 'Managing implosion', *Museum Management and Curatorship*, 15: 169–85.

—— (2002) 'Verisimilitude and Proffered Verisimilitude: Imagining and Imaging Tourist "North England"', CD-ROM file *2002/pt36.pdf* in Council of Australian University Tourism and Hospitality Education, *Ten Years of Tourism Research*, Council of Australian University Tourism and Hospitality Education.

—— (2005a) 'Tourists' Judgements and Recommendations: A Satisfaction Methodology', CD-ROM file, in P. Tremblay and A. Boyle (eds) *CAUTHE [Council of Australian University Tourism and Hospitality Education] Conference 2005, Alice Springs: Sharing Tourism Knowledge*, Darwin, NT: Charles Darwin University.

—— (2005b) *Denmark in the British Imagination: A Creative Tourism Perspective*, Official Opening of the Center for Tourism and Cultural Management, Copenhagen Business School. www.cbs.dk/content/download/18794/282166/file/RichardPrentice.pdf (accessed 3 July 2007).

—— (2006) 'Evocation and experiential seduction: updating choice-sets modelling', *Tourism Management*, 27(6): 1153–70.

Prentice, R.C. and Andersen, V.A. (2000) *Consumption Styles in Festival-Going*, paper to the Tourism and Leisure Research Network (TOLERN) and the Royal Anthropological Institute, 26 June, London: Department of Culture, Media and Sport.

Pretes, M. (1995). 'Postmodern tourism: the Santa Claus industry', *Annals of Tourism Research*, 22: 1–15.

Project for Public Spaces Association (2003) 'Hall of shame'. www.pps.org/great_public_spaces/one?public_place_id=624 (accessed 27 February 2007).

Puczkó, L. and Rátz, T. (2007) 'Trailing Goethe, Humbert, and Ulysses: Cultural Routes in Tourism', in G. Richards (ed.) *Cultural Tourism: Global and Local Perspectives*, Binghampton, NY: Haworth Press, pp. 131–48.

Punch, K.F. (2001) *Introduction to Social Research: Quantitative and qualitative approaches*, London: Sage.

Ramchander, P. (2007) 'Township tourism – Blessing or blight? The case of Soweto in South Africa', in G. Richards (ed.) *Cultural Tourism: Global and Local Perspectives*, Binghampton, NY: Haworth Press, pp. 39–67.

Rath, J. (2002) 'Immigrants and the tourist industry: The commodification of cultural resources', paper presented to 15th World Congress of Sociology, Brisbane, Australia, 7–13 July.

Ratzenböck, V., Demel, K., Harauer, R., Landsteiner, G., Falk, R., Hannes, L. and Schwarz, G. (2004) *An Analysis of the Economic Potential of the Creative Industries in Vienna*, Vienna: Kulturdokumentation.

Ray, C. (1998) 'Culture, intellectual property and territorial rural development', *Sociologia Ruralis*, 38: 3–20.

Ray, P.H. and Anderson, S.R. (2000) *The Cultural Creatives*, New York: Three Rivers Press.

Raymond, C. (2002) 'Creative Encounters', *Nelson Mail*, 29 May: 15–16. Reprinted on the CTNZ website: www.creativetourism.co.nz/press_releases/Creative_Encounters_Crispin_Raymond.rtf (accessed 6 August 2006).

Rehman, L.A. (2002) 'Recognizing the significance of culture and ethnicity: Exploring hidden assumptions of homogeneity', *Leisure Sciences*, 24(1): 43–57.

Relph, E. (1976) *Place and Placelessness*, London: Pion.

Republic of South Africa (1996) *White Paper on the Development and Promotion of Tourism in South Africa*, Pretoria: Department of Environmental Affairs and Tourism.

Richards, G. (1994) 'Cultural Tourism in Europe', in C.P. Cooper and A. Lockwood (eds) *Progress in Tourism, Recreation and Hospitality Management Vol. 5*, Chichester: Wiley Press, pp. 99–115.

—— (1996a) 'Skilled consumption and UK ski holidays', *Tourism Management*, 17(1): 25–34.

—— (1996b, ed.) *Cultural Tourism in Europe*, Wallingford: CABI.

—— (1996c) 'Production and consumption of European cultural tourism', *Annals of Tourism Research*, 23(2): 261–83.

—— (1999) 'Vacations and quality of life: patterns and structures', *Journal of Business Research*, 44: 189–98.

—— (2001, ed.) *Cultural Attractions and European Tourism*, Wallingford: CABI.

—— (2004) 'Symbolising Catalunya', unpublished report to the Catalan Department of Higher Education.

—— (2005) 'Textile Tourists in the European Periphery: New Markets for Disadvantaged Areas?', *Tourism Review International*, 8: 323–38.

—— (2006a) 'Les Actituds en relació al Turisme per part dels habitants de Barcelona: Resultats de Baròmetre 2006', unpublished report, Barcelona: Ajuntament de Barcelona, Direcció de Serveis de Turisme i Qualitat de Vida.

—— (2006b) 'ISTC/UNWTO survey on student and youth tourism among National Tourism Adminstrations/Organizations', in UNWTO (ed.) *Tourism Market Trends 2005 Edition*, Madrid: UNWTO, pp. 95–123.

—— (2007) *Cultural Tourism: Global and Local Perspectives*, Binghampton, NY: Haworth Press.

Richards, G. and Raymond, C. (2000) 'Creative tourism', *ATLAS News*, 23: 16–20.

Richards, G. and Wilson, J. (2003) *Today's youth travellers: tomorrow's global nomads: New horizons in independent youth and student travel*, Amsterdam: International Student Travel Confederation.

—— (2004a) 'The Impact of Cultural Events on City Image: Rotterdam Cultural Capital of Europe 2001', *Urban Studies*, 41(10): 1931–51.

—— (2004b, eds) *The Global Nomad: Theory and Practice in Backpacker Travel*, Clevedon, UK: Channel View.

—— (2006) 'Developing creativity in tourist experiences: a solution to the social reproduction of culture', *Tourism Management*, 27(6): 1209–23.

—— (2007) 'The Creative Turn in Regeneration: Creative Spaces, Spectacles and Tourism in Cities', in M. Smith (ed.) *Tourism, Culture and Regeneration*, Wallingford: CABI, pp. 12–24.

Ritchie, A. (2002) Planning and Community Liaison Officer, LB Tower Hamlets, personal interview with author, Bow Road, London, 27 September.

Ritzer, G. (1996) *The McDonaldization of Society: An Investigation into the Changing Character of Contemporary Social Life*, Thousand Oaks. CA: Pine Forge Press.

—— (1999) *Enchanting a disenchanted world: revolutionizing the means of consumption*, Thousand Oaks, CA: Pine Forge Press.

Ritzer, G. and Liska, A. (1997) '"McDisneyization" and "post-tourism": Complementary perspectives on contemporary tourism', in C. Rojek and J. Urry (eds) *Touring Cultures: Transformations of Travel and Theory*, London: Routledge, pp. 96–109.

Roberts, L. (2004) *New Directions in Rural Tourism*, Aldershot: Ashgate.

Roberts, L. and Hall, C. (2001) *Rural Tourism and Recreation*, Wallingford: CABI.

Rogerson C.M. (2002) 'Urban tourism in the developing world: the case of Johannesburg', *Development Southern Africa* 19(1): 169–90.

—— (2003) 'Tourism and transformation: small enterprise development in South Africa', *Africa Insight*, 33(1/2): 108–15.

—— (2004a) 'Regional tourism in South Africa: a case of "mass tourism of the South"', *GeoJournal*, 60: 229–37.

—— (2004b) 'Urban tourism and economic regeneration: the example of Johannesburg', in C.M. Rogerson and G. Visser (eds) *Tourism and Development Issues in Contemporary South Africa*, Pretoria: Africa Institute of South Africa, pp. 466–87.

—— (2004c) 'Urban tourism and small tourism enterprise development in Johannesburg: the case of township tourism', *GeoJournal*, 60: 247–57.

—— (2004d) 'Transforming the South African tourism industry: the emerging black-owned bed and breakfast economy', *GeoJournal*, 60: 273–81.

—— (2004e) 'Pro-poor local economic development in post-apartheid South Africa: the Johannesburg fashion district', *International Development Planning Review*, 26: 401–29.

—— (2005a) 'Conference and exhibition tourism in the developing world: the South African experience', *Urban Forum*, 16: 176–95.

—— (2005b) 'Globalization, economic restructuring and local response in Johannesburg – the most isolated "world city"', in K. Segbers, S. Raiser and K. Volkmann (eds) *Public Problems – Private Solutions?: Globalizing cities in the South*, Aldershot: Ashgate, pp. 17–34.

—— (2006) 'Local economic development in post-apartheid South Africa: a ten-year research review', in V. Padayachee (ed.), *The Development Decade?: Economic and social change in South Africa, 1994–2004*, Cape Town: HSRC Press, pp. 227–53.

Rogerson, C.M. and Visser, G. (2004, eds) *Tourism and Development Issues in Contemporary South Africa*, Pretoria: Africa Institute of South Africa.

Rogerson, C.M. and Lisa, Z. (2005) '"Sho't left": promoting domestic tourism in South Africa', *Urban Forum*, 16: 88–111.

Rogerson, C.M. and Visser, G. (2004, eds) *Tourism and Development Issues in Contemporary South Africa*, Pretoria: Africa Institute of South Africa.

Rogerson, C.M. and Visser, G. (2005) 'Tourism in urban Africa: the South African experience', *Urban Forum*, 16: 63–87.

—— (2007, eds) *Urban Tourism in the Developing World: The South African Experience*, New Brunswick, NJ: Transaction Press.

Rojal, F. (2005) 'Cultural Districts and the Role of Intellectual Property (Distinctive Signs)', EBLA Centre Working Paper 02/2005, University of Turin.

Rojek, C. (1993) *Ways of Escape: Modern Transformations in Leisure and Travel*, Basingstoke: Macmillan.

—— (2000) *Touring Cultures; Transformations of Travel and Theory*, London: Routledge.

Roodhouse, S. and Mokre, M.O. (2004) 'The MuseumsQuartier, Vienna: An Austrian Cultural Experiment', *International Journal of Heritage Studies*, 10: 193–207.

Rooijakkers, G. (1999) 'Identity factory southeast – Towards a flexible cultural leisure infrastructure', Finnica Seminar Paper, 8 April 1999. www.finnica.fi/seminaari/99/luennot/rooijakk.htm (accessed 3 July 2007).

Routenberg, M. (2003) 'l'Intervention ethnologique', in P. Porrier and L. Vadelorge (eds) *Pour Une Histoire des Politiques du Patrimoine*, Paris: Fondation Maison des sciences de l'Homme, pp. 469–89.

Rural Regeneration Cumbria (2006) *Desire Lines – Cultural Assets Investment Strategy for Cumbria*, Penrith: Rural Regeneration Cumbria.

Rushdie, S. (1992) *Imaginary Homelands*, London: Granta Books.

Russo, A.P. (2002) 'The "vicious circle" of tourism development in heritage destinations', *Annals of Tourism Research*, 29(1): 165–82.

—— (2006) 'A re-foundation of the talc for heritage cities', in R. Butler (ed.) *The Tourism Area Life Cycle, Vol. 1: Applications and Modifications*, Clevedon, UK: Channel View, pp. 139–61.

Russo, A.P. and Segre, G. (2005) 'Collective Property Rights for Glass Manufacturing in Murano: where Culture Makes or Breaks Local Economic Development', EBLA Centre Working Paper 05/2005, University of Turin.

Sandercock, L. (2006) 'Cosmopolitan urbanism: a love song to our mongrel cities', in J. Binnie, J. Holloway, S. Millington and C. Young (eds) *Cosmopolitan Urbanism*, London: Routledge, pp. 37–52.

Sant, M. and Waitt, G. (2000) 'Sydney: All day long, all night long', in J. Connell (ed.) *Sydney: The emergence of a world city*, South Melbourne: Oxford University Press, pp. 189–221.

Santagata, W. (2002) *Creativity, Fashion, and Market Behavior*, Working Paper 5/2002, International Centre for Research on the Economics of Culture, Institutions and Creativity, Department of Economics, University of Turin.

—— (2004) *Cultural Districts and Economic Development*, Working Paper 1/2004, International Centre for Research on the Economics of Culture, Institutions and Creativity, Department of Economics, University of Turin.

Santagata, W. and Cuccia, T. (2003) 'Collective Property Rights and Cultural Districts: The Case Study of Caltagirone Pottery in Sicily', in E. Colombatto (ed.) *Companion to Property Right Economics*, Cheltenham: Elgar, pp. 473–88.

Sassen, S. (2001) *The Global City: New York, London, Tokyo*, 2nd edn, Princeton, NJ: Princeton University Press.

Schein, E. (1996) *Strategic Pragmaticism: The Culture of Singapore's Economic Development Board*, Cambridge, MA: MIT Press.

Schnell, S.M. (2003) 'The ambiguities of authenticity in Little Sweden, USA', *Journal of Cultural Geography*, 20(2): 43–68.

Schor, J.B. (1998) *The Overspent American: Upscaling, downshifting and the new consumer*, New York: Basic Books.

Schouten, F. (2003) 'About the quality of life, and nothing less', in *Creating a fascinating world*, Breda: Breda University of Professional Education (NHTV), pp. 9–14.

Schwartz, D. (2003) *Suburban Xanadu: the Casino Resort on the Las Vegas Strip and Beyond*, New York: Routledge.

Scitovsky, T. (1976) *The Joyless Economy*, Oxford: Oxford University Press.

Scott, A.J. (1996) 'The craft, fashion and cultural-products industries of Los Angeles: competitive dynamics and policy dimensions in a multisectoral image-producing complex', *Annals of the Association of American Geographers*, 86(2): 306–23.

—— (2000) *The Cultural Economy of Cities*, London: Sage.

—— (2001) 'Capitalism, cities, and the production of symbolic forms', *Transactions: Institute of British Geographers*, 26: 11–23.

—— (2004) 'Cultural-products industries and urban economic development: prospects for growth and market contestation in a global context', *Urban Affairs Review*, 39: 461–90.

Seaton, A.V. and Lennon, J.J. (2004) 'Thanatourism in the Early 21st Century: Moral Panics, Ulterior Motives and Alterior Desires', in T.V.Singh (ed.) *New Horizons in Tourism: Strange Experiences and Stranger Practices*, Wallingford: CABI, pp. 63–82.

Selby, M. (2004) *Understanding Urban Tourism: Image, culture and experience*, New York: I.B. Tauris.

Sennett, R. (1994) *Flesh and Stone, The body and the city in western civilization*, New York: Norton.

Sesikhona Services and ECI Africa (2005) 'Tourism strategic plan and media launch for Johannesburg Inner City and Park Station', unpublished report for Gauteng Tourism Authority, Johannesburg.

Shapiro, C. (1983) 'Premiums for high quality as returns to reputations', *The Quarterly Journal of Economics*, 98: 659–79.

Shaw, G. and Williams, A.M. (1994) *Critical Issues in Tourism: A geographical perspective*, Oxford: Blackwell.

—— (1998) 'Entrepreneurship, small business culture and tourism development', in D. Ioannides and K. Debbage (eds) *The Economic Geography of the Tourist Industry*, London: Routledge, pp. 235–55.

—— (2002) *Critical Issues in Tourism: A geographical perspective*, 2nd edn, Oxford: Blackwell.

—— (2004) *Tourism and Tourism Spaces*, London: Sage.

Shaw, K. (2005) 'The place of alternative culture and the politics of its protection in Berlin, Amsterdam and Melbourne', *Planning Theory and Practice*, 6: 149–69.

Shaw, S. (2003) 'The Canadian world city and sustainable downtown revitalisation: messages from Montréal 1962–2002', *British Journal of Canadian Studies*, 16(2): 363–77.

—— (2006) 'Cities and national success: rapporteur's report of the Canada–United Kingdom Colloquium November 2005', Kingston, Ontario: School of Policy Studies, Queen's University.

Shaw, S. and MacLeod, N. (2000) 'Creativity and Conflict: Cultural Tourism in London's City Fringe', *Tourism, Culture and Communication*, 2(3): 165–75.

Shaw, S., Bagwell, S. and Karmowska, J. (2004) 'Ethnoscapes as spectacle: reimaging multicultural districts as new destinations for leisure and tourism consumption', *Urban Studies*, 41(10): 1983–2000.

Sheller, M. and Urry, J. (2004) *Tourism Mobilities: Places to play, places in play*, London: Routledge.

Shields, R. (1999) *Lefebvre, Love and Struggle: Spatial dialectics*, London: Routledge.

Short, J.R. (2004) *Global Metropolitan: Globalizing Cities in a Capitalist World*, London: Routledge.

Siddique, S. (1990) 'The phenomenology of ethnicity: A Singapore case study', in J.H. Ong, C.K. Tong and E.S. Tan (eds) *Understanding Singapore Society*, Singapore: Times Academic Press, pp. 107–24.

Silver, D., Clark, T.N. and Rothfield, L. (2006) *A Theory of Scenes*, Montreal: Urban Affairs Association.

Silver, N. (1994) *The Making of Beauborg*, Cambridge, MA: MIT Press.

Simmie, J. (2001) *Innovative Cities*, London: Spon.

Singapore Parliamentary Hansard (13 March 2004) Vol. 77, Session 1, Parliament sitting.

Singh, T.V. (2004) *New Horizons in Tourism: Strange Experiences and Stranger Practices*, Wallingford: CABI.

Sitgestiucultura '05 (2005) *Museums de Sitges. Passeja't per la cultura!*, Sitges.

Slater, T., Curran, W. and Lees, L. (2004) 'Guest editorial: Gentrification research – new directions and critical scholarship', *Environment and Planning A*, 36: 1141–50.

Smith, M. (2007) 'Space, Place and Placelessness in the Culturally Regenerated City', in G. Richards (ed.) *Cultural Tourism: Global and Local Perspectives*, Binghampton, NY: Haworth Press, pp. 91–112.

Smith, M. and Carnegie, E. (2004) 'Bollywood Dreams? The Rise of the Asian Mela as a Global Cultural Phenomenon', paper presented at Journeys of Expression III conference Innsbruck, Austria, 5–7 May.

Smith, M., Carnegie, E. and Robertson, M. (2005) 'Juxtaposing the Timeless and the Ephemeral: Staging Festivals and Events at World Heritage Sites', in A. Leask and A. Fyall (eds) *Managing World Heritage Sites*, Oxford: Butterworth-Heinemann, pp. 110–24.

Soisalon-Soininen, Tuovi and Lindroth, Kaija (2007) 'Networks as Generators of Creativity of Tourism SMEs?', in G. Richards and J. Wilson (eds) *Changing structures of collaboration in cultural tourism*, Arnhem: ATLAS.

Solling, M. and Reynolds, P. (1997) *Leichhardt: On the margins of the city – a social history of Leichhardt and the former municipalities of Annandale, Balmain and Glebe*, North Sydney: Allen & Unwin.

Statistics Canada (2006) 2001 Community Profiles. www12.statcan.ca/english/Profil01/CP01/Index.cfm?Lang=E (accessed 25 March 2007).

STB (11 January 2005) *Singapore Sets Out To Triple Tourism Receipts To S$30 Billion by 2015*, Singapore Tourism Board. http://app.stb.com/asp/new/new02a.asp (accessed 13 June 2006).

—— (20 January 2005) *STB Sets 2015 Targets – 8.9 Million Visitors and $10.4 Billion in Tourism Receipt*, Singapore Tourism Board. http://app.stb.com/asp/new/new02a.asp (accessed 13 June 2006).

—— (18 January 2006) *STB Exceeds 2005 Targets with 8.94 Million Visitor Arrivals and S$10.8 Billion Tourism Receipts*, Singapore Tourism Board. http://app.stb.com.sg/asp/new/new03a.asp (accessed 13 June 2006).

—— (23 February 2006) *Singapore Tourism Board And MTV Networks Asia In Benchmark $50 Million Alliance*, Singapore Tourism Board. http://app.stb.com.sg/asp/new/new03a.asp (accessed 13 June 2006).

Stebbins, R. (1997) 'Serious Leisure and Well-being', in J.T. Haworth (ed.) *Work, Leisure and Well-being*, London: Routledge, pp. 117–30.

Storm, E. (2002) 'Converting pork into porcelain: Cultural institutions and downtown development', *Urban Affairs Review*, 38(1): 3–21.

Sudjic, D. (2005) 'Can we still build iconic buildings?', *Prospect*, 111: 22–26.

Sweeting, D. (2002) 'Leadership in Urban Governance: The Mayor of London', *Local Government Studies*, 28: 3–20.

SWRDA (2003) *Regional Economic Strategy 2003–2012*, Bristol: SWRDA.

Szerszynski, B., Heim, W. and Waterton, C. (eds) (2003) *Nature Performed: Environment, culture and performance*, London: Sage.

Tallon, A.R. and Bromley, R.D.F. (2004) 'Exploring the attractions of city centre living: evidence and policy implications in British cities', *Geoforum*, 35: 771–87.

Taylor, B. (2003) 'Little Italy – very Little Italy: As other neighbourhoods burn out, College St. refuses to fade away', *Toronto Star*, 2 June: B.

Teedon, P. (2001) 'Designing a place called Bankside: On defining an unknown space in London', *European Planning Studies*, 9(4): 459–81.

Ter Borg, M. (2003) *De Zineconomie: De samenleving van de overtreffende trap*, Schiedam: Scriptum.

Terhorst, P., van de Ven, J. and Deben, L. (2003) 'Amsterdam: It's all in the mix', in L.M. Hoffman, S. Fainstein and D.R. Judd (eds) *Cities and Visitors: Regulating People, Markets and City Space*, Oxford: Blackwell, pp. 75–90.

Thomas, T. (1995) 'When "they" is "we": movements, municipal parties, and participatory politics', in J. Lightbody (ed.) *Canadian Metropolitics: Governing our Cities*, Toronto: Copp Clark, pp. 115–36.

Thrane, C. (2001) 'Everyday Life and Cultural Tourism in Scandinavia: Examining the Spillover Hypothesis', *Loisir et Societe/Society and Leisure*, 23(1): 217–34.

Timothy, D. (2002) 'Tourism and the growth of ethnic islands', in C.M. Hall and A. M. Williams (eds) *Tourism and Migration: New relationships between production and consumption*, Dordrecht: Kluwer Academic Publishers, pp. 135–51.

TNSW (2002) *New South Wales Tourism Profile Year End June 2001*, Sydney: Tourism New South Wales.

—— (2004a) *Adventure #01: Snazzy Sydney – Chinatown*, Tourism New South Wales. http://backpacker.visitnsw.com.au/scripts/runisa.dll?VISITNSWLIVE.85 2084:BP2COL:1088552725:pp=BPLNAV1D,pc=BPLNAV1DA (accessed 2 June 2004).

—— (2004b) *Adventure #01: Snazzy Sydney – Leichhardt*, Tourism New South Wales. http://backpacker.visitnsw.com.au/scripts/runisa.dll?VISITNSWLIVE.85 2084:BP2COL:1088552725:pp=BPLNAV1N,pc=BPLNAV1NA (accessed 2 June 2004).

—— (2004c) *New South Wales Tourism Profile Year End June 2003*, Sydney: Tourism New South Wales.

Tourism New Zealand (2003) *The Interactive Traveller*. www.tourismnewzealand. com/tourism_info/industry-resources/getting-started/the-interactive-traveller. cfm (accessed 6 August 2006).

Tourism Research Council New Zealand (2006) *New Zealand Regional Tourism Forecasts 2005 – 2011 Nelson RTO*. www.trcnz.govt.nz/NZ+Regions/South+ Island/Nelson+RTO (accessed 6 August 2006).

Turisme de Barcelona (2003) *Estadístiques de turisme a Barcelona*, Barcelona: Turisme de Barcelona.

Turnbull, L. (1999) *Sydney: Biography of a city*, North Sydney: Random House.

Turner, V. (1973) 'The center out there: Pilgrim's goal', *History of religions*, 12(3): 191–230.

Turok, I. (2003) 'Cities, clusters and creative industries: the case of film and television in Scotland', *European Planning Studies*, 11: 549–65.

UK Creative Industries Task Force (CITF) (2001) *Creative Industries Mapping Document 2001*, London: Department for Culture, Media and Sport.

United Nations Conference on Trade and Development (2004) 'Creative Industries and Development'. www.unctad.org/templates/Event___33.aspx (accessed 9 July 2007).

Urban Task Force (2005) 'Towards a Strong Urban Renaissance', an independent report by members of the Urban Task Force, chaired by Lord Rogers. www. urbantaskforce.org (accessed 27 February 2007).

Urry, J. (1990) *The Tourist Gaze: Leisure and Travel in Contemporary Societies*, London: Sage.

—— (1994) 'Cultural change and contemporary tourism', *Leisure Studies*, 13(4): 233–38.

—— (1995) *The Consumption of Place*, London: Routledge.

Vallbona, M. and Richards, G. (2007) 'The Meaning of Cultural Festivals: Stakeholder perspectives', *International Journal of Cultural Policy*, 27: 103–22.

Vallgrasa (2005) *Nature, the festival of ecotourism and protected natural spaces*, Sitges: Park of Garraf.

Van Dale (2007) www.vandale.nl/opzoeken/woordenboek (accessed 1 March 2007).

Van den Berg, L. and Russo, A.P. (2004) *The Student City: Strategic planning for students' communities in EU cities*, Aldershot: Ashgate.

Van den Berg, L., Van der Borg, J. and Van der Meer, J. (1995) *Urban Tourism: Performance and strategies in eight European cities*, Aldershot: Avebury.

Van den Berg, L., Van der Borg, J. and Russo, A.P. (2003) 'The Infrastructure for Urban Tourism: a European Model? A Comparative Analysis of Mega-Projects in Four Eurocities', in D.R. Judd (ed.) *The Infrastructure of Play*. Chicago, IL: M.E. Sharpe, pp. 296–320.

Van der Duim, R. (2007) 'Tourism and innovation: an actor-network approach', in G. Richards and J. Wilson, (eds) *Changing structures of collaboration in cultural tourism*, Arnhem: ATLAS.

Van der Duim, V.R. (2005) *Tourismscapes: An actor-network perspective on sustainable tourism development*, PhD Thesis, Wageningen University, the Netherlands.

Van der Poel, H. (1993) *De modularisering van het dagelijkse leven: Vrijetijd in structuratietheoretisch perspectief*, Amsterdam: Thesis Publishers.

—— (1997) 'Leisure and the Modularization of Daily Life', *Time and Society*, 6(2/3): 171–94.

—— (1999) *Tijd voor vrijheid. Inleiding tot de studie van de vrijetijd*, Amsterdam: Uitgeverij Boom.

Vanolo, A. (2006) 'The image of the creative city', Dipartimento Interateneo Territorio, Politecnico e Università di Torino. http://web.econ.unito.it/vanolo/index_file/paper/tiotcc.pdf (accessed 15 February 2007).

Vassal, S. (1987) 'L'Europe des Universités', *Cahiers de Sociologie Economique et Culturelle. Ethnopsychologie*, 8: 149–57.

Verdú, V. (2003) *El Estilo del Mundo: la Vida en el Capitalismo de Ficción*, Barcelona: Anagrama.

Ville de Montréal (1992) *City Plan: Master Plan of the Ville-Marie District*, Montréal: Ville de Montréal Service de l'Urbanisme.

—— (1998) *Chinatown Development Plan*, Montréal: Ville de Montréal Service de l'Urbanisme.

VisitBritain (2006) *Cultural Tourism Advisory Guide*, London: VisitBritain.

—— (2007) Literary Landscapes. www1.visitbritain.com/VB3-de-DE/experiences/Tour/inspirational_landscapes/writers_poets_introduction.aspx (accessed 9 January 2007).

Visser, G. 2003: 'Gay men, tourism and urban space: reflections on Africa's "gay capital"', *Tourism Geographies*, 5(2): 168–89.

Viviani, N., Coughlan, T. and Rowland, T. (1993) *Indo-Chinese in Australia: The issues of unemployment and residential concentration*, Melbourne: Bureau of Immigration Research.

VVAA (2001) 'Université: droit de cité', *Urbanisme*, 317 (3/4).

Waitt, G. and Markwell, K. (2006) *Gay Tourism: Culture and Context*, Binghampton, NY: Haworth Press.

Walberg, D. (2006) 'Political Party Boy', *Xtra! Magazine*, in: *Creative Places + Spaces News Journal*, Issue 2, Toronto, Artscape, February.

318   *References*

Waldinger, R. (1986) *Through the Eye of the Needle: Immigrants and Enterprise in New York's garment trades*, New York: New York University Press.

Wang, N. (1999) 'Rethinking authenticity in tourism experience', *Annals of Tourism Research*, 26(2): 349–70.

Whatmore, S. (2002) *Hybrid Geographies: Natures, cultures, spaces*, London: Sage.

Whitworth, M. (2006) 'London fashion is back in vogue as Britmania grips US', *Daily Telegraph*, 4 February, p. 5.

Wickens, E. (2002) 'The sacred and the profane: A tourist typology', *Annals of Tourism Research*, 29(3): 834–51.

Willis, K., Fakhri, S.M.A.K. and Yeoh, B. (2002) 'Introduction: Transnational Elites', *Geoforum*, 33: 505–7.

Wilson, E. (2003) *Bohemians: The Glamorous Outcasts*, London: Tauris Parke.

Wood, P. and Taylor, C. (2004) 'Big ideas for a small town: the Huddersfield creative town initiative', *Local Economy*, 19: 380–95.

World Travel and Tourism Council (1998) *South Africa's Travel and Tourism: Economic Driver for the 21st Century*, London: World Travel and Tourism Council.

Wu, F. (2000) 'The global and local dimensions of place-making: remaking Shanghai as a world city', *Urban Studies*, 37: 1359–77.

Wu, W. (2005) *Dynamic Cities and Creative Clusters*, World Bank Policy Research Working Paper 3509, World Bank, Washington D.C.

Yan, A. (2002) *Revitalization Challenges for Vancouver's Chinatown*, Vancouver: Carnegie Community Action Project.

Yin, R. (2003) *Case Study Research: Design and methods*, 3rd edn, Thousand Oaks, CA: Sage.

Youl Lee, S., Florida, R. and Acs, Z.J. (2004) 'Creativity and entrepreneurship: a regional analysis of new firm formation', *Regional Studies*, 38(8): 879–91.

Yusuf, S. and Nabeshima, K. (2005) 'Creative industries in East Asia', *Cities*. 22: 109–22.

Zukin, S. (1990) 'Socio-spatial prototypes of a new organization of consumption: the role of real cultural capital', *Sociology*, 24(1): 37–56.

—— (1991) *Landscapes of Power, From Detroit to Disneyworld*, Berkley: University of California Press.

—— (1995) *The Cultures of Cities*, Oxford: Blackwell.

—— (1998) 'Urban lifestyles: diversity and standardisation in spaces of consumption', *Urban Studies*, 35(5–6): 825–39.

—— (2001) 'How to create a Culture Capital: Reflections on Urban Markets and Places', in I. Blazwick (ed.) *Century City: Art and Culture in the Modern Metropolis*, London: Tate Publishing, pp. 259–64.

Index